The Darwinian Tradition in Context

Richard G. Delisle
Editor

The Darwinian Tradition in Context

Research Programs in Evolutionary Biology

Springer

Editor
Richard G. Delisle
Departments of Liberal Education and Philosophy
University of Lethbridge
Lethbridge, AB, Canada

ISBN 978-3-319-69121-3   ISBN 978-3-319-69123-7 (eBook)
https://doi.org/10.1007/978-3-319-69123-7

Library of Congress Control Number: 2017960194

Printed on acid-free paper

This Springer imprint is published by Springer Nature
The registered company is Springer International Publishing AG
The registered company address is: Gewerbestrasse 11, 6330 Cham, Switzerland

Preface

In the wake of the Evolutionary Synthesis constituted in the 1930s, 1940s, and 1950s, historians and philosophers of biology have devoted considerable attention to the Darwinian tradition linking Charles Darwin to mid-twentieth-century developments in evolutionary biology. This historiographical focus may not be wholly coincidental, given the professionalization of the fields of history of science and philosophy of biology that accompanied the post-1960 era. Since then, more recent developments in evolutionary biology challenged the heritage of the Darwinian tradition as a whole or in part. Predictably, perhaps, this was followed by a historiographical "recalibration" by historians and philosophers toward other research programs and traditions since Darwin's time.

As this recalibration is going on, it is difficult not to have the impression of confusion or dismay regarding what exactly happened in evolutionary biology. In order to dispel some of this confusion, it seems timely to reunite in this volume synthetic contributions concerned with historical, philosophical, and scientific issues. It is the main goal of this volume to contextualize the Darwinian tradition by raising such questions as: How should it be defined? Did it interact with other research programs? Were there any research programs whose developments were conducted largely independently of the Darwinian tradition? Authors of this volume explicitly reflect upon the nature of the relationship between the Darwinian tradition and other parallel research traditions.

A more traditional approach to the topic might have required organizing the volume's contributions along themes like the "main Darwinian tradition," "non-Darwinian theories," "evolutionary biology in national traditions," "pre-synthetic developments," the "Evolutionary Synthesis," or "post-synthetic developments." As much as this was the editor's original intention, many contributions collected here suggested to him that historiographical studies are currently moving beyond this more traditional outlook, pointing at other intellectual avenues. In order to acknowledge this historiographical shift and foster new thinking on these matters, the papers are organized in a sequence that highlights how the boundaries of the various research programs within evolutionary biology are apparently more porous than often assumed. The papers can be meaningfully arranged into two main threads:

1. Part I: The view that sees Darwinism as either originally pluralistic or acquiring such a pluralism through modifications and borrowings over time.
2. Part II: The view blurring the boundaries between non-Darwinian and Darwinian traditions, either by holding that Darwinism itself was never quite as Darwinian as previously thought or that non-Darwinian traditions took on board some Darwinian components, when not fertilizing Darwinism directly.

Between a Darwinism reaching out to other research programs and non-Darwinian programs reaching out to Darwinism, the least that can be said is that this crisscrossing of intellectual threads blurs the historiographical field.

In Part I of this volume, Timothy Shanahan argues in "Selfish Genes and Lucky Breaks: Richard Dawkins' and Stephen Jay Gould's Divergent Darwinian Agendas" that Darwin's Darwinism was polymorphic or pluralistic enough to legitimately accommodate future developments as divergent as those opposing Dawkins's genetic reductionism and Gould's holistic hierarchical thinking. In a similar vein, John Alcock's "The Behavioral Sciences and Sociobiology: A Darwinian Approach" holds that Darwin's strong adaptationist stance has been successfully maintained in the behavioral sciences but by applying it to new phenomena, this time involving both genetic entities and individual organisms (and excluding higher entities), as seen in scholars like N. Tinbergen, W. Hamilton, E. O. Wilson, and R. Dawkins, among others. Embracing the same historiographical view, but simultaneously allowing for an expansion of Darwinism, David Depew argues in his "Darwinism in the Twentieth Century: Productive Encounters with Saltation, Acquired Characteristics, and Development" that Darwinism continually and successfully met the challenges of evolutionary developmentalism, the inheritance of acquired characteristics, and saltationism by taking on board new explanatory components but within its own ways of doing things. This evolving and flexible Darwinian tradition is presented in Massimo Pigliucci's "Darwinism after the Modern Synthesis" as having permitted the transition from the Evolutionary Synthesis to the Extended Evolutionary Synthesis—incorporating new phenomena, mechanisms, and concepts—yet without moving beyond the confines of the same paradigm. This view is also shared by Adam Van Arsdale in his "Human Evolution as a Theoretical Model for an Extended Evolutionary Synthesis," who uses the case of human evolution to reflect upon the nature of this theoretical expansion when it comes to integrating unique features such as encephalization, as well as nongenetic and flexible behaviors.

In Part II of this volume, Richard Delisle holds in "From Charles Darwin to the Evolutionary Synthesis: Weak and Diffused Connections Only" that key Darwinian scholars (including Darwin himself) and some proponents of the Evolutionary Synthesis were also simultaneously committed to ideas that were not particularly Darwinian, making the boundary between Darwinian and non-Darwinian ideas porous. Indeed, Georgy Levit and Uwe Hossfeld argue in "Major Research Traditions in Twentieth-Century Evolutionary Biology: The Relations of Germany's

Darwinism with Them" how evolutionary biology in German-speaking countries which centered around notions like "type," "monism," and "holism" variously integrated some Darwinian elements, especially as seen in E. Haeckel, L. Plate, and B. Rensch.

What used to be seen in the traditional historiography as past blind intellectual alleys are increasingly seen as possible early insights now in need of some sort of revival. In "Alternatives to Darwinism in the Early Twentieth Century," Peter Bowler expands on how important Lamarckism, Orthogenesis, and Saltationism had been for evolutionism in the late nineteenth and early twentieth centuries, some of these ideas now being reconsidered. In a similar vein, Maurizio Esposito holds in "The Organismal Synthesis: Holistic Science and Developmental Evolution in the English-Speaking World, 1915–1954," how a fairly robust tradition founded on the centrality of organismic biology persisted in the English-speaking world throughout the first half of the twentieth century—largely independently of what was perceived as a reductionistic and mechanistic neo-Darwinism—a tradition revived today in Evo-Devo under different guises. And precisely because these research programs overlap, one strain of the multifaceted, robust, and long-lasting Lamarckian movement in France paved the way for important innovations in molecular biology in the 1950s and 1960s, as argued in the "Lamarckian Research Programs in French Biology (1900–1970)" of Laurent Loison and Emily Herring. This situation erected a bridge between two movements usually opposed over the divide of "hard inheritance" (molecular biology) and "soft inheritance" (Lamarckism), Darwinism being traditionally associated more closely to the former than the latter. In "Molecularizing Evolutionary Biology," Michel Morange further reflects upon the nature of the interrelationship between molecular biology and evolutionary biology, arguing that the former has insinuated itself ever more profoundly into evolutionary questions since the 1960s, to the point of significantly modifying the character of the so-called Modern Synthesis.

Whether or not the Darwinian/non-Darwinian divide is judged to have been more porous than often assumed, some research programs managed to grow without much contact with Darwinism, until recent bridges were established. In "Cells, Development, and Evolution: Teeth Studies at the Intersection of Fields," Kate MacCord and Jane Maienschein offer an alternative to the narrative of a "gene-centered" evolutionary biology by recounting how development, evolution, and cells were brought together throughout the twentieth century. In a different case study, Ulrich Kutschera's "Symbiogenesis and Cell Evolution: an Anti-Darwinian Research Agenda?" explains how the research program on the rise of more complex cells in the early history of life (the symbiogenesis theory) was for too long conducted from the viewpoint of an anti-Darwinian agenda.

Just as the first two contributions to this volume argued that Darwinism's original pluralism was sufficient to explain a wide scope of evolutionary phenomena, so the volume closes with Derek Turner's analysis in "Paleobiology's Uneasy Relationship with the Darwinian Tradition: Stasis as Data" in which he holds that Darwinism today has been destabilized by what paleobiology brought to evolutionary studies since the 1970s.

Irrespective of how one understands the Darwinian tradition, most contributions to this volume show the extent to which the various research programs in evolutionary biology are deeply pluralistic, often being composed of many overlapping or semi-distinct intellectual strains, suggesting an overall picture of a tight and complex network of ideas across evolutionary biology.

Lethbridge, Canada Richard G. Delisle

Contents

Introduction: Darwinism or a Kaleidoscope of Research Programs and Ideas?

Richard G. Delisle

Who controls the past controls the future: who controls the present controls the past.
George Orwell, Nineteen Eighty-Four (1949)

The Preface to this volume has provided an exposition of the project, the rationale for the order of presentation of the papers, and a brief description of each paper. In this introductory chapter, I would like to embrace a more personal view about what seems to have emerged in this research area with the assistance of insights provided by some of the contributions included here.[1]

We know too well that history is continually being rewritten in light of a changing present. The current turmoil in evolutionary biology about evo-devo, epigenetic inheritance, holistic manifestations, and stochastic modes of change, for example, can only serve as a stimulus for revisiting the past in search of antecedents. Yet, ironically, this salutary quest under way reveals that what we thought were novel claims may not be entirely so. Indeed, if the currently sought pluralism for accommodating recent developments in evolutionary biology is found to have existed in the past, then the past and the present of evolutionary biology cannot be entirely incommensurable with one another. I would argue that the current phase in evolutionary biology may be characterized as follows: after initial claims of novelties proclaimed during the last decades, historians and philosophers are in the process of researching the past in order to see how the pieces of the overall puzzle fit together. Suddenly, it seems, the past comes hunting for the present, generating complex interactions between the two. The jury is still out on what the outcome of these reflections will be. Darwinism, today, has never been more of a "moving target" (Burian 1988: 250).

[1]The view presented here is solely my own. The reader is strongly encouraged to discover what other authors have had to say for themselves in this regard beyond the brief summary already presented in the Preface.

R.G. Delisle (✉)
Departments of Liberal Education and Philosophy, University of Lethbridge, 4401 University Drive, Lethbridge, AB, Canada, T1K 3M4
e-mail: richard.delisle@uleth.ca

R.G. Delisle (ed.), *The Darwinian Tradition in Context*,
https://doi.org/10.1007/978-3-319-69123-7_1

In a sense, evolutionary biology today fell victim to events occurring between the 1930s and the early 1960s, known collectively as the Evolutionary Synthesis. The synthesists proclaimed themselves to have reached a theoretical unity of unprecedented breadth, and, serendipitously, the newly rising professionalization of history and philosophy of biology offered the synthesists an amplifier for their voice. As Mark Largent (2009: 4, 8) writes:

> The architects of the modern evolutionary synthesis and the historians who followed them constructed a discontinuous history of their discipline... By unquestionably interpreting early twentieth-century evolutionary biology through the lenses of the triumphalist synthesizers, [historians of biology] distort our understanding... and commit a historiographical blunder... We need to go back to the generation prior to synthesis... and free ourselves from the propagandistic claims made by the mid-twentieth century synthesizers.

The Evolutionary Synthesis stands in the middle of twentieth century like a horizontal cut, defining what came before and after. In fact, the horizontality of the standard view is such that distinct yet overlapping intellectual layers have been proposed: the first synthesis (mathematical population genetics plus selectionism in the 1920s and 1930s), the second synthesis (adding organismic biology in the late 1930s and 1940s), and the hardening of the synthesis (focusing nearly exclusively on selectionism in the 1950s and after). The first and second phases are sometimes reunited together under the epithet of the "pluralistic phase of the synthesis," in opposition to its hardened and later phase (Provine 1985; Gould 1983; Mayr 1982).

It is not my aim to criticize the historians and philosophers of the first generation: after all, we are largely building upon their work. Rather, I am merely raising the obvious historiographical point that if the Evolutionary Synthesis is not what has been advertised, it is only natural that what has been left out of the official story needs to be reinserted into the narrative. As much as the Evolutionary Synthesis seems to be an established reality in the mind of many scholars, it should be noted that a small but significant number of them question its historiographical reality. The following quotes will make this point more obvious:

> There was more to the synthesis project than work done by the so-called architects or the supposed merger of Mendelian genetics and selection theory via mathematical theory. The evolution books of the revived Columbia Biological Series have been far too dominant in the synthesis historiography. There was more—much more—to evolutionary studies in the 1920s and 1930s than is suggested in the mainline narratives of the period... I propose we abandon the unit concept of 'the evolutionary synthesis'... What do we find when we remove the organization this master narrative imposes? What is gained if we presume the unit concept 'the evolutionary synthesis' obscures more than it clarifies (Cain 2009: 621, 622, 625).

> [O]ur analysis of Rensch's theoretical work can be seen as a case study of the heterogeneity of the Modern Synthesis. The scale of this heterogeneity is, in fact, so significant that the picture of the Synthesis as a unified movement needs to be deconstructed. The idea that the Synthesis is an interdependent body of beliefs covering not only all major branches of empirical biology, but also the general questions of methodology, history, and philosophy of science, collapses in front of such hardly compatible world views as Rensch's and Mayr's. It is also important to remark that all parts of their theoretical constructions were equally important for their arguments in favor of Darwinism... Rensch's holistic

theoretical system is in almost direct opposition to Mayr's philosophy, coinciding with it only on the purely phenomenological level and in empirically testable explanations. Indeed, beyond the elementary level of accepting mutation, recombination, geographical isolation, and natural selection as the most important factors of evolution, there is little that unites them considering deep philosophical differences between their systems. This makes the picture of the Synthesis as an amalgamation of closely interrelated theoretical systems very questionable (Levit et al. 2008: 320–321).

[B]eneath a superficial appearance of unity among neo-Darwinians [like J.S. Huxley, Th. Dobzhansky, B. Rensch, G.G. Simpson, and E. Mayr] lies genuine *foundational* oppositions in the epistemological and metaphysical choices made by them... (1) oppositions in the predominate epistemology (descriptive/synthetic, ontological monism, or etiological); (2) oppositions in the interpretation given to the direction of evolution; (3) oppositions in the scope of application of the evolutionary principles (to the biological realm only or to the entire cosmos); (4) oppositions in the nature of the evolutionary process in time (open-ended, cyclical, or stagnating)... [N]eo-Darwinism... is not a movement from which all neo-Darwinians sprung, but is rather a meeting place from which each drew evolutionary mechanisms in order to insert them in distinct and quasi-incommensurable... frameworks (Delisle 2009a: 120).

In addition to what has already been said, one will find among the various implicit and explicit points made by Mark Largent (2009), Joe Cain (2009), George Levit, Michal Simunek, and Uwe Hossfeld (2008), and Richard Delisle (2008, 2009a, b, 2011, 2017) the following:

1. Substantial synthetic work was already conducted before the Evolutionary Synthesis (ES) by evolutionists, although these were excluded from the ES by its official promoters.
2. The synthesists retroactively created an intellectual vacuum and a discontinuity before the Evolutionary Synthesis, presenting many pre-1930 ideas as misguided.
3. Architects of the Evolutionary Synthesis were actively engaged in rhetorical arguments, a kind of self-promotion: they advanced a certain narrative regarding what they believed themselves to have achieved, which was later taken up and echoed by subsequent historians and philosophers.
4. The Evolutionary Synthesis may well be better conceived as a political or sociological event than as a conceptual one.
5. The Evolutionary Synthesis became a straw man for its post-1960 opponents, thus reinforcing its political utility. It is indeed always useful to have an enemy against whom one can formulate opposite ideas, even if that enemy does not exist.

In short, skeptics[2] of the Evolutionary Synthesis have raised doubts about both its internal/conceptual coherence and its external/contextual isolation from the rest

[2]By referring to my colleagues Mark Largent, Joe Cain, George Levit, Michal Simunek, and Uwe Hossfeld, as "skeptics" of the Evolutionary Synthesis, I am not trying to co-opt their views or subsume them under my own understanding of the issues. Nor do I pretend there exists a single and unified front against the traditional historiography. Ultimately, only additional concerted work

of the field of evolutionary biology. Whether or not one is prepared to follow these analyses to their logical conclusion—the abandonment of the notion of an "Evolutionary Synthesis"—it seems to this author that a call for a historiographical shift is at least not only reasonable but also an emergent reality of recent studies. This proposed shift would move us from an evolutionary biology conceived around temporal horizontal cuts or layers (Darwinian revolution, eclipse of Darwinism, phases of the Evolutionary Synthesis, post-synthetic developments) to an analysis of vertical intellectual movements and ideas evolving in parallel and interacting in complex ways. Indeed, this shift toward historiographical continuity is supported by a fascinating crisscrossing of intellectual threads in evolutionary studies: while some scholars argue that Darwinism evolves by co-opting ideas from competing research programs, others hold that non-Darwinian programs have availed themselves of Darwinian explanatory components. For instance, David Depew (2017) argues in this volume that:

> Darwinism's continued dominance in evolutionary science reflects its proven ability to interact productively with these other traditions, an ability impressed on it by its founder's example. Evolution by sudden leaps (saltations) is alien to the spirit of Darwinism, but Darwinism advanced its own agenda by incorporating and subverting saltationist themes. Similarly, Lamarckism's belief in the heritability of acquired characteristics has been discredited, but some of the facts to which it seems congenial reappear in genetic Darwinism as phenotypic plasticity and niche construction.

The contribution of Massimo Pigliucci (2017) to this volume is of a similar spirit, promoting what he calls an Extended Evolutionary Synthesis which incorporates within Darwinism explanations about processes such as epigenetic inheritance, self-organizing biological phenomena, and self-emergent properties.

Now, looking at things from the viewpoint of non-Darwinian theories, Peter Bowler (2017) writes in this volume:

> When the author of this chapter first began to study the 'eclipse of Darwinism' in the 1980s the triumph of the modern Darwinian theory made it easy to dismiss the alternatives as blind alleys into which scientists had been led temporarily... In recent decades our interpretation of this episode has been transformed by the emergence of evolutionary developmental biology. This has reopened issues once marginalized by genetics and the modern Darwinian synthesis. Some enthusiasts see 'evo-devo' as reintroducing a role for non-selectionist factors such as Lamarckism, while even those skeptical of this view acknowledge that the older theories were not as wide of the mark as was once claimed... We can now appreciate that this concern was not merely a distraction from the main business of evolutionary biology... The historians who look back at these early non-Darwinian theories can, perhaps, see evidence of ideas being explored that may once again come to play a role in evolution theory.

---

along these lines could clarify this question; organizing a symposium in a near future may be a timely idea. This being said, it is interesting to note that, to my knowledge at least, many "skeptics" have arrived at similar conclusions independently of each other. Apparently, the time is ripe for a questioning of the historiography.

Similarly, documenting the existence of a robust research tradition concerned with organismic biology in the first half of the twentieth century, Maurizio Esposito (2017) notes in this volume:

> [T]he developmentalist perspective advanced by these "romanticists" was neither equivalent with Ernst Haeckel's biogenetic law nor represented a variant of orthogenetic evolution, but rather endorsed Walter Garstang's idea according to which ontogeny does not recapitulate phylogeny, it creates it... In that sense, this perspective is closer to contemporary evo-devo than 19th century recapitulationist theories... Now, if after the 1940s, the "romanticists'" views gradually fell in the background, in favor of neo-Darwinian hypotheses... then the question is: why did that happen? The question is particularly relevant because, in the last few decades, the consensus about the neo-Darwinian synthesis has been eroded in favor of a new form of developmental evolutionism (evo-devo) and novel versions of organicist philosophies have again entered onto the stage... The overreaching perspective offered by the developmental system theory, or by the complex epigenetical models of gene expression, are certainly closer to the systemic view of the "romanticists" than the adaptationist models of the modern synthesizers.

Far from always being segregated, competing research programs are intellectually fertilized by each other. Laurent Loison and Emily Herring (2017) describe in this volume how this happened during the transition from Lamarckism to Darwinism:

> We describe how Teissier and L'Héritier's interests, ideas and conjectures, despite their Darwinian inclinations, were influenced by the Lamarckian atmosphere of French biology. This example perfectly shows how non-Darwinian ideas influenced the development of the Modern Synthesis... Despite their indisputable commitment to Darwinism, Teissier and L'Héritier also showed interest in certain aspects of inheritance and evolution that did not belong to the classical Mendelian-Darwinian account of evolution. Here, we would like to briefly sketch these unorthodox dimensions of their work and emphasize their connections with the predominantly Lamarckian atmosphere of French zoology during the 1930s and 1940s... L'Hérititier's and Teissier's heterodox position underlines the specificity of the French context: at the time of the Synthesis, French biology was under the domination of Lamarckian-Bergsonian thought which prioritized the separation between adaptation and true evolution and which tended to favor non-Mendelian modes of heredity: these two main characteristics were central to L'Héritier's and Teissier's rethinking of the structure of the Evolutionary Synthesis.

As we dig deeper into the annals of evolutionary biology, the task of clearly distinguishing between the various research programs does not get any easier. Presenting in this volume what was once called "Old-Darwinism," Georgy Levit and Uwe Hossfeld (2017) write (see also Levit and Hossfeld 2006):

> 'Old-Darwinism' in its fully established and explicit form cannot be reduced to any other theoretical school. The specificity of this theory lay to combine the 'standard' Darwinian factors of evolution (mutation, recombination, geographic isolation, natural selection) with the neo-Lamarckian and orthogenetic mechanisms in order to define the exact role of all these mechanisms in evolutionary process proceeding from the whole complex of biosciences including genetics. Old-Darwinians legitimately insisted that they follow the initial ideas of Darwin, who assumed some roles for Lamarckian mechanisms as well as for the auxiliary hypothesis of constraints. The very idea of combining various evolutionary mechanisms was wide spread at that time within various cultural contexts... In addition to Darwin, Haeckel and himself, Plate counted Richard Semon (1859–1919), Wilhelm Roux (1850–1924), Richard von Hertwig (1850–1937), Fritz v. Wettstein (1895–1945), Berthold Hatschek (1854–1941), Jan Paulus Lotsy (1867–1941), Franz Weidenreich (1873–1948)

> and the future "co-architect" of the Evolutionary Synthesis, Bernhard Rensch, among the old-Darwinians... In Plate's later works (Plate 1932–1938) we find all the basic factors of evolution later adapted by the Evolutionary Synthesis. Thus, Plate claimed that random mutations and recombination deliver the bulk of raw material for evolution. Natural selection and geographical isolation perform a major role in evolution... Also, what is now known as 'population thinking' is of great importance for Plate as he analyses the 'laws of populations' with some mathematics... Yet Plate also admitted other evolutionary mechanisms going beyond the basic tenets of the Synthetic Theory of Evolution. Plate accepted both macro- and directed- mutations, orthogenesis and the inheritance of acquired characters.

The closer we study the development of evolutionary biology, the more the overall picture seems to be blurred and networklike, rendering the neat separation of research programs and historical periods somewhat difficult (see also Levit and Hossfeld 2011). The obvious question to ask at this stage is whether or not distinguishing research programs from one another is even possible. I, myself, do not know the answer, given the current state of our knowledge. However, the apparently obvious solution, that of identifying research programs by their distinct "hard cores," is no panacea. Assuming that research in any particular area consists in a continually evolving quest—research programs are historical entities, to use familiar terminology—must we assume also that such an area is accompanied by an explanatory hard core? Let us take Darwinism as an example. If one answers "yes," then one is committed to the "multilevel model" of science which promotes an explanatory structure positioning the cause (i.e., natural selection) in a privileged hierarchical position relative to other components or fields being explained by that cause or hard core.

There exists, however, an alternative view of science and Darwinism: it consists in arguing that Darwinism was never quite as advertised in the traditional historiography. Already, John C. Greene (1981, 1999) argued for a Darwinism that is intimately connected with so many ontological and metaphysical issues as to make the traditional view centered around selective mechanisms a pale representation of its multidimensional complexion. No surprise, then, that Ernst Mayr (1986) expressed frustrations at Greene's depiction of Darwinism, busy as the former was in representing it as a pure product of positivistic science (Delisle 2009c). Moving away from the "mechanism-centered" bias of the historiography allows for the uncovering of self-proclaimed rhetorical arguments in favor of Darwinism during its two main phases: Darwin's *Origin of Species* and the Evolutionary Synthesis (Delisle 2008, 2009b, c, 2011, 2014, 2017). Several of these so-called Darwinians (including Darwin himself) are committed to so many distinct and competing empirical, conceptual, ontological, and metaphysical choices that they are unable to use the concept of natural selection without significant distortion. On this view, "Darwinians" are not truly bound together under a strong and common intellectual thread; rather, they each exploit some Darwinian components in a piecemeal fashion only by inserting them in distinct research entities. The dissolution of Darwinism, under this thesis, expresses itself through the weak and diffused spread of "Darwinian" explanatory components among a wide scope of evolutionary views, irrespective of how we call them. This alternative view appeals to a "reticulate model" of science founded on a diffused and flexible explanatory structure, with no privileging of causal components over more descriptive ones, and no hierarchical organization of disciplines over others (explanatory core versus explained periphery), in contradiction to the multilevel model of science. In other words, the reticulate model holds that Darwinism has no real hard core, and without it, the task of segregating research programs from one another becomes nearly impossible.

Without a doubt, the debate is an open and fascinating one. Irrespective of which thesis one favors, studies in evolutionary biology tend to show that this area is a kaleidoscope of research entities and ideas, an ever thicker and complex intellectual network. One thing is sure, though: much work remains to be done regarding the development of evolutionary biology. I, for one, once thought that the main events of its development had already been fairly well established. I no longer think this to be the case. New research perspectives and new material are what the future will be made of. "Who controls the past controls the future: who controls the present controls the past," wrote George Orwell in his 1949 novel. I would like to close by extending a plea to historians and historically inclined philosophers: let us prove Orwell's motto wrong in its second half, and true in its first. New viewpoints about the development of evolutionary biology need not always wait for the Whig judgment of a changing present.

Acknowledgments I thank James Tierney (Yale University) for assistance in improving the English version of this paper.

References

Bowler PJ (2017) Alternatives to Darwinism in the early twentieth century. In: Delisle RG (ed) The Darwinian tradition in context: research programs in evolutionary biology. Springer, Cham, pp 195–218

Burian R (1988) Challenges to the evolutionary synthesis. Evol Biol 23:247–269

Cain J (2009) Rethinking the synthesis period in evolutionary studies. J Hist Biol 42:621–648

Delisle RG (2008) Expanding the framework of the holism/reductionism debate in neo-Darwinism: the case of Theodosius Dobzkansky and Bernhard Rensch. Hist Philos Life Sci 30:207–226

Delisle RG (2009a) The uncertain foundation of neo-Darwinism: metaphysical and epistemological pluralism in the evolutionary synthesis. Stud Hist Phil Biol Biomed Sci 40:119–132

Delisle RG (2009b) Les philosophies du néo-darwinisme. Presses Universitaires de France, Paris

Delisle RG (2009c) Ernst Mayr's philosophy of science: its connections with logical empiricism, the unity of science movement, and the scientific revolution (unpublished manuscript available on internet at: http://philbioihpst.free.fr/DelisleArticle.pdf)

Delisle RG (2011) What was really synthesized during the evolutionary synthesis? A historiographic proposal. Stud Hist Phil Biol Biomed Sci 42:50–59

Delisle RG (2014) Evolution in a fully constituted world: Charles Darwin's debts towards a static world in the *Origin of Species* (1859). Endeavour 38:197–210

Delisle RG (2017) From Charles Darwin to the evolutionary synthesis: weak and diffused connections only. In: Delisle RG (ed) The Darwinian tradition in context: research programs in evolutionary biology. Springer, Cham, pp 133–168
Depew DJ (2017) Darwinism in the 20th century: productive encounters with saltation, acquired characteristics, and development. In: Delisle RG (ed) The Darwinian tradition in context: research programs in evolutionary biology. Springer, Cham, pp 61–88
Esposito M (2017) The organismal synthesis: holistic science and developmental evolution in the English-speaking world, 1915–1954. In: Delisle RG (ed) The Darwinian tradition in context: research programs in evolutionary biology. Springer, Cham, pp 219–242
Gould SJ (1983) The hardening of the modern synthesis. In: Grene M (ed) Dimensions of Darwinism. Cambridge University Press, Cambridge, pp 71–93
Greene JC (1981) Science, ideology, and world view. University of California Press, Berkeley
Greene JC (1999) Debating Darwin. Regina Books, Claremont
Largent M (2009) The so-called eclipse of Darwinism. Trans Am Philos Soc (New Ser) 99(1):3–21
Levit GS, Hossfeld U (2006) The forgotten "Old-Darwinian" synthesis: the evolutionary theory of Ludwig H. Plate (1862–1937). NTM 14:9–25
Levit GS, Hossfeld U (2011) Darwin without borders? Looking at "generalised Darwinism" through the prism of the "hourglass model". Theor Biosci 130:299–312
Levit GS, Hossfeld U (2017) Major research traditions in 20th century evolutionary biology: the relations of Germany's Darwinism with them. In: Delisle RG (ed) The Darwinian tradition in context: research programs in evolutionary biology. Springer, Cham, pp 169–194
Levit GS, Simunek M, Hossfeld U (2008) Psychoontogeny and psychophylogeny: Bernhard Rensch's (1900–1990) selectionist turn through the prism of panpsychistic identism. Theor Biosci 127:297–322
Loison L, Herring E (2017) Lamarckian research programs in French biology (1900–1970). In: Delisle RG (ed) The Darwinian tradition in context: research programs in evolutionary biology. Springer, Cham, pp 243–270
Mayr E (1982) The growth of biological thought. Belknap Press, Cambridge
Mayr E (1986) The death of Darwin? Rev Synth (3):229–236
Pigliucci M (2017) Darwinism after the modern synthesis. In: Delisle RG (ed) The Darwinian tradition in context: research programs in evolutionary biology. Springer, Cham, pp 89–104
Provine WB (1985) Adaptation and mechanisms of evolution after Darwin: a study in persistent controversies. In: Kohn D (ed) The darwinian heritage. Princeton University Press, Princeton, pp 825–866

Part I
From a Pluralistic Darwinism to an Ever More Inclusive Darwinism

Selfish Genes and Lucky Breaks: Richard Dawkins' and Stephen Jay Gould's Divergent Darwinian Agendas

Timothy Shanahan

Abstract Darwin expressed alternative theoretical perspectives on a range of issues fundamental to our understanding of evolution, thereby making it possible for his intellectual descendants to develop his ideas in markedly different and even incompatible directions while still promoting their views as authentically "Darwinian." The long-running and well-publicized scientific rivalry between Richard Dawkins and Stephen Jay Gould is a striking case in point. In elegantly written books and essays spanning the last quarter of the twentieth century, they developed and defended diametrically opposed views on the units of selection, the scope and depth of adaptation, the significance of chance events, and the reality and meaning of evolutionary progress—each explicitly juxtaposing his own views against those of the other while insisting that his own conclusions represent the genuinely "Darwinian" view. These skirmishes raise many questions. If there is just one world, why do they reach such different conclusions about it? Does each have an equally good claim to represent authentic "Darwinism"? Are they best viewed as defending different interpretations of a *single* Darwinian tradition, or as representing alternative (e.g., competing) Darwinian *traditions*? More generally, is a scientific tradition best characterized by a set of propositions that define its essence, or by causal interactions providing cohesiveness in terms of self-identification, social relations, and historical continuity? An analysis of the Dawkins–Gould rivalry provides a fertile opportunity to address these and other questions concerning "the Darwinian tradition" in the twentieth century.

Keywords Charles Darwin • Richard Dawkins • Stephen Jay Gould • Evolution • Natural selection • Adaptation • Constraints • Convergence • Contingency • Progress • Darwinism • Research programs

---

T. Shanahan (✉)
Department of Philosophy, Loyola Marymount University, Los Angeles, CA 90045, USA
e-mail: tshanahan@lmu.edu

R.G. Delisle (ed.), *The Darwinian Tradition in Context*,
https://doi.org/10.1007/978-3-319-69123-7_2

1 Introduction

Richard Dawkins (b. 1941) and Stephen Jay Gould (1941–2002) are among the best-known evolutionists of the last half-century, each having produced an impressive stream of scholarly and popular works intended to educate readers about the nature of science and to persuade them to accept their respective interpretations of evolution. Although they agree on many issues, they disagree in significant ways on a range of issues fundamental to our understanding of evolution.[1] A critical comparison of their strikingly different views promises to illuminate not only the character of the Darwinian tradition (or traditions) in the twentieth century but also the interpretive nature of scientific knowledge more generally.

Understanding Dawkins' and Gould's divergent Darwinian agendas requires situating them in relation to a pair of parallel, culturally inflected research traditions descended from Darwin's own polymorphic evolutionary theorizing. Darwin expressed his understanding of evolution in ways that (like species diverging from a common ancestor) permitted subsequent theorists to develop his ideas in markedly different directions while viewing themselves as remaining within the Darwinian clade. As Delisle (2017) observes, "Darwin does not provide for the evolutionists of the future a unified view of evolution, but instead offers a whole range of tools and concepts from which one can individually pick." Consequently, identifying some of the theoretical branching points in Darwin's view (in Sect. 2) will prove useful for comparing, contrasting, and explaining their differential expressions in the work of Dawkins and Gould (Sects. 3, 4, 5 and 6). We can then draw upon these comparative analyses to assess the significance of the Dawkins–Gould dispute for understanding the nature of the Darwinian tradition in the twentieth century and for the interpretive nature of scientific knowledge more generally (Sect. 7). I will argue that the Darwinian tradition has a distinctive "hard core" that differentiates it from other approaches to understanding life but also possesses ample conceptual resources to permit biologists to develop this tradition in divergent ways while legitimately representing themselves as carrying on and extending Darwin's seminal work, thereby endowing "Darwinism" with a remarkable capacity to continually adapt and evolve.

2 Darwin's Polymorphic Theorizing

Depending upon how generously one understands the extension of the word "evolution," theories of biological evolution predate publication of *On the Origin of Species* (1859) anywhere from decades to millennia. By the mid-nineteenth century, a belief in the *fact* of evolution, in some form, was common. Darwin's most important contribution was the idea of *natural selection* and his detailed argument, supported by facts

[1] Although Gould died in 2002, for consistency I will continue to refer to both biologists in the present tense.

culled from diverse domains, that it offers the best explanation for organisms' remarkable appearance of having been intelligently designed (and, significantly, for *deviations* from perfection) and for the tendency of new species to arise from preexisting species via a gradual process of "descent with modification." The basic idea is simple enough (in retrospect). Living things tend to differ slightly from one another in ways that confer on some a small advantage in the struggle for survival and reproduction. Some of these characteristics are heritable and are passed on to offspring, who in turn exhibit differential fitness with respect to their own (often slightly different) environments. Over time, kinds of living things become better adapted to their diverse environments and tend to further diverge from one another. Adaptation and diversification are thereby explained by appeal to natural causes alone.

That bare-bones outline is accepted by all Darwinians, yet it embodies many unresolved puzzles, the pursuit of solutions to which has been the driving force in the development of evolutionary biology since Darwin. Among these puzzles are fundamental questions concerning the units of natural selection, the scope of adaptation, the significance of chance, and the reality of evolutionary progress (see Shanahan 2004). A brief review of Darwin's views on these issues is essential for understanding their subsequent differential development in the work of Richard Dawkins and Stephen Jay Gould.

2.1 Darwin on Natural Selection

First, consider Darwin's characterization of *natural selection*. In all six editions of the *Origin*, he maintains that "natural selection works solely by and for the good of each being" (Darwin 1859: 489; 1959: 758). But for the good of *which being(s)* does natural selection work? There are many kinds of biological entities, from cells to organisms to species to ecosystems. Darwin generally thought of natural selection as discriminating among, and thereby ultimately being for the good of, *individual organisms*. In a pack of wolves, for example, the swiftest and slimmest will be the most effective predators, and hence selection will favor individual wolves possessing such characteristics (Darwin 1859: 90). But Darwin realized that explanations in terms of individual advantage alone are limited. For example, in Chapter VII of the *Origin*, he considers "one special difficulty, which at first appeared to me insuperable, and actually fatal to my whole theory. I allude to the neuters or sterile females in insect-communities" (Darwin 1859: 236). Why this should be a problem for Darwin's theory is clear. Sterile individuals, by definition, do not reproduce. Instead, they appear to sacrifice their reproductive interests to serve the interests of the hive or colony. If natural selection can operate only on individuals that pass on their characteristics, it is difficult to see how sterile castes can be products of evolution. Yet eusocial insects, with their sterile castes, are among the most widespread and successful living systems on earth—a great puzzle, indeed.

Despite the serious threat it appeared to pose to his theory, Darwin thought that the problem of sterile castes could be handled rather easily: "[I]f such insects had been social, and it had been *profitable to the community* that a number should have been annually born capable of work, but incapable of procreation, I can see no very great difficulty in this being effected by natural selection" (Darwin 1859: 236; emphasis added). Here, at least, Darwin was willing to entertain the idea that there could be selection for characteristics beneficial to the community, even though they were of no use (and actually detrimental) to the fitness of the individuals possessing those characteristics. Whether this process involved selection operating at the individual level, or a special form of selection operating on more inclusive organizational levels, remained unclear (perhaps even to Darwin himself) and was left for others to work out.

2.2 *Darwin on Adaptation*

Second, consider Darwin's treatment of *adaptation*. Natural selection is said by him to work "for the good of each being." But as resulting from a blind, unguided process, *how good* should one expect the products of such adaptation to be? On the one hand, Darwin was fond of describing adaptations as "perfect" when he wanted to emphasize "the beauty and infinite complexity of the coadaptations between all organic beings, one with another and with their physical conditions of life, which may be effected in the long course of time by nature's power of selection" (Darwin 1859: 109). Indeed, sometimes when he used the word "perfection" he meant it *literally*. In the *Origin*'s chapter on "Instinct," he devotes twelve pages to providing a speculative reconstruction of the evolution of the cell-making instinct of hive-bees. Such bees have succeeded in solving a difficult mathematical problem—that of constructing a hive that will hold the greatest quantity of honey while using the least amount of wax. They solved the problem by constructing hexagonal cells that fit together with no wasted intercellular spaces. As Darwin (1859: 235) remarks, "Beyond this stage of perfection in architecture, natural selection could not lead; for the comb of the hive-bee, as far as we can see, is absolutely perfect in economizing wax." On the other hand, he was aware that living things generally will *not* attain biological perfection and indeed in many instances fall far short of this high standard. Vestigial and rudimentary organs (e.g., the human appendix and male nipples) are classic examples. Indeed, "Organs or parts in this strange condition, bearing the stamp of inutility, are extremely common throughout nature" (Darwin 1859: 450). Therein lay the puzzle: Why does selection produce absolute perfection in some cases but not in others? What degree of perfection should we expect, and what factors prevent some living things from achieving perfection? Again, Darwin begat the problem but ultimately left it unresolved.

2.3 Darwin on Chance

Third, consider Darwin's understanding of the role of *chance* in evolution. What many of his contemporaries found most objectionable about his theory was not evolution *per se* or even natural selection, but rather the idea that the entire process depends on *chance* variations, thus leaving evolution bereft of a preordained goal or even an inherent direction. Darwin seemed to make evolution more haphazard than anyone before him had dared to imagine (Shanahan 1991).

"Chance" also enters his theory in another important way, one that underscores the *historical* nature of evolution. As he inferred from his biogeographical studies, present-day organisms bear the marks of contingent historical events. That long ago one or a few birds were blown off course during a storm and were stranded on a remote island was a purely contingent event; no law of nature dictates that this must happen. But given the right conditions and sufficient time, such accidental colonizers may evolve into distinct species. Thus, the origin of new species will be *governed* by natural laws, but will not be *predictable* from the knowledge of such laws, as Darwin explained using a striking simile: "Throw up a handful of feathers, and all must fall to the ground according to definite laws; but how simple is the problem where each shall fall compared to the action and reaction of the innumerable plants and animals which have determined, in the course of centuries, the proportional numbers and kinds of trees now growing on the old Indian ruins!" (Darwin 1959: 75). What is true for those trees growing on the old Indian ruins is true in spades for species over millions of years of undirected evolution. Evolutionary change is both lawlike *and* subject to innumerable historical, chance events. Yet, although the notion of chance is fundamental to Darwin's theory, by his own admission he had difficulty grasping its precise role. In a 22 May 1860 letter to the American botanist Asa Gray, he confided: "I am inclined to look at everything as resulting from designed laws, with the details, whether good or bad, left to the working out of what we may call chance. Not that this notion *at all* satisfies me. I feel most deeply that the whole subject is too profound for the human intellect. A dog might as well speculate on the mind of Newton" (Darwin 1993, vol. 8: 224). Darwin recognized this basic property of evolution but never fully explained *which* features of the evolutionary process are predictable and which are contingent and in principle unpredictable.

2.4 Darwin on Evolutionary Progress

Finally, consider *evolutionary progress*. On the one hand, Darwin again and again expresses confidence that "natural selection is ... silently and insensibly working, whenever and wherever opportunity offers, at the *improvement* of each organic being in relation to its organic and inorganic conditions of life" (Darwin 1859: 84; emphasis added). Indeed, "The inhabitants of each successive period in the world's history have beaten their predecessors in the race for life, and are, in so far, *higher* in the scale of nature"—a fact which accounts for "that . . . sentiment, felt by many paleontologists, that organization on the whole has progressed" (Darwin 1859: 345; emphasis added). On the other hand, he also seems to categorically *reject* talk of "higher" and "lower." In the third edition of the *Origin* (1861), he rhetorically asks: "[W]ho will decide whether a cuttle-fish be higher than a bee?" (Darwin 1959: 550). By the sixth edition (1872), he was prepared to answer that question with a degree of confidence that seems to leave no doubt about his position: "To attempt to compare members of distinct types in the scale of highness seems hopeless; who will decide whether a cuttle-fish be higher than a bee, that insect which the great Von Baer believed to be 'in fact more highly organized than a fish, although upon another type'?" (Darwin 1959: 550) Moreover, he was very much concerned to distance his view from Lamarck's "law of progressive development." In an 11 January 1844 letter to Joseph Hooker, he wrote: "Forfend me from Lamarck nonsense of a 'tendency to progression'! But the conclusions I am led to are not widely different from his; though the means of change are wholly so" (Darwin and Seward 1903, vol. I: 41). Statements like these clearly illustrate the problem concerning evolutionary progress bequeathed by Darwin to later biologists. Progress is real (in some hard-to-define sense), but its nature and causes are wholly different from those previously attributed to it.

2.5 Darwinian Puzzles

All of the unresolved theoretical issues just briefly discussed are summed up in Darwin's remarkable claim, expressed *verbatim* in all six editions of the *Origin*, that "As natural selection works solely by and for the good of each being, all corporeal and mental endowments will tend to progress towards perfection" (Darwin 1859: 489, 1959: 758). This is a stirring summary statement of astounding scope and significance. But it leaves many questions of fundamental importance unresolved. *For the good of which being(s)* does natural selection work? *How perfectly adapted* should we expect these beings to be? How should we understand the relationship between lawlike and chance *tendencies* in evolutionary change? How, if at all, should *evolutionary progress* be characterized? To point out that there are unresolved issues in Darwin's view is not to criticize his magnificent accomplishment. On the contrary, it reflects the fact that in forging a novel perspective, some of his ideas were bound to be inchoate. Moreover, the fact that biologists continue to debate these issues suggests that nature itself speaks ambiguously on them. As we shall see, Dawkins' and Gould's disagreements about each of these issues reflect divergent interpretations of Darwin's polymorphic theorizing.

3 Dawkins and Gould on Natural Selection

3.1 *Selfish Genes*

Evolutionists since Darwin generally have followed him in viewing natural selection as operating primarily on individual organisms, and *perhaps* occasionally on groups of organisms as well, with a few biologists (e.g., Wynne-Edwards 1962) taking group selection to be both common and important. Richard Dawkins argues that there is a more penetrating and powerful view, namely, that *genes*—not organisms, and certainly not groups or species—are the "beings" (to use Darwin's term) for whose good natural selection works. As he memorably puts it in one essay: "Birds' wings are obviously 'for' flying, spider webs are for catching insects, chlorophyll molecules are for photosynthesis, DNA molecules are for... What *are* DNA molecules for? .... [This] is the forbidden question. DNA is not 'for' anything.... all adaptations are for the preservation of DNA; DNA just *is*" (Dawkins 1982a: 45). Previously some biologists (e.g., Williams 1966) had explicitly proposed such a view, and it was perhaps implicit in the seminal work of R. A. Fisher (1930), but in *The Selfish Gene* (1989a) Dawkins made it into a powerful organizing first principle for addressing a range of biological puzzles, from the origin of life to altruism to the social behaviors of animals (see also Alcock 2017). He deployed two kinds of arguments in support of the "selfish gene" view.

First, according to Dawkins, only genes have the requisite properties to function as "units of selection" and thereby to be the ultimate beneficiaries of natural selection. Genes (usually) replicate faithfully, exist in large numbers in virtue of their many copies in a population, and persist for long periods of time. Genotypes, organisms, and groups, by contrast, are ephemeral, short-lived entities whose components are repeatedly reshuffled, exist in far fewer numbers, and can be said to replicate in only a very loose sense. According to Dawkins (1989a: 34), "[T]he individual [organism] is too large and too temporary a genetic unit to qualify as a unit of natural selection. The group of individuals is an even larger unit. Genetically speaking, individuals and groups are like clouds in the sky or duststorms in the desert. They are temporary aggregations or federations." Only genes are preserved intact from one generation to the next; hence, only genes have the properties necessary to be the units of selection.

Second, the selfish gene view has unrivaled explanatory *power* and *scope*. Darwin struggled to explain the existence of sterile castes in the eusocial insects by a vague appeal to what would be "profitable to the community." But William D. Hamilton (1964), one of Dawkins' intellectual heroes, showed how sterile insect castes could evolve and be maintained in terms of selection operating at the level of shared genes within the peculiar haplo-diploid reproductive systems of eusocial insects. Hamilton's key insight was that these sterile individuals are unusually closely related to fertile members of the colony. Although themselves reproductively sterile, by helping their fertile relatives to survive and reproduce they assist in the propagation of copies of their own *genes*, many of which are shared with close relatives. Such a process [later dubbed "kin selection" by John Maynard Smith (1964)] obviates the need to postulate selection at some higher biological level. Dawkins' insight was to realize that this striking explanatory success has far-reaching implications. Whereas only *some* biological phenomena can be explained in terms of selection operating at the level of organisms, *every* such phenomenon, Dawkins contends, can be explained in terms of selection operating at the level of genes. The selfish gene view therefore provides a *deeper explanation* and a *more general theoretical perspective* than any of its theoretical alternatives (see Shanahan 1997).

3.2 The Invisibility of Genes

Across the Atlantic, Gould was not convinced. He claimed to find an elementary flaw in the selfish gene theory: "No matter how much power Dawkins wishes to assign to genes, there is one thing he cannot give them—direct visibility to natural selection. Selection simply cannot see genes and pick among them directly. It must use bodies as an intermediary. A gene is a bit of DNA hidden within a cell. Selection views bodies" (Gould 1980a: 90). Moreover, Gould claimed that the selfish gene view grossly misconstrues the relationship between genes and bodies: "Bodies cannot be atomized into parts, each constructed by an individual gene" (Gould 1980a: 91).[2] Even if the one gene/one body part view *were* true, the selfish gene view would still be flawed, Gould contended, because it is the whole organism, rather than the individual gene, that is naturally selected. Gould attributed the fascination generated by Dawkins' view to "some bad habits of Western scientific thought—from attitudes ... that we call atomism, reductionism, and determinism" (Gould 1980a: 91–92). By contrast, his own evolutionary perspective is proudly hierarchical: "The world of objects can be ordered into a hierarchy of ascending levels.... Different forces work at different levels" (Gould 1980a: 85). Insofar as Darwin (usually) thought of selection as operating on individual organisms rather than on discrete units of heredity (of which he knew nothing), Gould could claim to be more "Darwinian" than Dawkins on this point. Indeed, Gould saw himself as restoring the organism to the central role assigned to it by "the orthodox, Darwinian view" (Gould 1980a: 85). Endorsing David Hull's (1976) pithy formulation, he declared that "genes mutate, organisms are selected, and species evolve" (Gould 1980a: 85). Fifteen years later, Gould was still chastising Dawkins as a "strict Darwinian zealot ... who's convinced that everything out there is adaptive and a function of genes struggling. That's just plain wrong, for a whole variety of complex reasons" (Brockman 1995: 63). The battle between "orthodox" and "zealous" [latter dubbed by Gould (1997a) "fundamentalist"] Darwinian visions was well under way.

[2]See MacCord and Maienschein (2017) for a contemporary critique of the overemphasis on the role of genes as the locus of explanation for development and evolution.

3.3 *Replicators and Vehicles*

It did not take long for Dawkins (1982a: 47) to strike back, emphasizing that insisting on the causal primacy of genes "does not mean, of course, that genes ... literally face the cutting edge of natural selection. It is their phenotypic effects that are the proximal subjects of selection." Differences in genes give rise to differences at the phenotypic level, resulting in the differential propagation of the genes responsible for those phenotypes. Natural selection operates directly on "vehicles" (i.e., phenotypes), but it is the indirect effects on the differential fate of "replicators" (i.e., genes) that is crucial for understanding evolutionary change. Evolution is essentially a contest in which genetic replicators vie with each other by constructing bodies by which they lever themselves into subsequent generations. Moreover, Dawkins disavowed the idea that the selfish gene theory requires that there be a simplistic one-to-one mapping of genes to phenotypic characteristics. It is quite enough, he pointed out, that *differences* among genes be responsible for *differences* at the phenotypic level.

4 Dawkins and Gould on Adaptation

4.1 *Spandrels and the Panglossian Paradigm*

Darwin was convinced that natural selection is a perfecting agent, yet left unresolved the issue of *how* perfect one should expect the products of natural selection to be. At least two questions in this regard need to be distinguished, pertaining to the *scope* and the *depth* of adaptation. First, should *every* phenotypic characteristic be considered an adaptation? Second, is every bona fide adaptation *optimal*?[3] In a widely cited paper, "The Spandrels of San Marco and the Panglossian Paradigm: A Critique of the Adaptationist Programme" (1979) (coauthored with his Harvard colleague Richard Lewontin), Gould answers both questions with a resounding "No." The first part of the paper's title comes from a comparison of some organismal traits to certain architectural features of St. Mark's Basilica in Venice. Spandrels are described by Gould as the tapering triangular spaces that arise as the necessary architectural by-products of mounting a dome on rounded arches meeting at right angles. Each of the spandrels in St. Mark's is decorated with a Christian motif. One ignorant of architectural necessity might suppose that the spandrels exist in order to provide spaces for the depiction of religious themes. But according to Gould, one would be dead wrong. The spandrels came into existence for inescapable architectural reasons and were *then* pressed into service for religious purposes; the fact that they provide suitable surfaces for religious iconography in no way *explains* their existence. Gould claims that biologists make an

[3]Other questions include whether biological entities above or below the level of the individual organism can be, and sometimes are, the *bearers* or "owners" of adaptations.

analogous mistake in their analysis of organisms when they uncritically assume that every phenotypic characteristic exists because it serves some adaptive purpose, thereby ignoring the "architectural constraints" that delimit the structures of organisms. By simply assuming that all characteristics are adaptive, "ultra-adaptationists" (like Dawkins) fail to distinguish between the current utility of a phenotypic characteristic and the real evolutionary reasons for that characteristic's existence in the first place.

The second part of the title of the "Spandrels" paper refers to Dr. Pangloss in Voltaire's satire, *Candide*, who assumed that whatever exists (e.g., earthquakes and all the rest) does so because it is for the best. So too, Gould maintains, evolutionary biologists are prone to exhibit unlimited "faith in natural selection as an optimizing agent" (Gould and Lewontin 1979: 147). The only brake ever admitted on the perfection of each trait consists in trade-offs among competing selection pressures: "Any suboptimality of a part is explained as its contribution to the best possible design for the whole. The notion that suboptimality might represent anything other than the immediate work of natural selection is usually not entertained" (ibid: 151). Even non-optimality is thereby accounted for in terms of selection-driven adaptation. Moreover, "This program regards natural selection as so powerful and the constraints upon it so few that direct production of adaptation through its operation becomes the primary cause of nearly all organic form, function, and behavior" (ibid: 150–151). A telltale symptom of this unquestioned assumption is the failure to even consider various non-adaptationist explanations for biological structures. Gould also hints at his preferred alternative approach, one with a distinguished European pedigree (Levit and Hossfeld 2017). Instead of viewing organisms as suites of interchangeable, atomized characteristics, he maintains that "organisms must be analyzed as integrated wholes, with *Baupläne* (fundamental body plans) so constrained by phyletic heritage, pathways of development, and general architecture that the constraints themselves become more interesting and more important in delimiting pathways of change than the selective force that may mediate change when it occurs" (ibid: 147). Significantly for the broader concerns of the present paper, Gould explicitly associates this perspective with "Darwin's own pluralistic approach to identifying the agents of evolutionary change" (ibid: 147).

4.2 Adaptationism Reasserted

Dawkins is not cited in the Spandrels paper, but he may well have taken his own approach to be among the primary targets of its pointed criticisms. Only a few years after that paper appeared, he explicitly addressed the issue of "Constraints on Perfection" in his book *The Extended Phenotype* (1982b), mentioning the authors of the Spandrels paper in the very first paragraph and then responding to them, singularly and together, throughout. He argues on theoretical grounds that we should *not* expect optimal adaptations, nor is such optimality empirically confirmed. Living things are, after all, products of blind processes. Although Darwin is not explicitly referenced, Dawkins' conclusion is exactly the same as *one* of Darwin's, with which he was surely familiar: "Natural selection will not produce absolute perfection, nor do we always meet, as far as we can judge, with this high standard under nature" (Darwin 1859: 202). (For further discussion, see Shanahan 2008.)

Having explained why one should *not* embrace the form of ultra-adaptationism critiqued by Gould, Dawkins nevertheless emphasizes in subsequent works that the adaptations of living things are, far more often than is generally appreciated, incredibly well designed. For example, the chapter entitled "Good Design" in *The Blind Watchmaker* (1986) is a *tour de force* in conveying the stupefyingly impressive adaptations that permit insectivorous bats to locate and capture prey. Natural theologians like the Rev. William Paley, author of *Natural Theology, or Evidences of the Existence and Attributes of the Deity* (1802), sought to show that a careful examination of living things provides indisputable proof of a divine Designer. Dawkins, of course, rejects Paley's specific explanation for the appearance of design. But he nonetheless thinks that Paley was right to emphasize living things' *appearance* of having been intelligently designed. The emphasis throughout the chapter and indeed the entire book is on the fact that living things have the sort of astonishingly complex "design" (i.e., adaptations) that an intelligent designer *would* impart if such a being was trying to make a nearly perfect machine of that sort; yet such astounding results have been achieved without any conscious agency whatsoever.[4]

4.3 Odd Arrangements and Funny Solutions

Whereas for Dawkins complex organic "design" is *the* preeminent biological datum requiring scientific explanation, Gould finds biological *oddity* and *poor design* to be far more significant for understanding the nature of Darwinian evolution. His essay "The Panda's Thumb" is a striking case study in historically constrained biological *imperfection* that is said to provide powerful evidence for Darwinian evolution—precisely *because* the panda's "thumb" (an extension of the radial sesamoid bone) manifests biological imperfection. In stark contrast to Dawkins' perspective, Gould writes that: "[I]deal design is a lousy argument for evolution, for it mimics the postulated action of an omnipotent creator. Odd arrangements and funny solutions are the proof of evolution—paths that a sensible God would never tread but that a natural process, constrained by history, follows perforce" (Gould 1980a: 20–21). In another essay, he explains: "[Y]ou cannot demonstrate evolution with perfection because perfection need not have a history" (Gould 1980a: 28). For Gould, historical factors trump functional factors in explaining the most interesting aspects of life.

[4]Segerstråle (2006, p. 88) interprets *The Blind Watchmaker* as a whole as Dawkins' response to Gould's critique of adaptationism. This may be going too far, but Gould is certainly *a* target.

The differential importance Gould and Dawkins attach to the "historical" manifests itself in other ways as well. Dawkins is impressed by living things' seemingly limitless ability to adapt to new challenges, especially those posed by other living things, remarking: "I believe that there's not a lot that genes can't achieve in the way of small-scale, gradual, step-by-step change from what's already there" (Brockman 1995: 81). By contrast, Gould is impressed by constraints that place limits on evolutionary change, maintaining that: "There are certain pathways that are more probable, and there are certain ones that aren't accessible, even though they might be adaptively advantageous. It really behooves us to study the influence of these structural constraints upon Darwinian and functional adaptation; these are very different views" (Brockman 1995: 53).

5 Dawkins and Gould on Chance

5.1 A Minor Ingredient in the Darwinian Recipe

The notion of *chance* is fundamental to Darwin's conception of evolution, yet by his own admission he found it difficult to explain its precise role, thereby rendering his theory vulnerable to endless misunderstanding and misrepresentation. For example, creationists argue that "random evolution" could never explain the beautifully designed features of living things, to say nothing of uniquely human characteristics. They are right, of course, but their facile mistake, as Dawkins points out with undisguised exasperation, is "to believe that Darwinism explains living organization in terms of chance ... alone. This belief, that Darwinian evolution is 'random', is not merely false. It is the exact opposite of the truth. Chance is a minor ingredient in the Darwinian recipe" (Dawkins 1986: 49). By contrast, "the most important ingredient" of Darwinian evolution, in Dawkins' view, is *cumulative selection*, "which is quintessentially *non*random" (ibid: 49; emphases in original). Cumulative selection is simply the iterated operation of natural selection whereby the accumulation of small changes over time results in significant evolutionary change. Always armed against the doubters of Darwinism, Dawkins is concerned to show that slight, chance improvements in functionality can accumulate to produce the astoundingly complex adaptations of living things we observe. Chance variations are crucial to this process, but all the heavy lifting involved in forging adaptations is done by natural selection, a nonrandom process.

5.2 Lucky Breaks

Gould's primary concerns lay elsewhere, in the vast expanse of the history of life, a history that is characterized by unpredictable twists and turns. In his book *Wonderful Life: The Burgess Shale and the Nature of History* (1989), he encourages readers to think of life on earth as shot through with *contingency*.[5] Fossil remains in the Burgess Shale of British Columbia reveal a bonanza of long-extinct *phyla*, the likes of which have not existed for half a billion years since the mysterious Cambrian Explosion, dubbed evolution's "big bang." Why did these bizarre body plans flourish and then suddenly vanish? No one knows. But according to Gould (1989: 47), it was a genuine *decimation* in the sense that those that left descendants were a minute, *random* sample of those that had previously flourished. An ultra-adaptationist, Gould points out, would interpret this pruning of the tree of life as yet another example of natural selection in action, no doubt insisting that "all but a small percentage of Burgess possibilities succumbed, but the losers were chaff, and predictably doomed. Survivors won for cause—and cause includes a crucial edge in anatomical complexity and competitive ability" (ibid: 48). Against this ultra-adaptationist interpretation, Gould insists, those that survived were just the beneficiaries of *lucky breaks*; consequently, their distant descendants (including us) are merely the products of "a thousand ... happy accidents" (ibid: 48). The survival of entire phyla often depends more on *luck* than on fitness. Were it possible to restart the evolutionary process from its beginning, there is every reason to conclude that an entirely different biota would evolve. Contingency rules over Darwinian evolution.

5.3 Convergence

In response, Dawkins essentially accused Gould of grossly exaggerating the significance of some well-known facts. In a withering review of *Wonderful Life*, first published in 1990, Dawkins (2003: 205) writes: "Since, for Gould, the Cambrian was peopled with a greater cast of phyla than now exist, we must be wonderfully lucky survivors. It could have been our ancestors who went extinct.... We came 'that close' to not being here. Gould expects us to be surprised. Why? The view that he is attacking—that evolution marches inexorably towards a pinnacle such as man—has not been believed for years." Elsewhere Dawkins (1986) had already considered, and rejected, the claim that, were the evolutionary process to be restarted from its beginning, an *entirely* different biota would evolve. On the contrary, he noted: "It is ... a striking testimony to the power of natural selection ... that numerous examples can be found in real nature, in which independent lines of evolution appear to have converged, from very different starting points, on what looks like the same endpoint" (Dawkins 1986: 94). In *Climbing Mount Improbable* (1996: 19–22), Dawkins argues that eyes have evolved independently a number of times because organs for seeing are likely to be useful under a wide array of

[5]Later, in his final major work, Gould (2002: 47) defines "contingency" as "the tendency of complex systems with substantial stochastic components, and intricate nonlinear interactions among components, to be unpredictable in principle from full knowledge of antecedent conditions, but fully explainable after time's actual unfoldings" [*sic*].

recurring circumstances. Replay life's tape and it is indeed unlikely that the same *species* would evolve again, but it is overwhelmingly likely that evolution would again produce organisms with organs for detecting light—*and* functional types such as autotrophs, herbivores, carnivores, scavengers, parasites, etc. In his view, Gould fails to understand that the fundamental nature of Darwinian evolution makes it likely that organisms playing the same ecological roles would invariably arise again because selection channels chance variations into broadly predictable paths. In Dawkins' view, the conclusions that Gould draws from his "replaying the tape of life" thought experiment simply do not follow from, and indeed are contradicted by, the basic principles of Darwinian evolution.

6 Dawkins and Gould on Progress

6.1 A Noxious Idea

Darwin's view of evolutionary progress is best described as *guarded.* He was confident that natural selection improves the beings on which it operates, making organisms that appear later in an evolving lineage "higher" in the scale of nature than their predecessors in the same lineage. But he was contemptuous of a Lamarckian "tendency to progression" and consequently dismissive of any attempt to rank as higher or lower organisms of different "types." Still, he believed that he discerned a real, if qualified, sense in which evolution manifests progress (Shanahan 2000). Gould entertains no such qualifications. As he explains in the first sentence of one essay, "Progress is a noxious, culturally embedded, untestable, nonoperational, intractable idea that must be replaced if we wish to understand the patterns of history" (Gould 1988: 319). Writing eight years later, he adamantly denies "that progress characterizes the history of life as a whole, or even represents an orienting force in evolution at all" (Gould 1996: 3). At least five distinct arguments for these claims can be extracted from the latter work.

First, we humans have a lamentable, albeit understandable, tendency to place ourselves atop nature's hierarchy and to arrange all other living things somewhere down the evolutionary ladder. The very fact that we are so predisposed to believe in progress, and to place ourselves at evolution's pinnacle, should render this belief deeply suspect. Second, there is nothing about the evolutionary process *per se* that would make progress inevitable, or even likely. Instead, the history of life is rife with chance, contingency, and historicity, making each stage in the process utterly unpredictable given what came before. Third, because life necessarily began in a simple, relatively uncomplicated form, the only regions of morphospace available for colonization were those for more complex organisms. Organisms became more complex, not because increased complexity was "better," but just because there was nothing else to do *but* to become (on average) more complex. Fourth, evolutionary progress is an illusion because bacteria and insects far outnumber mammals.

Finally, evolution should be viewed as "a history of change as the increase or contraction of variation in an entire system (a 'full house'), rather than as a 'thing' moving somewhere" (Gould 1996: 146). In short, progress is an illusion, albeit a seductive one.

6.2 Dyed-in-the-Wool Progress

In a scathing review of Gould's *Full House*, Dawkins agrees that "complexity, braininess and other particular qualities dear to the human ego should not necessarily be expected to increase progressively in a majority of lineages" (Dawkins 1997: 1018), but nonetheless finds fault with Gould's broader critique of evolutionary progress: "Why should any thoughtful Darwinian have expected a majority of lineages to increase in anatomical complexity? Certainly it is not clear that anybody inspired by adaptationist philosophy would" (ibid: 1017). In his view, "Gould is wrong to say that the appearance of progress in evolution is a statistical illusion" (ibid: 1018) because there is an alternative, and far more plausible, way of construing evolutionary progress, namely, as "a tendency for lineages to improve cumulatively their adaptive fit to their particular way of life, by increasing the numbers of features which combine together in adaptive complexes" (ibid: 1016). "By this definition," Dawkins writes, "adaptive evolution is not just incidentally progressive, it is deeply, dyed-in-the-wool, indispensably progressive" (ibid: 1017). For example, "The evolution of the vertebrate eye *must* have been progressivee.... Without stirring from our armchair, we can see that it must be so" (ibid: 1018; emphasis in original). Evolutionary progress, which does not require the baggage Gould attempts to saddle it with, is thus quite real.

6.3 Not Evolution's Defining Feature

The Gould–Dawkins debate over evolutionary progress may be a classic case of interpreting the same facts through the lenses of two different conceptual frameworks. For his part, Gould (1996: 197) grudgingly acknowledges the fact of increasing *complexity* in the history of life, but insists that this should not be regarded as evolution's "defining feature," for two reasons. First, although increasing complexity (on average) *is* an undeniable a feature of evolution, it is not a pervasive feature of most lineages. Second, increasing complexity, where it occurs, arises as an incidental by-product of processes whose causes do not include a mechanism for progress or increased complexity. Dawkins, likewise, believes that complexity (on average) has increased over time but interprets this increase as an inevitable consequence of a mechanism, natural selection, which may bias evolution in that direction. Consequently, while agreeing on many of the *facts*, Dawkins and Gould nonetheless fundamentally disagree on the *significance* of these facts for understanding Darwinian evolution. Perhaps more clearly than in any other area, their dispute over the reality of evolutionary progress demonstrates that simple appeals in science to "the evidence" are sometimes insufficient to resolve fundamental theoretical issues because it is precisely the *interpretation* of the evidence that is at issue.

7 Dawkins, Gould, and Darwinian Traditions in the Twentieth Century

So far we have considered Dawkins' and Gould's alternative, and often diametrically opposed, views on a range of fundamental issues concerning evolution, along with their stated reasons for holding such views. Without in the least minimizing the importance of those reasons, we can also delve more deeply into the different contextual factors and associated methodological agendas that contribute to such divergent interpretations and applications of Darwinism.[6] Chief among these factors are different disciplinary priorities and culturally inflected research agendas.

7.1 Disciplinary Priorities and Culturally Inflected Research Agendas

In the 1960s, Dawkins was a student in Oxford of Niko Tinbergen (1907–1988), one of the founders of ethology, a biological subdiscipline that aims to understand the adaptive significance of animal behavior in the context of an animal's natural environment, and hence a field of inquiry that takes adaptationism as a central organizing principle. Its limitations (e.g., as pointed out by Gould and Lewontin) notwithstanding, adaptationism is unarguably a powerful heuristic in the study of animal behavior—one that Dawkins thoroughly absorbed in his scientific training. He was also ideally situated to inherit an exciting new set of ideas strongly associated with late mid-century British evolutionary theorizing. He credits William D. Hamilton (1936–2000) and John Maynard Smith (1920–2004), in particular, for introducing him to the ideas of inclusive fitness and evolutionarily stable strategies, respectively—ideas around which much of *The Selfish Gene* is organized. Dawkins' work also reflects key ideas and ideals associated with fellow Englishman Ronald A. Fisher (1890–1962), whom Dawkins once lauded as "the greatest biologist since Darwin."[7] Fisher's "Fundamental Theorem of Natural Selection" states that, in an infinite population, "The rate of increase in fitness of any organism at any time is equal to its genetic variance in fitness at that time" (Fisher 1930: 35).

[6]For a more detailed analysis of such factors, see Shanahan (2001).

[7]http://edge.org/conversation/who-is-the-greatest-biologist-of-all-time

Although there are (as Fisher recognized) conditions under which this prediction will not be borne out (since all real biological populations are finite), the theorem nonetheless provides a basis for an inherent directionality in evolution. Not coincidentally, Fisher was also a staunch believer in evolutionary progress. Indeed, Ruse (2006: 147) describes Fisher's *The Genetical Theory Natural Selection* (1930) as "a hymn to evolutionary progress." Dawkins' belief that evolution *must* be progressive has a strong Fisherian flavor. Like Fisher, Dawkins begins with an idealized conception of the evolutionary process as adaptive change powered by natural selection and logically deduces the necessary consequence: organisms will become progressively better adapted to their specific conditions of life.

Other seminal influences on Dawkins are less direct but no less consequential. Julian Huxley (1887–1975) managed to surpass even Fisher as an enthusiast for evolutionary progress. Like Dawkins, he studied and then taught at Oxford University. From his earliest writings (Huxley 1912: 114–115) straight through to his later writings (Huxley 1953: 31), he emphasized the objective reality of evolutionary progress and the importance of co-evolutionary arms races for understanding progressive evolution—a topic on which Dawkins would later conduct original research (Dawkins and Krebs 1979). Eventually, Huxley (1954: 11) defined evolutionary progress as consisting in the appearance of biological innovations that make possible further progress—an idea that strikingly presages Dawkins' (1989b) idea of the "evolution of evolvability." Dawkins' emphasis on arms races, adaptation, progress, and the evolution of evolvability, as well as his highly public role in the promotion of science, are all themes with striking Huxlean precedents.[8] In myriad ways, Dawkins sports a distinctively English neo-Darwinian pedigree (Kohn 2004).

Gould's Darwinian pedigree is strikingly different. In 1967, he completed a doctorate at Columbia University in evolutionary biology and paleontology—the latter a discipline that aims to understand patterns of change and diversification among (overwhelmingly extinct) biological lineages during the last 550 million years. Gould became a paleontologist at a time when paleontology still labored under a second-class professional status within evolutionary biology, being overshadowed first by population genetics in the 1940s and then by molecular biology in the 1950s. The former situation had begun to be rectified during Gould's childhood by the American paleontologist George Gaylord Simpson (1902–1984) who, in *Tempo and Mode in Evolution* (1944), sought to integrate paleontology into the congealing "modern synthesis." Simpson also combatted what he saw as the naïve anthropocentricism of evolutionary progressionists like Huxley by arguing in *The Meaning of Evolution* (1949) that "The [fossil] record has demonstrated that evolution is not some over-all cosmic influence that has been changing all living things in a regular way throughout the periods of the earth's history" (Simpson 1949: 97).

[8]What has been said about Huxley could with equal justice be said about Dawkins: "Huxley's contributions of new knowledge were far less important than his infectious enthusiasm and encouragement, as well as his ability to combine scattered concepts or ideas into general principles and meaningful visions" (Cain 2009a: 649).

Supposed instances of progression in the fossil record are merely artifacts of selective and faulty analysis of the paleontological data. In his view, the "tempo" of evolution is characterized by a diversity of evolutionary rhythms varying from one evolutionary branch and geological period to another, with contingent historical factors playing a crucial role.

Simpson's influence on Gould was profound. Like Simpson, Gould spent most of his career at the American Museum of Natural History in New York and at the Museum of Comparative Zoology at Harvard. (Prior to joining the American Museum, Simpson was a professor at Columbia University, Gould's alma mater.) Like Simpson, Gould rails against popular but (in his view) mistaken progressionist conceptions of evolution and aims to demonstrate that a critical interpretation of the fossil record renders such popular beliefs empirically untenable. Gould's deep admiration for the work of Simpson is clearly evident in his assessment of Simpson's contribution to the modern synthesis (e.g., in Gould 1980b: 120). Also, "This View of Life"—the title of Gould's long-running monthly column in *Natural History* magazine—is the title of one of Simpson's books (Simpson 1964). Indeed, at times Gould's prose is virtually indistinguishable from Simpson's. Compare Gould's denial "that progress characterizes the history of life as a whole, or even represents an orienting force in evolution at all" (Gould 1996: 3) with Simpson's nearly identical claim that "evolution is not invariably accompanied by progress, nor does it really seem to be characterized by progress as an essential feature" (Simpson 1949: 262). In crucial respects, Gould trod in Simpson's influential footsteps.[9]

Gould's understanding of evolution also owes a powerful debt to the American population geneticist Sewall Wright (1889–1988). Wright was suspicious of mathematical models that treat populations as infinite and as lacking significant internal structure and that treat chance events as relatively unimportant (Provine 1986). In Wright's view, random genetic drift—a process that characterizes all real, finite populations—may underlie the ability of biological populations to cross genetic valleys and thereby to ascend higher adaptive peaks. Gould followed Wright in his suspicion of models that fail to acknowledge the multitude of complicating factors to which real biological systems are always subject, that fail to consider the evolutionary history of evolved entities, and that downplay the pervasiveness of chance factors in evolution. The theory of punctuated equilibrium—the scientific idea for which Gould is best known—owes much to Wright's notion (later given greater prominence by Ernst Mayr via his model of allopatric speciation) that speciation may be favored by the subdivision of populations by random genetic drift into reproductive isolates that continue to diverge until new species formation is complete (Turner 2017). In these and other ways (e.g., his frequent allusions to baseball to drive home key points), Gould is a product of distinctly American influences.

[9]In time, however, Gould sought to distance his views from those of Simpson. As Cain (2009b) discusses, Gould later embarked on a campaign of "ritual patricide" against his one-time hero.

7.2 *Dawkins and Gould as "Darwinians"*

Despite their methodological and substantive disagreements, Dawkins and Gould each sees himself as representing authentic "Darwinism" and each enthusiastically (albeit selectively) appropriates Darwin for his own purposes—a rhetorical strategy that Darwin facilitated through his polymorphic theorizing. Recall the two great principles that Darwin explains and defends in the *Origin*: natural selection and descent with modification. To a first approximation, Dawkins and Gould each can be viewed as prioritizing one of Darwin's great principles over the other. For Dawkins, the most striking feature of living things requiring cogent explanation—namely, their complex organization—requires understanding how natural selection could have forged such remarkable design: "The problem is one of complex design.... Complicated things, everywhere, deserve a very special kind of explanation. We want to know how they came into existence and why they are so complicated" (Dawkins 1986: ix, 1). The answer, of course, is natural selection. For Gould, by contrast, the most important features of living things requiring explanation are patterns of similarity and diversity over immense periods of time, e.g., as evident in the fossil record. These are characteristics of biological lineages, not individual organisms, and therefore require first and foremost understanding historical patterns and processes of descent with modification: "In our Darwinian traditions, we focus too narrowly on the adaptive nature of organic form, and too little on the quirks and oddities encoded into every animal by history" (Gould 1995: 371). (See Shanahan 2011 for how Darwin attempted to reconcile these themes.)

Like observers attending to different aspects of the same Gestalt image, Dawkins and Gould naturally privilege different elements of Darwin's theory, with consequences for their respective self-identifications with "the Darwinian tradition." Dawkins, it is fair to say, always sees himself as carrying on the scientific tradition inaugurated by Darwin. He wears the "Darwinian" mantle with obvious pride while recognizing that biologists have learned much that Darwin necessarily could not have known. In his view, the most important piece missing from Darwin's understanding of evolution is modern genetics: "If only Darwin had read Mendel! A gigantic piece of the jigsaw puzzle would have clicked into place.... Darwin would have been delighted and astounded by the population genetics, the neo-Darwinism of the 1930s. It's also nice to think that he might have been pleased about kin selection and selfish genes as well" (Brockman 1995: 75). Kin selection and selfish genes are, of course, central to Dawkins' own interpretation of evolutionary theory. By judging that Darwin would have approved of these ideas, Dawkins thereby situates himself as heir to a Darwinian tradition stretching back to, and deriving authentication from, the great man himself.

By contrast, Gould's self-conception in relation to "the Darwinian tradition" is more ambiguous. Early in his career he declared that "the essence of Darwinism lies in its claim that natural selection creates the fit. Variation is ubiquitous and random in direction. It supplies the raw material only" (Gould 1977: 44). Later he came to characterize "strict Darwinism" as a rigid ideology according to which natural selection is regarded as the only important cause of evolutionary change, organisms are infinitely malleable under the influence of natural selection, micro-evolutionary processes can be extrapolated to explain all macro-evolutionary phenomena, and the history of life as a whole can be defined by a drive toward better, more complex organisms. Indeed, it is vital for "strict Darwinism" that selection operating on individual bodies explains "all major patterning forces in the history of life.... unless you can argue that Darwinian selection on bodies is, by extrapolation, the cause of evolutionary trends and of the major waxing and waning of groups through time, then you don't have a fully Darwinian explanation for life's history" (Brockman 1995: 63). Gould obviously does not consider himself a Darwinian in this "strict" (i.e., constricted) sense. Indeed, in this constricted sense, he says, even "Darwin is not a strict Darwinian" (Brockman 1995: 53). At various times, Gould (1980b, 1997b, respectively) has prophesized the demise of strict Darwinism and contrasted it with a more open, pluralistic attitude toward evolutionary principles that, he says characterized Darwin's own work.

Gould considers himself a "Darwinian" in this more expansive, pluralistic sense and speculates that were Darwin to learn of asteroid impacts, mass extinctions, and even punctuated equilibrium, he would be open to such ideas (Brockman 1995: 64). With respect to his own distinctive views on contingency and biological oddities, however, he opines that Darwin would be *fully* on board: "Darwin invoked contingency in a fascinating way as his *primary support* for the fact of evolution.... One might think that the best evidence for evolution would reside in those exquisite examples of optimal adaptation presumably wrought by natural selection.... Yet Darwin recognized that ... *the primary evidence for evolution* must be sought in quirks, oddities, and imperfections that lay bare the pathways of history" (Gould 1989: 300; emphases added).

Likewise, Gould interprets Darwin's view of evolutionary progress as virtually indistinguishable from his own, although given the social milieu in which he lived Darwin was forced to disguise his doubts about the inevitability of progress. Therefore, when Darwin expresses progressionist sentiments, they should not be understood to represent his *real* views, but rather as concessions to the then-prevailing *Zeitgeist* that had enshrined "progress" as an inevitable social law. In Gould's view, although Darwin categorically rejected any notion of evolutionary progress, he nonetheless sometimes weakened and included progressionist language in his writings so as to not upset the status quo of which he was such an indisputable beneficiary: "Darwin, the social conservative, could not undermine the defining principle of a culture ... to which he felt such loyalty, and in which he dwelt with such comfort" (Gould 1996: 141).

7.3 What Is "Darwinism"?

The fact that Dawkins and Gould can each think of himself as a "Darwinian," and that each can justify such self-identification by citing Darwin himself, while nonetheless holding such different views from one another, raises more general questions about the nature of "Darwinism" and "the Darwinian tradition." What *is*

"Darwinism"? Does it have defining features? If so, what are they? What constitutes "the Darwinian tradition"? Is it uniform or it is, like Darwin's own theorizing, polymorphic? Are "Darwinism" and "the Darwinian tradition" co-extensive, or distinct?

Some scholars maintain that "Darwinism" has something like an essential nature that distinguishes it from other understandings of evolution, e.g., those promulgated in the years following Darwin's death and right into the early twentieth century (see Bowler 2017). James Lennox (2015), for example, maintains that "Darwinism" consists in a distinctive set of concepts, principles, and methodological maxims concerning the history and diversity of life on earth, centering on five themes: (1) probability and chance; (2) the nature, power, and scope of selection; (3) adaptation and teleology; (4) nominalism vs. essentialism about species; and (5) the tempo and mode of evolutionary change. According to Lennox, it is possible to identify the Darwinian position with regard to each of the foregoing issues; Darwin and his contemporaries recognized the distinctiveness of Darwin's position on each of these topics; and these elements continue to differentiate Darwinism from rival views of evolution. Such an approach aims to distill the essence of Darwinism in all its fullness into a comprehensive but finite set of theses.

A comparatively stripped-down but still essentialist approach is taken by David Depew (2017), who takes "Darwinism" to refer to "Darwin's claim that gradual natural selection is the primary (but not the only) cause of evolutionary diversification." Absent from this spare conception is any reference to chance, the units of selection, adaptation, the nature of species, and whether evolution itself (as distinct from natural selection) is gradual. What makes something "Darwinian" on this view is just the central importance attributed to natural selection in accounting for life's diversity. As Depew recognizes, in his view T. H. Huxley, Darwin's most formidable advocate in the years following the *Origin*, yet who always doubted the paramount power of natural selection, would fail to qualify as a "Darwinian." Presumably all biologists who consider natural selection to be "*the* fundamental idea in biological evolution" (Pigliucci 2017), despite their other differences, would qualify as Darwinians in the fullest sense of that term. Dawkins almost certainly would be included; Gould, most likely, would not.

An even more liberal approach is taken by Richard Delisle (2011: 57) who treats "Darwinism broadly construed [as] any evolutionary approach that appeals to natural selection." Here, natural selection need not even be the primary explanatory concept. This inclusivist strategy permits biologists as diverse in their understandings of the evolutionary process and its implications as Julian Huxley and George Gaylord Simpson, or Richard Dawkins and Stephen Jay Gould, to equally represent "Darwinism" while differing on many fundamental issues. Likewise, various neo-Lamarckian, neo-vitalist, and "romanticist" biological theories that flourished in the early years of the twentieth century would qualify as *fully* "Darwinian" on this liberal account inasmuch as their proponents generally attributed *some* role to natural selection (Esposito 2017). Only those approaches that deny or fail to mention *any* role for natural selection would remain outside, e.g., those forms of Lamarckism that flourished in France right through the mid-twentieth century (Loison and Herring 2017). More problematic cases include creationists who grudgingly accept some role for natural selection (typically restricted to microevolution) and extraterrestrial biologists who (we might suppose) have never heard of Charles Darwin, yet nonetheless embrace the *principle* of natural selection without, presumably, calling it that.[10]

An even more accommodating approach would be to include within the Darwinian fold all those biological theories that are *compatible* with Darwin's emphasis on the importance of natural selection, even if the proponents of those theories did not see it that way. Understood in this way, "Darwinism" might encompass some theories explicitly put forth as *anti*-Darwinian (Kutschera 2017) so long as, with hindsight, logical compatibility can be established.

The foregoing approaches all seek to characterize "Darwinism" conceptually, sometimes treating "Darwinian" and "Darwinism" as logically co-extensive. David Hull (1985: 809) distinguishes between "the Darwinians" as a *social group* and "Darwinism" as a *conceptual system* and maintains that a scientist can be a Darwinian without accepting all or even a large proportion of tenets identified with Darwinism; conversely, a scientist can accept the tenets of Darwinism without being a Darwinian. For example, in various ways Thomas Henry Huxley, Asa Gray, Alfred Russel Wallace, Ernst Haeckel, Charles Lyell, and Herbert Spencer could be considered "Darwinians" inasmuch as each accepted and promoted elements of Darwin's theory. But each also *rejected* important elements of Darwin's views. Huxley preferred saltationism to Darwin's gradualist perspective. Gray reserved a place for divine guidance in the evolutionary process. Lyell could never bring himself to extend evolutionary theory to include human beings. Even Wallace, the co-discoverer of natural selection, eventually came to doubt the power of selection to account for man's spiritual nature. Michael Ruse (1979: 203) had earlier suggested that a Darwinian is "someone who identifies with Darwin, not necessarily someone who accepted all of Darwin's ideas." In this view, one can be a "Darwinian" without accepting even key elements of "Darwinism" (whatever those may be).

Hull's bifurcation distinguishes "the Darwinians" as a social group from "Darwinism" as a conceptual system. Given some of the difficulties of defining "Darwinism" conceptually, it may be tempting to collapse Hull's distinction by treating "Darwinism" as whatever it is that unites Darwinians into a cohesive social group, thereby obviating the need to define "Darwinism." As Richard Delisle (2011: 50) observes, however, the dominant historiography of evolutionary biology since Darwin classifies biologists as belonging to one or the other side of a Darwinian versus non-Darwinian divide, thereby requiring historians of biology to wield *some* principle, explicitly or implicitly, for deciding who belongs in which camp—which returns us once again to the question of what is distinctive of "Darwinism."

[10]It is worth noting that Delisle (2017) expresses skepticism about the "extreme pluralism" that Darwin presents in the *Origin* as "being reducible to a sort of neat, compact, and abstract theoretical construct."

A view that seems to capture what is usually meant by "Darwinism," without leading to counterintuitive consequences (e.g., extraterrestrial Darwinians), treats it as a scientific research program roughly as described by Imre Lakatos (1970) as consisting of an incorrigible "hard core" that distinguishes that program from competing programs, surrounded by a malleable "protective belt" that permits considerable modification of the theory's "auxiliary hypotheses." The hard core of Darwinism is the central importance accorded to natural selection. That was Darwin's most novel, influential, and enduring contribution to evolutionary theorizing. But Darwinism as a research program consists of more than that. It consists of those evolutionists, their professional affiliations, research activities, products, and beliefs that constitute a nexus of causal interactions centered on a shared recognition of the fundamental importance of the seminal scientific ideas of Charles Darwin. In this view, the evolutionary theorizing and research activities of almost all mid- to late-twentieth-century biologists, Dawkins and Gould included, constitute "Darwinism." It also includes the theorizing and research activities of virtually all contemporary evolutionists, but not that of creationists nor (presumably) that of extraterrestrial biologists. "Darwinism" in this sense can be understood as a historically evolving approach to understanding life that takes Darwin's emphasis on natural selection as its origin and point of departure, but that given Darwin's pluralistic theorizing can be, has been, and presumably will continue to be, developed in significantly different ways.

8 Conclusions

Construing Darwinism as a scientific research program leaves open the question of precisely how many Darwinian traditions there are. As in biological systematics, so, too, in the history of science, there are "lumpers" and "splitters." Lumpers who emphasize commonalities will see just one, albeit multiform, Darwinian tradition. Splitters who emphasize differences may see two or more divergent Darwinian traditions. What our discussion of the Dawkins–Gould rivalry should make clear is the fact that *scientists* often care a great deal about whether their view is, or seen to be, part of a specific scientific tradition. This fact signals something important about the power of the *idea* of such traditions to shape scientific rhetoric and research agendas. "Darwinism" as a pluralistic scientific research program that can encompass a number of identifiable Darwinian traditions is flexible enough to undergo significant additions, alterations, and adjustments while retaining its distinctive identity. Consequently, reports of the *de facto* or imminent "dissolution of Darwinism" (a phrase which, shorn of its scholarly qualifications, can easily be exploited by those promoting an anti-science agenda) at present seem premature. If the past is any guide, then barring any truly revolutionary developments, Darwinism will continue to evolve in response to the multiplicity of demands placed upon it.

References

Alcock J (2017) The behavioral sciences and sociobiology: a Darwinian approach. In: Delisle RG (ed) The Darwinian tradition in context: research programs in evolutionary biology. Springer, Cham, pp 37–60

Bowler PJ (2017) Alternatives to Darwinism in the early twentieth century. In: Delisle RG (ed) The Darwinian tradition in context: research programs in evolutionary biology. Springer, Cham, pp 195–218

Brockman J (1995) The third culture. Simon & Schuster, New York

Cain J (2009a) Huxley, Julian S. (1887–1975). In: Ruse M, Travis J (eds) Evolution: the first four billion years. The Belknap Press of Harvard University Press, Cambridge, pp 645–649

Cain J (2009b) Ritual patricide: why Stephen Jay Gould assassinated George Gaylord Simpson. In: Sepkoski D, Ruse M (eds) The paleobiological revolution: essays in the growth of modern paleontology. University of Chicago, Chicago, pp 346–363

Darwin C (1859) On the origin of species, 1st edn. John Murray, London

Darwin C (1959) The origin of species. In: Peckham M (ed) A variorum text. University of Pennsylvania Press, Philadelphia

Darwin C (1993) The correspondence of Charles Darwin. Cambridge University Press, Cambridge

Darwin F, Seward AC (eds) (1903) More letters of Charles Darwin, vol 2. John Murray, London

Dawkins R (1982a) Replicators and vehicles. In: King's College Sociobiology Group (ed) Current problems in sociobiology. Cambridge University Press, Cambridge, pp 45–64

Dawkins R (1982b) The extended phenotype. W H Freeman, Oxford

Dawkins R (1986) The blind watchmaker: why the evidence of evolution reveals a world without design. W W Norton, New York

Dawkins R (1989a) The selfish gene, revised edition. Oxford University Press, New York

Dawkins R (1989b) The evolution of evolvability. In: Langton C (ed) Artificial life. Addison-Wesley, Santa Fe, pp 201–220

Dawkins R (1996) Climbing mount improbable. W W Norton, New York

Dawkins R (1997) Human chauvinism. Evolution 51:1015–1020

Dawkins R (2003) Hallucigenia, wiwaxia and friends. In: Dawkins R (ed) A devil's chaplain: reflections on hope, lies, science, and love. Houghton Mifflin, Boston, pp 203–205

Dawkins R, Krebs JR (1979) Arms races within and between species. Proc R Soc Lond Ser B 205 (1161):489–511

Delisle RG (2011) What was really synthesized during the evolutionary synthesis? A historiographic proposal. Stud Hist Philos Biol Biomed Sci 42:50–59

Delisle RG (2017) From Charles Darwin to the evolutionary synthesis: weak and diffused connections only. In: Delisle RG (ed) The Darwinian tradition in context: research programs in evolutionary biology. Springer, Cham, pp 133–168

Depew DJ (2017) Darwinism in the 20th century: productive encounters with saltation, acquired characteristics, and development. In: Delisle RG (ed) The Darwinian tradition in context: research programs in evolutionary biology. Springer, Cham, pp 61–88

Esposito M (2017) The organismal synthesis: holistic science and developmental evolution in the English-speaking world, 1915–1954. In: Delisle RG (ed) The Darwinian tradition in context: research programs in evolutionary biology. Springer, Cham, pp 219–242

Fisher RA (1930) The genetical theory of natural selection. Clarendon Press, Oxford

Gould SJ (1977) Ever since Darwin: reflections in natural history. W W Norton, New York

Gould SJ (1980a) The panda's thumb: more reflections in natural history. W W Norton, New York

Gould SJ (1980b) Is a new and general theory of evolution emerging? Paleobiology 6(1):119–130

Gould SJ (1988) On replacing the idea of progress with an operational notion of directionality. In: Nitecki MH (ed) Evolutionary progress. University of Chicago Press, Chicago, pp 319–338

Gould SJ (1989) Wonderful life: the Burgess Shale and the nature of history. W W Norton, New York

Gould SJ (1995) Dinosaur in a haystack. Harmony Books, New York

Gould SJ (1996) Full house: the spread of excellence from Plato to Darwin. Harmony Books, New York
Gould SJ (1997a) Darwinian fundamentalism. http://www.nybooks.com/articles/1997/06/12/darwinian-fundamentalism/. Accessed 25 Feb 2017
Gould SJ (1997b) The pleasures of pluralism. http://www.nybooks.com/articles/1997/06/26/evolution-the-pleasures-of-pluralism/. Accessed 25 Feb 2017
Gould SJ (2002) The structure of evolutionary theory. The Belknap Press of Harvard University Press, Cambridge
Gould SJ, Lewontin RC (1979) The spandrels of San Marco and the panglossian paradigm: a critique of the adaptationist programme. Proc R Soc Lond Ser B 205:581–598
Hamilton WD (1964) The genetical evolution of social behavior. J Theor Biol 7:1–52
Hull DL (1976) Are species really individuals? Syst Zool 25:174–191
Hull DL (1985) Darwinism as a historical entity: a historiographic proposal. In: Kohn D (ed) The Darwinian heritage. Princeton University Press, Princeton, pp 773–812
Huxley JS (1912) The individual in the animal kingdom. Cambridge University Press, Cambridge
Huxley JS (1953) Evolution in action. Chatto & Windus, London
Huxley JS (1954) The evolutionary process. In: Huxley JS, Hardy AC, Ford EB (eds) Evolution as a process. George Allen & Unwin, London, pp 1–23
Kohn M (2004) A reason for everything: natural selection and the English imagination. Faber and Faber, London
Kutschera U (2017) Symbiogenesis and cell evolution: an anti-Darwinian research agenda? In: Delisle RG (ed) The Darwinian tradition in context: research programs in evolutionary biology. Springer, Cham, pp 302–332
Lakatos I (1970) Falsification and the methodology of scientific research programmes. In: Lakatos I, Musgrave A (eds) Criticism and the growth of knowledge. Cambridge University Press, Cambridge, pp 91–196
Lennox J (2015) Darwinism. http://plato.stanford.edu/entries/darwinism/. Accessed 25 Feb 2017
Levit GS, Hossfeld U (2017) Major research traditions in 20th century evolutionary biology: the relations of Germany's Darwinism with them. In: Delisle RG (ed) The Darwinian tradition in context: research programs in evolutionary biology. Springer, Cham, pp 169–194
Loison L, Herring E (2017) Lamarckian research programs in French biology (1900–1970). In: Delisle RG (ed) The Darwinian tradition in context: research programs in evolutionary biology. Springer, Cham, pp 243–270
MacCord K, Maienschein J (2017) Cells, development, and evolution: teeth studies at the intersection of fields. In: Delisle RG (ed) The Darwinian tradition in context: research programs in evolutionary biology. Springer, Cham, pp 289–308
Maynard Smith J (1964) Group selection and kin selection. Nature 201:1145–1147
Paley W (1802) Natural theology, or evidences of the existence and attributes of the deity, collected from the appearances of nature, 1st edn. Faulder, London
Pigliucci M (2017) Darwinism after the modern synthesis. In: Delisle RG (ed) The Darwinian tradition in context: research programs in evolutionary biology. Springer, Cham, pp 89–104
Provine WB (1986) Sewall Wright and evolutionary biology. University of Chicago Press, Chicago
Ruse M (1979) The Darwinian revolution: science red in tooth and claw. University of Chicago Press, Chicago
Ruse M (2006) Richard Dawkins and the problem of progress. In: Grafen A, Ridley M (eds) Richard Dawkins: how a scientist changed the way we think. Oxford University Press, Oxford, pp 145–163
Segerstråle U (2006) An eye on the core: Dawkins and sociobiology. In: Grafen A, Ridley M (eds) Richard Dawkins: how a scientist changed the way we think. Oxford University Press, Oxford, pp 75–97
Shanahan T (1991) Chance as an explanatory factor in evolutionary biology. Hist Philos Life Sci 13:249–269

Shanahan T (1997) Pluralism, antirealism, and the units of selection. Acta Biotheor 45:117–126
Shanahan T (2000) Evolutionary progress. BioScience 50(5):451–459
Shanahan T (2001) Methodological and contextual factors in the Dawkins/Gould dispute over evolutionary progress. Stud Hist Philos Biol Biomed Sci 32(1):127–151
Shanahan T (2004) The evolution of Darwinism: Selection, adaptation, and progress in evolutionary biology. Cambridge University Press, New York
Shanahan T (2008) Why don't zebras have machine guns? Adaptation, selection, and constraints in evolutionary theory. Stud Hist Philos Biol Biomed Sci 39(1):135–146
Shanahan T (2011) Phylogenetic inertia and Darwin's higher law. Stud Hist Philos Biol Biomed Sci 42:60–68
Simpson GG (1944) Tempo and mode in evolution. Columbia University Press, New York
Simpson GG (1949) The meaning of evolution: a study of the history of life and of its significance for man. Yale University Press, New Haven
Simpson GG (1964) This view of life: the world of an evolutionist. Harcourt Brace & World, New York
Turner DD (2017) Paleobiology's uneasy relationship with the Darwinian tradition: stasis as data. In: Delisle RG (ed) The Darwinian tradition in context: research programs in evolutionary biology. Springer, Cham, pp 333–352
Williams GC (1966) Adaptation and natural selection: a critique of some current evolutionary thought. Princeton University Press, Princeton
Wynne-Edwards VC (1962) Animal dispersion in relation to social behaviour. Oliver & Boyd, Edinburgh

The Behavioral Sciences and Sociobiology: A Darwinian Approach

John Alcock

Abstract Darwinian natural selection theory is the theory of choice for the overwhelming majority of behavioral biologists interested in the evolution of adaptive social behavior despite many challenges to the discipline of sociobiology over the years. The initial criticisms came from a group of ideologues who argued that sociobiology was based on a form of genetic determinism. This view was dismissed by working sociobiologists for several reasons but primarily because the criticism fundamentally misrepresents the goal of the discipline, which is to test hypotheses about the evolution and adaptive value of social traits and not to examine the proximate causes that influence the development of an individual's social behavior during its lifetime. Among the additional opposing views to the discipline that have been presented over the years are those held by academics who believe (1) that evolution occurs because of differences among groups (not individuals), (2) that kin selection theory (an amendment to Darwinian natural selection theory) has failed and should be dropped, and (3) that Darwinian sexual selection (a subset of natural selection theory) should be replaced by a more modern and inclusive theory of the evolutionary effects of the interactions between the sexes. This chapter debunks each of these competitors. Darwinian theory as modified over the last 50 years continues to be the basis for evolutionary research into the interactions between the sexes, helpful behavior in its various forms, especially the evolution of altruistic behavior, and all aspects of human sociality.

Keywords Adaptation • Altruism • Cooperation • Darwinism • Evolutionary psychology • Group selection theory • Natural selection theory • Selfish genes • Sexual selection theory • Sociobiology

---

J. Alcock (✉)
School of Life Sciences, Arizona State University, Tempe, AZ 85287-4501, USA
e-mail: j.alcock@asu.edu

R.G. Delisle (ed.), *The Darwinian Tradition in Context*,
https://doi.org/10.1007/978-3-319-69123-7_3

1 The History of Sociobiology

Sociobiology or the study of the adaptive value of social behavior has obvious implications for persons interested in human sociality, some of whom believe strongly that human social behavior cannot be studied properly from the perspective of the sociobiologist. And what is this supposedly improper perspective? It is that the evolution of the social behavior of all organisms, humans included, deserves to be analyzed from a modern Darwinian viewpoint [see Shanahan (2017) for why it is that behavioral biologists so often follow the "selfish gene" model of evolutionary theory]. A goal of the Darwinian adaptationist [as opposed to those who study the origin of species as brilliantly outlined by Pigliucci (2017)] is to determine whether and how an organism's social behavior contributes to the reproductive or genetic success of individuals of that species (generally referred to by evolutionists as individual "fitness").

The Darwinian approach to adaptations is rarely controversial currently [although many critiques were raised by those operating primarily before or during the premodern evolutionary synthesis—see Bowler (2017), Esposito (2017), and Kutschera (2017)]. The positive reaction to the application of Darwinian adaptationism to creatures other than humans can be illustrated by the early response to Edward O. Wilson's book *Sociobiology, The New Synthesis*. Wilson (1975b) assembled a great many findings in his massive (697 pages) treatise on the social behavior of a very broad range of animals. His contribution lay in his thorough review of this research, discussing (for example) sociality in Florida scrub jays, siphonophores, bulldog ants, and wildebeest. Wilson's synthesis depended upon the work of many pre-sociobiologists who analyzed the social behavior of their research species in the light of natural selection theory, Darwin's great contribution to biology but one whose usefulness for research scientists is not emphasized by those focusing on such things as symbiogenesis in the early evolution of eukaryotes (Kutschera 2017) and the role played in Darwinism by a modern understanding of genes (Delisle 2017). There was general agreement among behavioral biologists in 1975 that it was good to have a complete review of evolutionary research into a broad spectrum of animal social behaviors and the mechanisms that provided an adaptive foundation for these behaviors.

Indeed, Wilson's summary of what evolutionary biologists had discovered about social behavior was largely accepted—except for his comments on human social behavior in a 28-page chapter at the end of a total text of 575 pages. These comments aroused intense objections on the part of many led by a group called Science for the People, which included some colleagues of Wilson at Harvard, namely, the evolutionist Stephen Jay Gould and the geneticist Richard Lewontin. Science for the People claimed that Wilson had in *Sociobiology* revived some old and discredited positions advocated by the Nazis with their murderous "scientific" racism and the eugenicists with their attempts to "improve" the human species through restrictions on reproduction by supposedly undesirable types (Sociobiology Study Group 1976).

Wilson quickly replied to his critics (Wilson 1976) and so did others (e.g., Ruse 1979), but the strong criticism of an apparently new discipline colored the response to *Sociobiology*, the book, and sociobiology, the entire field of study. This criticism was based largely on a misreading, willful or otherwise, of the nature of the discipline, which the opponents said was a dangerous reversion to outmoded ideas about the relationship between genes and behavior. In reality, sociobiology is a Darwinian field that focuses on the role of natural selection in shaping the behavioral traits of the members of a species over evolutionary time, *not* a discipline that attempts to study the immediate or proximate underlying causes of social behavior, such as the genetic or experiential factors that shape the behavioral development of an individual during its lifetime. The early critics chose to ignore this point, taking advantage of the general view that sociobiologists employed a "biological" approach to social behavior. This tactic enabled the critics to blur the distinction between the evolutionary (ultimate or long-term) and the immediate (proximate or short-term) causes of behavior, thereby linking sociobiology to false claims about genetic determinism and human behavior.

In 1975, all evolutionists knew, as did everyone else, that human social behavior is developmentally flexible. Sociobiologists then and now were and are not trying to identify particular genes for specific social behaviors but instead would like to know how a behavior, flexible or not, might be adaptive, that is, show the effects of past natural selection. So for example, the ability of very young babies to acquire the essential features of their population's language is highly flexible but reflects past selection favoring infants whose brains can acquire information on the particular sounds and grammatical rules that will enable the baby as an older child and as an adult to communicate effectively with others of his or her language group.

That sociobiologists were not trying to determine the proximate effects of genes on human behavior can be shown by examining how sociobiologists studied nonhuman animals in the 1970s. No one was trying to find a gene for the social behavior of Florida scrub jays or myrmicine ants. Instead, sociobiologists tested the prediction that the social actions of jays, ants, or what have you would enable individuals to pass on their genes successfully to the next generation. Yes, genes are involved in sociobiological research but not in the sense of requiring sociobiologists to figure out how particular genes control the metabolic activities occurring in a developing organism. Indeed, sociobiology is flourishing currently despite a general lack of knowledge about the proximate bases of behavioral development in the overwhelming majority of animal species.

1.1 Why Gould and Lewontin's Views Have Had Little Effect on Sociobiological Researchers

Gould and Lewontin (particularly Gould) realized that to further depreciate the sociobiological study of humans, they needed to undercut the entire discipline which they attempted to do well after launching their initial salvos at Wilson.

This effort took the form of an article that they jointly published in a prestigious journal produced by Royal Society of London (Gould and Lewontin 1979) and also in many shorter pieces written by Gould alone and designed for the general public in the magazine *Natural History* (Gould 1976, 1981, 1982, 1984). Both the academic article and those in *Natural History* almost certainly contributed to the general feeling by persons other than evolutionary biologists that there was something wrong with sociobiology. But for evolutionary biologists, the problem lay with Gould and Lewontin's arguments (Alcock 1998).

First, it was common knowledge in biological circles that Gould and Lewontin were avowed Marxists. As such, they permitted their ideological persuasion to dictate aspects of their academic conclusions as the evolutionist Robert Trivers (2015) describes in an assessment of their achievements. Although Gould and allies labeled their opponents as closet ideologues whose research supported their political views, in reality it was Gould and company who permitted their unconcealed political positions to affect their conclusions. Indeed, Lewontin, in the company of Richard Levins (1985, p. 165), wrote, "As working scientists in the field of evolutionary genetics and ecology, we have been attempting with some success to guide our own research by a conscious application of Marxist philosophy." Thus, the views of Lewontin, Gould, and others of their allies were suspect for those aware of the environmentalist bias of Marxists, who believe that all human social behavior can be altered by the correct set of cultural influences.

Second, evolutionary biologists wrote direct rebuttals of Gould and Lewontin's famous joint paper as well as analyses of Gould's articles in *Natural History* and elsewhere (Borgia 1994; Queller 1995; Alcock 1998). Among the many flaws these critics identified was the notion that Darwin was a pluralist when it came to thinking about the causes of evolutionary change so that, according to Gould and Lewontin, he gave as much weight to other factors (such as the random effects of genetic drift) as to natural selection as an agent of evolutionary change. He did not. Darwin's *On the Origin of Species* (1859) is, to paraphrase the words of the evolutionary biologist Ernst Mayr, one long argument about natural selection, not about any other element that might influence the way in which a species evolves. Natural selection occurs when individuals in a species differ in their heredity in ways that affect individual reproductive output, or as we would say today, the genetic success of individuals or their fitness as measured by the number of genes they propagate. In other words, Darwin was an adaptationist, someone who proposes that the complex traits exhibited by organisms have been naturally selected on the basis of their ability to promote the survival of individuals that possess these attributes, which is linked to their genetic success. He did not write a book about randomness or contingency or competition among groups or any other possible feature of life that might conceivably influence the path of evolution and thus the attributes of a species. The theory that he constructed guides the research done by the vast majority of evolutionary scientists today.

Gould and Lewontin (1979) also claimed that adaptationists tended to accept speculative positions as soon as they were proposed. This too is manifestly false as can be seen by reviewing the literature in evolutionary biology. Students of adaptive value propose speculative hypotheses for the same reason that all scientists hypothesize, as a starting point for the development of testable predictions, whose tests give reason to accept or reject a particular hypothesis (Brown 1982). If it were found, for example, that very young babies exposed to Mandarin Chinese were no more able to discriminate between the word sounds (phonemes) associated with this language than between the phonemes that they had never heard in other languages, then the adaptationist would reject the speculative hypothesis that past selection had shaped the language learning abilities of infants.

Opponents of adaptationist thinking, whether applied to humans or to other species, by devaluing the importance of natural selection theory might well convince persons who were not evolutionarily oriented to be skeptical of the theory and the scientific way it has been used. To some degree, this effect of sociobiological criticism still persists in the writing of non-evolutionists about human behavior (see Gottleib 2012; Slater 2013).

But what about evolutionary biologists? The intensity of the initial criticism and the presence among the critics of well-known biologists like Gould and Lewontin made sociobiologists and others interested in the evolution of social behavior nervous, so much so that sociobiology as a label is rarely used today. Instead, the term has been replaced by behavioral ecology (for the sociobiology of nonhuman animals) and evolutionary psychology (for the sociobiology of humans). But sociobiologists, broadly defined, have in effect voted on the competing claims of Wilson versus the early opponents of sociobiology by adopting in an overwhelming way the adaptationist approach inherent in the discipline even while largely dropping the label "sociobiology."

1.2 The Origins of the Adaptationist Approach to Social Behavior

As noted earlier, the Darwinian approach to the study of adaptations was well developed before Wilson wrote *Sociobiology*. Among the notable practitioners of adaptationism were not just the individual researchers whose work formed the basis of Wilson's book but theoreticians who updated the underlying formulation of natural selection theory for the sociobiologists of the era (who were usually called ethologists). Indeed, one of the founders of modern behavioral biology, the ethologist Niko Tinbergen, identified four basic questions about the behavior of all animals (Tinbergen 1963). One of these questions dealt with the adaptive significance of a behavior for which an understanding of natural selection was critical. Tinbergen employed Darwinian theory when he and his colleagues did what was to become famous work on egg shell removal by gulls from the nest after the young bird had hatched (Tinbergen et al. 1962). They showed convincingly that this behavior had evolved by natural selection because it made the nestlings harder to locate by predators that would otherwise use the conspicuous white interior color of egg shells to find nests with edible gull chicks, thereby reducing the fitness (i.e., reproductive success) of parents that did not remove the broken egg shells after their nestlings had hatched.

W. D. Hamilton in 1964 expanded and modernized Darwinian theory by showing that reproductive success was not the only means by which individuals could pass on their genes and thus influence the course of evolution [see Shanahan (2017) for Darwin's initial solution to the problem of cooperation among "community" members]. This work (Hamilton 1964), which is directly relevant to so much social behavior in nonhumans and humans alike, made behavioral researchers aware that one of the major puzzles of social behavior had a potential solution. If, as Darwinians propose, individuals evolve traits that maximize their reproductive success (as measured by the number of surviving offspring produced), then cases in which individuals sacrifice their reproductive success to help others are highly puzzling. Hamilton demonstrated, however, that if the altruists that did the sacrificing did so in ways that increased the reproductive success of their relatives, *with whom they shared genes in common*, then the puzzle might be resolved. In other words, selection favoring altruism toward relatives (*aka* kin selection) could sometimes enable individuals to gain more genetic success than they could by attempting to reproduce personally.

If not equal to Hamilton, then George C. Williams is a close second in terms of importance in updating Darwinian thought as it relates to behavior, especially social behavior. In *Adaptation and Natural Selection*, also written well before *Sociobiology*, Williams (1966) argued brilliantly that differences in the genetic success of individuals were the key to understanding the evolution of social behavior. Despite the work of Darwin and Hamilton, many biologists by the mid-1960s had become used to invoking benefits to the group as a whole as the supposed basis for the evolution of adaptations. Indeed, V. C. Wynne-Edwards (1962) published a massive book in which he interpreted various social behaviors as adaptations at the group level, designed to help prevent the extinction of the group or species. Williams demolished this idea in the course of explaining that if a trait benefited the group as a whole but harmed the individual's ability to pass on its genes, then no matter how beneficial the trait to the species, the loss of the altruist's genes would guarantee the eventual disappearance of the characteristic. The power of this simple but largely ignored argument (at the time) convinced almost all behavioral biologists to abandon facile hypotheses about how such and such a social characteristic might help groups or entire species survive and to instead propose (and then test) alternative hypotheses on how the trait in question might lead to greater representation of the genes of the social individuals in subsequent generations.

Williams's ideas were in substantial measure represented in Richard Dawkins's *The Selfish Gene* (1976) in which he used the title as a metaphor to explain why individuals behaved the way they did, which is as if they were controlled by genes that "selfishly" attempted to maximize their transmission at the expense of competing versions of those genes. This book makes no reference to *Sociobiology* by E. O. Wilson, presumably because it was largely written before Wilson's book appeared.

The point here is that many well-known social theoreticians and evolutionists were practicing adaptationists before Wilson coined the term sociobiology to refer to adaptationist research on social behavior. Indeed, Wilson could not have written his book were it not for the fact that prior to 1975 the development of an expanded Darwinian approach to social behavior had already occurred and had been widely adopted by behavioral researchers interested in ultimate (evolutionary) questions about the attributes of living things.

2 Remaining Criticisms of Sociobiology

One still encounters claims that the adaptationist or selfish gene approach to human social behavior is defective for reasons presented by Gould and Lewontin years ago, as if these claims were novel, which they are not (Alcock 2001; Hagen 2005). In any event, we can safely dismiss repeats of the just-so story criticism and the view that sociobiology is a form of genetic determinism.

In a similar fashion, the creationist criticism of all things evolutionary continues to be broadcast in certain circles even today despite an abundance of rebuttals from non-creationists. The critique that evolutionary theory violates biblical accounts of creation and therefore must be wrong is so obviously ideological in nature that it neither needs nor deserves a detailed response, this having been done many times before (e.g., Berra 1990; Coyne 2009). Richard Dawkins (1981) in particular has shown how complexity, as exhibited in the social behavior of many animals, can arise from natural selection acting on naturally occurring mutational variation given sufficient time. And the literature on the evolution of human morality is also large and convincing, showing that human impulses toward what we call moral behavior motivate us to behave in ways that are advantageous to our genes (Alexander 1987; Ridley 1996). To claim that our morality is inexplicable in any terms other than the supernatural, as the head of the National Institutes of Health, Frances Collins (2006), does, demonstrates a complete failure to become familiar with the evolutionary literature on this subject.

More interesting is another objection to sociobiology launched by academics who argue that Darwinian natural selection is insufficient to deal with highly complex social organisms, such as the social insects and humans. These scientists have been led by David Sloan Wilson (1975a, 1977) but include the philosopher of science Elliot Sober and the founder of sociobiology Edward O. Wilson himself (Sober and Wilson 1998; Wilson and Wilson 2007). As noted earlier, the idea that evolution could lead to group-benefiting adaptations was largely demolished by George C. Williams but D. S. Wilson, no relative of E. O. Wilson, has tried for years to resuscitate group selection theory largely by developing the idea that there are multiple levels of selection, one of which involves competition among groups.

And it is true, as Darwin (1859) acknowledged, that groups can and do sometimes compete in ways that affect their survival and thus affect, in modern terms, the collective propagation of their genes. However, despite the enthusiasm of multilevel selection advocates (Wilson and Wilson 2007), no one has convincingly shown that selection at the group level has overridden selection at the level of the individual or its genes. When the argument has been made that such and such a trait requires group selection to evolve, additional analysis has shown that within-group competition driven by natural selection acting at the level of the individual is responsible for the trait in question (Biernaskie and Foster 2016). More importantly perhaps, several workers have shown mathematically that the genetic outcome of multilevel selection would be the same as the result from genetic competition at the level of individuals, only requiring different calculations (Reeve 2000; Marshall 2011). In essence, to employ the group level approach one has to know the genetic contribution of the group as a whole to the next generation, often an impractical demand. In the absence of a demonstration that group selection can effectively counter individual level selection, and given the mathematical difficulties involved in using the "new" form of group selection for hypothesis generation, the vast majority of behavioral researchers have ignored group selection theory altogether even as supposedly modified under the multilevel label. They do so largely because of the difficulties of using the theory and the confusions that it stimulates (West et al. 2011) rather than because of some ulterior motive as is sometimes argued.

2.1 The Validity of Kin Selection

In fact, sociobiologists have also voted via their research decisions against another criticism of a component of sociobiology, namely that Hamiltonian kin selection (as noted, selection for altruism directed toward relatives) is defective as a theory. Here again E. O. Wilson has recently joined others (Nowak et al. 2010) in writing a forceful critique of an important element of sociobiology, namely, kin selection theory, an idea based on the realization that differences among individuals in the ability to pass on their genes are the foundation for all evolutionary change. Darwin's *Origin* focused on the *direct* means of achieving genetic success, which is to reproduce personally since offspring possess half the genes of their parents. Natural selection favors traits that enable parents to have surviving offspring, which will inject the genes of their parents into the next generation.

Hamilton convinced biologists that there was an *indirect* wayin which individuals might transmit their genes to the next generation, namely by helping relatives survive and reproduce successfully thereby passing on some of the same genes as their helper as a result of having a common ancestor. Thus, nephews and nieces have one-quarter of the genes of their aunt or uncle so that if an aunt or uncle could boost the reproductive success of these relatives they would in effect be creating more copies of the genes they share in common with their nephews or nieces. Individuals with a tendency to help their relatives live and reproduce could in theory increase the production of related individuals sufficiently to overcome any losses in the direct route to genetic success caused by their helpfulness to relatives other than their offspring. Kin selection is predicted therefore to favor helping out close relatives rather than distant relatives or non-kin, a prediction that has been checked many times with positive results as will be noted later.

But Nowak and company claimed that kin selection theory should be replaced by a combination of natural selection (with the reproductive benefits to individuals arising from membership in a group that controls a key resource) and multilevel selection in which group competition determines the survival and reproductive success of entire groups. Kin selection was invalidated in their view by their mathematical calculations and by the fact that sterile castes had evolved in some groups, such as termites, which lacked the haplo-diploid system of sex determination, a supposedly critical piece of evidence for kin selection theory (according to Nowak et al. 2010). Why? Because when males are haploid and females diploid as they are in the bees, ants, and wasps, daughters of mothers that have mated with only one haploid male will have received the same set of genes from their father, and so will be especially closely related. Therefore, the Hymenoptera are predicted by kin selection theory to be especially likely to sacrifice personal reproduction in favor of behaving altruistically toward their female siblings. The existence of sterile workers in the Hymenoptera matches expectation, but a worker caste in species with other systems of sex determination means, according to the critics, that kin selection theory is inadequate.

The response was strongly negative from most evolutionary biologists who responded to this criticism. The many signers of a letter to *Nature* (Abbot et al. 2011) noted that Hamilton was fully aware of the diplo-diploid sex determining system of termites and that he had pointed out that the degree of relatedness (the proportion of genes shared as a result of having a common ancestor) was only one of several factors determining whether altruism would be adaptive for individuals in a social setting. Another element was the cost of the action in terms of reducing the individual's potential reproductive success. If an animal had little chance of reproducing personally, as is true for workers in colonies of social insects, then altruism could still generate a net benefit for the helpful individual. In other words, kin selection theory does not fail because of the existence of sterile castes in species with a method of sex determination other than one found in the bees, ants, and wasps. In addition, the critics' assertion that kin selection theory had not been useful to researchers in sociobiology was rebutted by those who noted that the theory had contributed in a major way to our understanding of a host of phenomena, including why it is that altruism is directed primarily toward close kin, why worker social insects sometimes lay eggs, why other workers sometimes prevent their fellow workers from reproducing, why eusociality has consistently evolved in the Hymenoptera whose queens mate with a single male (Strassmann et al. 2011), as well as such things as adoption, conflict resolution, kin discrimination, multicellularity, and more (Gardner and West 2014). Finally, Liao et al. (2015) have critically reexamined the mathematical model that Nowak et al. (2010) developed. This team of researchers showed that the model simply assumes close relatedness among the members of a group. As a result, the math actually supports the kin selectionist view of the evolution of sociality.

That sociobiologists have not abandoned kin selection theory in favor of the alternative presented by Nowak et al. (2010) is made clear with an example taken from the recent literature, the work of Scott and Nancy Creel on the sociality of the wild dogs of Africa (Creel and Creel 2015). Wild dogs have long been known to live in groups in which only the dominant male and female breed, a point that led to the suggestion that the other members of the group were related helpers whose protection and feeding of the young had evolved by kin selection, a suggestion that Wilson (1975b) accepted when writing about wild dogs in *Sociobiology*. Subsequent research has shown that the reproductive success of the dominant reproducers in a group increases with increasing pack size; the more helpers, the more food for the offspring, and the greater number of pups that survive.

The Creels' paper asks whether helpers in large packs of wild dogs benefit personally or pay a price for living with others. The Creels found that nonreproducing group members had 25% shorter lives in large groups compared to the life span of wild dog helpers in the smallest packs. So here we have a case in which some group members sacrifice for others, giving up a longer life and the reproductive success associated with a longer life span. This fact coupled with evidence that members of larger groups do not leave their packs as well as the fact that helpers are related to at least one of the dominant breeders led the Creels to conclude that kin selection has favored helping behavior in this species. In their paper, there is no discussion of Nowak et al. (2010) nor any mention of multilevel selection *à la* David Sloan Wilson (1975a, 1977). The Creels clearly find kin selection theory a perfectly adequate tool for the evolutionary analysis of helpful behavior.

2.2 *Is Sexual Selection Theory Outmoded?*

Finally, another challenge comes from Joan Roughgarden (2004, 2012) and Erol Akçay (2010) who state that sexual selection theory, a major component of Darwinian theory, is invalid in part because the theory emphasizes conflict between the sexes rather than cooperation and in part because the theory employs outdated gender stereotypes of the nineteenth century.

To understand why these criticisms carry little weight with most students of sexual behavior, we need first to explain that Darwin (1871) developed the theory of sexual selection as an amendment to natural selection theory. For someone who viewed natural selection as acting to promote survival, as Darwin did, the existence of conspicuous body ornaments and displays, the willingness of individuals to fight for mates with others of the same sex, and the drive to reproduce even in the face of predation are puzzles because they clearly do not promote survival. But attributes that reduced an individual's survival chances while at the same time enabling the individual to acquire mates could still spread through a population if the mortality risk was offset by increased personal reproduction. Many sexually selected traits, like the remarkable displays of peacocks, which spread their enormous, decorative tails with many "eye spots" when courting females, appear to have evolved because females prefer sexual partners that possess these characteristics. Indeed, Petrie and Halliday (1994) have shown that when some eye spots are experimentally removed from the trains of peacocks, the mating success of the males so affected is reduced.

The theory of sexual selection has been widely used by researchers ever since Darwin but especially over the past five decades or so (Andersson 1994). Today, many evolutionists do not emphasize the differences between natural and sexual selection in the way that Darwin did because they realize that the two kinds of selection both operate on the basis of differences in the genetic success of individuals (Clutton-Brock 2010). Although it is true that relatively long lived individuals often have more opportunities to pass on their genes than short lived individuals, so do those individuals that have greater access to mates no matter how long they live. In the last analysis, both natural and sexual selection occur only when certain traits are associated with genetic propagation.

As we shall see shortly, sexual selection theory has resulted in major discoveries by evolutionary biologists investigating reproductive behavior (Andersson 1994). To dismiss the theory as fatally flawed would be to erase these discoveries, which is why most researchers are reluctant to abandon a theory that has proven highly useful as a research tool. Moreover, Roughgarden's views miss the mark (Kavanagh 2006; Clutton-Brock 2010). For one thing, the position that reproduction is essentially a cooperative venture is severely weakened by the abundant evidence of conflicts between sexual partners. For another, the idea that Darwin's social milieu led him to adopt old-fashioned stereotypes of male and female behavior is negated by the observation that he did not define sexual selection strictly in terms of male versus female behavior. Instead, his definition of selection of this sort was that differences in access to mates, either by males or by females, would generate evolutionary change in favor of attributes of *either* sex that increased an individual's ability to acquire sexual partners. As it turns out, females are usually a limited resource for males which is why males typically compete for females and attempt to overcome their resistance to mating. But there are some species in which males are the limited sex, because of such things as the valuable resources they provide their partners, and in these species, females often compete with one another for access to males, which exhibit mate choice (Edward and Chapman 2011). Indeed, as predicted by the theory, sex roles can change within a species when resources become difficult for males to acquire and pass on to females (Gwynne 1984, 1985; Gwynne and Simmons 1990). The theory of sexual selection is not dependent upon Victorian attitudes about male and female behavior. Roughgarden's views have been adopted by very few researchers.

3 Modern Sociobiology: Behavioral Ecology

Thus, despite the critics, sociobiologists have continued to use Darwinian natural selection and sexual selection theory with the modifications provided by W. D. Hamilton. Since the 1960s, sociobiologists have proposed that a wide range of social traits have evolved because of their contribution to either individual direct fitness (via personal reproduction) or indirect fitness (via aid given to genetically similar individuals), not groups. The sum of direct and indirect fitness is labeled inclusive fitness, which is why Hamilton is said to have developed inclusive fitness theory (Gardner and West 2014). In cases in which groups do compete for resources, the inclusive fitness or selfish gene approach is based on the premise that how individuals behave socially affects the genetic success of the animals that make up the group. Therefore, rather than presuming that individuals have evolved attributes that help the group as a whole, most students of animal social behavior find it more instructive to view these characteristics as ones that help individuals to live with others in ways that maximize their own inclusive fitness. In other words, any benefits to the group are the result of characteristics that have evolved because of their positive effects on the direct or indirect fitness of group members and not because they are advantageous to the group per se (West et al. 2011).

3.1 The Sociobiology of Reproductive Behavior

The results of sociobiological research founded on Darwinian principles are impressive as in demonstrating that males and females often act to thwart the reproductive success or fitness of their partners, when it is in their self-interest to do so. Consider the possible functions of courtship signals. Once these behaviors were thought to have evolved primarily as the means by which the two sexes coordinated their reproductive physiologies, the better to integrate their sexual activities. The fact that males often provided distinctive signals during the courtship process was viewed largely as a way in which females could insure that they were mating with a member of their own species, thereby avoiding the cost of producing inferior hybrid offspring. But more recent research founded on adaptationist thinking has led to a different interpretation of male and female sexual behavior.

Females of many species are now viewed as evaluators of what males have to offer, whether it be their genes or the resources under their control. The evaluation of male quality can be done sometimes via his courtship, according to the revised view. Thus, the courting individual, usually but not always a male, is attempting unconsciously to demonstrate either that he has the potential to provide females with genes that will enhance their direct fitness *or* that the male will provide better-than-average parental care or other resources that will also raise the reproductive success of his sexual partners. In other words, male courtship has evolved in the context of female mate choice with the choosy female discriminating among would-be mates not on the basis of species membership or physiological readiness to reproduce but on the basis of his potential contribution to her production of surviving offspring. For example, in the lizard *Anolis sagrei* the male's dewlap under his chin is expanded rigidly in courtship and the orange and yellow color of the extended dewlap conveys information about the health and immune condition of the male, information that conceivably helps females mate with genetically superior partners (Driessens et al. 2015). And males of the brown booby, a pelagic bird, display their bluish throat skin during courtship, with the color of the gular pouch indicating the degree to which the male will help feed the offspring of a pair (Montoya and Torres 2015).

3.2 *Extrapair Copulations*

Even after a female has permitted a male to mate with her, she may still be able to improve her reproductive success by copulating with other males despite the loss of fitness that this usually entails for her previous partner(s). This is merely one of several post-copulatory actions that affect the reproductive success of females and their partners (Birkhead 2010; Eberhard 1996). Copulation with several males occurs even in females that have a social mate, one that assists them in the rearing of a brood of offspring, as is the case for most species of birds. Although at one time, researchers took the social monogamy of birds with male parental care to be synonymous with genetic monogamy, we now know that extrapair offspring are found in about 90% of all bird species (Giffith et al. 2002). By determining the paternity of the offspring of individual females, something that can be done through molecular means unavailable in the pre-sociobiological era, researchers have established that a pairbond that leads a male to form a union of sorts with his partner is no guarantee that she will use only his sperm in fertilizing her eggs.

Of course, if a female is "unfaithful" in a species whose females acquire social partners, then some males will be "unfaithful" as well. Indeed, males that are socially monogamous often are willing and able to mate with females other than their social partner and in this way pass their genes on to some offspring that will be reared by other males (Westneat et al. 1990). Thus, reproduction, even in bird species that were once considered paragons of cooperative monogamy, can be fraught with conflict as individuals attempt to maximize their success in leaving copies of their genes, no matter what the effect on other individuals, even a current mate.

3.3 *Sperm Competition*

Therefore, the copulatory choices made by males and females can influence the chance that these individuals will pass copies of their genes to the next generation. But in addition, males can further affect their genetic success via a competition that takes place *within* the bodies of their partners, a point that demonstrates clearly that access to mates is not the only factor that shapes the intensity of sexual selection. Males do so through sperm competition (Birkhead 2002, Birkhead and Møller 1998) as they either behaviorally or physiologically employ stratagems that increase the odds that their sperm, and not those of a competitor, will be used by a sexual partner to fertilize her eggs.

Geoffrey Parker (1970) first realized that the number of mates affected male genetic success less than the number of eggs fertilized by a male. Although Darwin failed to recognize this point (Eberhard 2009), sociobiologists realize that if male or his sperm could increase the odds of successful egg fertilization, selection should favor the individual with these abilities especially in those species, a vast majority, in which females mate with more than one male during a reproductive cycle.

One adaptation that promotes a male's sperm competitiveness is mate guarding (Alcock 1994) in which the copulating male remains with his partner to prevent her from mating with another male soon thereafter. Mate guarding carries costs, of course, in that the time spent with one female cannot be spent searching for additional mates. Thus, mate guarding should be more pronounced when the likelihood of a female's remating (if left unguarded) is greater than the likelihood that the male will encounter an unguarded potential sexual partner. So, for example, the association between a male and his mate is particularly prolonged in the stick insect, *Micrarchus hystriculeus*, in which the sex ratio is consistently biased toward males, a fact that means that the likelihood of a male finding an unguarded female is very low (Kelly 2015).

Another sperm competition adaptation is the ability of the male to remove sperm donated to his mate by another male (Simmons 2001) as occurs dramatically in the damselfly, *Calopteryx maculata*. In this species, the male's intromittent organ has lobes and spines that can be used to remove previously acquired sperm from within the female's sperm storage organ prior to the release of the male's own gametes. Jon Waage (1979) showed that males were highly effective in emptying the female's storage organ so that if a female laid eggs after copulating, only sperm from the last partner were available for any egg fertilizations that she effected at this time. Sperm removal by males has evolved in any number of other organisms as well including cuttlefish (Wada et al. 2010) and birds (Davies 1983), although the role of the female in determining which male's sperm has a competitive edge must also be considered (Dean et al. 2011).

Sperm competition has also been demonstrated to occur in the humble fruit fly, *Drosophila melanogaster*, whose males supply their copulatory partners with chemical supplements in the ejaculate that reduce the sexual receptivity of the female and induce her to lay eggs (Chapman et al. 1995). By investing in these materials, the male donor lowers the chances that his mate will go on to mate with another male shortly and thereby receive sperm that she may use in the fertilization of her eggs.

The point is that once Parker had shown the potential importance of sperm competition to male genetic success, sociobiological researchers began to look for and find the means by which males gave their sperm an edge in the race to fertilize eggs even if these competitive devices reduced the number of females a male could inseminate. And males might even gain an advantage if their sperm competition mechanisms reduced the lifetime reproductive success of a given female (Edward et al. 2015).

3.4 Cryptic Female Choice

Sperm competition is not the only way in which egg fertilization might be skewed in favor of one male over another. Females also have a stake in which sperm reach their eggs (Thornhill 1983; Eberhard 1991), as noted above. So, for example, females of the spider *Pisaura mirabilis* that receive a nuptial gift in the form of a prey item given to them by a male store more sperm from their partner and lay eggs that are more likely to hatch than females that are mated for the same period of time but with a non-gift giving partner (Albo et al. 2013).

Thus, mate choice, extrapair copulations, sperm competition, and cryptic female choice are all discoveries made by persons willing to test Darwinian hypotheses. The idea, which one not infrequently encounters even today, that evolutionary hypotheses are easy to generate and easy to accept without testing is false. Yes, the adaptations that promote individual genetic success in this arena are the product of evolutionary speculations (hypotheses), but before any of the conclusions about the characteristics discussed in this section were ever published, the underlying hypotheses were tested to the satisfaction of scientific reviewers, which is why evolutionary biologists believe that sperm competition and cryptic female choice really do occur.

4 The Sociobiology of Cooperation and Reciprocity

The same point can be made by examining the history of the study of cooperation and reciprocity in nonhuman animals. There was a time when cooperation was considered part of the repertoire of most animals as a means by which a species preserved itself from extinction. As noted earlier, however, this group benefit approach to animal social behavior has been replaced with the sociobiological or Darwinian approach. Instead of expecting animals to sacrifice for the benefit of their species, evolutionary biologists now expect animals to behave in ways that will promote the propagation of their genes. The adaptationist approach, however, creates a potential puzzle, which is how can animals evolve cooperative behavior if each individual in a group is acting in its own self-interest. The answer is that cooperation can be in the genetic interests of all involved if the costs to the helpful individuals are outweighed by the benefits they derive from their cooperative actions. In some cases, it has been shown that cooperation does have the potential to provide direct reproductive benefits for helpful individuals (Dìaz-Muñoz et al. 2014). For example, male chimpanzees may form coalitions for the purpose of attacking other males; Gilby et al. (2013) document that males able to work cooperatively in this fashion are more likely to rise in dominance rank in their groups and more likely to father offspring. Thus, cooperation has benefits for cooperators in this and in other species as well, even when the helpful individuals are unrelated to one another as sometimes happens in lions in which small groups of unrelated males may be able to dislodge other males from prides of receptive females (Packer et al. 1991).

However, just as in the analysis of reproductive behavior, the adaptationist approach leads one to predict possible conflicts among the would-be cooperators as individuals jockey for reproductive advantage in the system. Conflicts among cooperative societies are especially well documented among the social insects (Ratnieks et al. 2006). For example, conflict in these species often arises when self-sacrificing workers attempt to reproduce against the interests of the queen and certain other workers, leading to the harmed individuals to punish the non-cooperators (Singh and Boomsma 2015).

When it can be shown that some helpful animals receive no immediate benefits but do secure delayed rewards, then another explanation for cooperation is worth testing, namely, reciprocity. For example, consider the reciprocal interactions that may occur when pied flycatchers fly at and vocalize around potential predators, the better to encourage a hawk or owl to go elsewhere. Experimental methods have shown that pairs that received help from others when mobbing a potential predator are more likely to reciprocate when their helpers are in need of mobbing assistance (Krams et al. 2007). In this way, a helpful mobber is paid later for his or her current helpfulness.

4.1 Modern (Human) Sociobiology: Evolutionary Psychology

Cooperation and reciprocity are the cornerstones of human behavior, a point that has inspired an adaptationist analysis of our behavior, primarily by evolutionary psychologists (Barclay 2013; Roberts 2015). Although many of these academics are reluctant to be labeled sociobiologists (Buss 1995), they clearly employ the adaptationist approach as they explore the evolutionary basis for human behavior. However, the person who pioneered this approach was a biologist, Richard Alexander. He showed, well before the publication of *Sociobiology*, that natural and kin selection theories could be productively applied to the social behavior of nonhumans (Alexander 1974) and humans (Alexander 1971) alike. Alexander (1975) noted that when the adaptive value of human behavior had been discussed in the past, the focus had been on benefits to the group or species, not on the benefits to the individual, as required by a modern understanding of evolutionary theory.

Alexander went on to publish two books, one on human social behavior in 1979 and the other on human morality in 1987, in which he presented a host of adaptationist hypotheses and their tests with an emphasis on the role of kin selection. Subsequently, psychologists began to employ an adaptationist approach to human reproductive behavior, cooperation, reciprocity, and altruism. There is now a vast literature on these subjects in which psychologists aware of evolutionary theory attempt to test hypotheses on various aspects of our behavior the better to *explain* why we do the things we do, and not to claim that these behavioral traits are good, moral, and desirable. We can illustrate the explanatory goal of evolutionary psychologists by looking at some examples of their research into human sexual behavior, one of the primary areas of interest for these scientists (Buss 1998; Geary et al. 2004; Li and Kenrick 2006).

We begin with the observation, surely not controversial, that men consider older women less sexually attractive than women in their 20s. Indeed, a male sexual preference for younger women is universal, occurring across a wide range of cultures as shown by the cross-cultural research of Buss (1989). The sociocultural hypothesis that men are conditioned by the traditions of their particular culture, which in turn are the products of human imagination, generates the expectation that in at least some cultures, men ought to be taught to prefer older women with the result that they would learn to consider older females more sexually appealing. This does not happen. Anywhere.

So what is an evolutionary explanation for the male preference for younger women? If male sexual psychology is the product of natural selection, then our brains should help individual men seek out mates that are especially fertile, since an enthusiasm for sexual activity results in genetic success only to the extent that a man succeeds in copulating with a fertile woman who then becomes pregnant (Buss 2003; Geary et al. 2004). Evolutionary psychologists have used this hypothesis (an explanation for the psychological preferences of men) to produce testable predictions, one of which is that women in their 20s should be more fertile than older ones. Adult female age is of course negatively linked to fertility in our species (Dunson et al. 2002). In fact, women in their 30s are about three to five times as likely to be infertile as women in their 20s. Because older women are less fertile, indeed much less fertile, than younger ones, researchers have concluded that the evolutionary explanation for male preferences hypothesis had been supported. Note that these researchers were not trying to justify male sexual preferences, which can lead to cruel and unthinking sexism, but to explain why we do what we do in evolutionary terms.

Evolutionary psychologists are no more satisfied with one or two tests of a hypothesis than are physicists. The more tests, the better. The more challenging the prediction, the more convincing the test. Here is an example of another test of the evolutionary hypothesis that male mating preferences have been shaped by natural selection to increase the odds that a man will seek out fertile partners. Until very recently, biologists thought that neither men nor women could tell when a woman was in the ovulatory phase of her menstrual cycle (Gangestad and Thornhill 2008). However, several evolutionary psychologists predicted that men *should* have evolved the ability to detect subtle cues associated with ovulation when evaluating the sexual attractiveness of women (Miller et al. 2007; Miller and Maner 2011). One way to check this prediction is to construct two composite photographs of the same set of women, but with one image composed of these women when ovulating and the other of the same women in the non-ovulatory portion of the menstrual cycle (Roberts et al. 2004). The differences between the two composite images are slight and yet when men were asked which image they preferred, they significantly more often selected the one linked to ovulation—as predicted from the evolutionary hypothesis under examination. In other words, men apparently can tell unconsciously when a woman is ovulating, an ability that could motivate them to try to mate with a woman who was especially likely to become pregnant.

And here is another example. Lap dancers secure tips for their performances, which has been used to test the prediction that these women should receive larger tips if they were ovulating. The actual data collected by a team of evolutionary psychologists at the University of New Mexico (Miller et al. 2007) showed that ovulating lap dancers received more money per shift from their male clients than non-ovulating dancers. Either men could smell the subtle signals of ovulation or the behavior of the lap dancers provided cues of ovulation. The point is that this work constituted a test of an evolutionary hypothesis about the adaptive value of male sexual preferences.

5 Conclusion: The Sociobiology of Human Altruism

By looking at the evolution of human sexual preferences, evolutionary psychologists have greatly advanced our understanding of how natural selection has shaped elements of our behavior. But what about those cases of "genuine" altruism in which individuals give up some chances to reproduce in order to help others? Here, we have arguably the strongest challenge to modern Darwinism. As noted earlier, one powerful solution to the puzzle of altruism was provided by William D. Hamilton, who showed that in addition to personal reproduction, individuals can gain genetic success by helping genetically similar individuals survive and reproduce. Hamilton's kin selection theory rested on Hamilton's rule, namely that altruism would evolve and spread if the benefit to the recipient of altruism (multiplied by the degree of genetic similarity between the recipient and its helper) was greater than the cost to the altruist as measured by the loss of personal reproductive output resulting from the helpful act(s) of the altruist.

The key prediction from Hamilton's work is that altruism requires that the altruist share genes with those individuals that he or she helps. In practical terms, this almost always means that the altruist is predicted to assist relatives, usually close relatives that share a relatively large fraction of their genes as a result of common descent. This prediction has been tested a multitude of times with overwhelmingly positive results in animals as different as bacteria (Rumbaugh et al. 2012), aphids (Wu et al. 2010), ant larvae (Schultner et al. 2014), wild dogs (Creel and Creel 2015—see above), and humans (Burton-Chellew and Dunbar 2015).

So, for example, when undergraduates were asked to rank a set of individuals in terms of social distance (roughly, degrees of friendship) and then were asked to imagine how much money they were prepared to give up in order to transfer 75.00 to a particular person, the experimental subjects were more inclined to sacrifice a personal reward to help those with whom they were close friends. But for any given degree of social distance, the subjects were even more likely to pay the price to help a relative rather than a non-relative, as predicted by kin selection theory (Rachlin and Jones 2008). This work shows that people unconsciously employ Hamilton's rule when thinking of helping others.

Naturally not all cases of this sort are "truly" altruistic as in those instances in which persons donate to charities in order to impress others and thereby (perhaps) gain access to sexual partners, an example of a direct benefit from their charitable giving (Moore et al. 2002). Alternatively, the generous individual may acquire a reputation that results in (indirect) benefits for him or her via his or her relatives (Alexander 1987). The point is that these cases too are explicable in adaptationist terms. Many more examples of the beauty and power of a Darwinian approach can be found in textbooks about animal behavior and evolutionary psychology (Alcock 2013; Buss 2016; Gaulin and McBurney 2004). Adaptationism is very much alive and well thanks to sociobiologists in their modern forms as behavioral ecologists and evolutionary psychologists.

References

Abbot P, Abe J, Alcock J et al (2011) Inclusive fitness theory and eusociality. Nature 471:E1–E4. https://doi.org/10.1038/nature09831

Albo MJ, Bilde T, Uhl H (2013) Sperm storage mediated by cryptic female choice for nuptial gifts. Proc R Soc B 280. https://doi.org/10.1098/rspb.2013.1735

Alcock J (1994) Post-insemination associations between males and females in insects: the mate guarding hypothesis. Ann Rev Ent 39:1–21

Alcock J (1998) Unpunctuated equilibrium in the natural history essays of Stephen Jay Gould. Evol Hum Behav 19:321–336

Alcock J (2001) The triumph of sociobiology. Oxford University Press, New York

Alcock J (2013) Animal behavior: an evolutionary approach, 10th edn. Sinauer Associates, Sunderland

Alexander RD (1971) The search for an evolutionary philosophy of man. Proc R Soc Vic 84:99–120

Alexander RD (1974) The evolution of social behaviour. Ann Rev Ecol Syst 5:325–383

Alexander RD (1975) The search for a general theory of behavior. Behav Sci 20:77–100

Alexander RD (1979) Darwinism and human affairs. University of Washington Press, Seattle

Alexander RD (1987) The biology of moral systems. Aldine de Gruyter, New York

Andersson M (1994) Sexual selection. Princeton University Press, Princeton

Barclay P (2013) Strategies for cooperation in biological markets, especially for humans. Evol Hum Behav 34:164–175

Berra TM (1990) Evolution and the myth of creationism. A basic guide to the facts in the evolution debate. Stanford, Stanford University Press

Biernaskie JM, Foster KR (2016) Ecology and multilevel selection explain aggression in spider colonies. Ecol Let 19:873–879

Birkhead TR (2002) Promiscuity: an evolutionary history of sperm competition. Harvard University Press, Cambridge

Birkhead TR (2010) How stupid not to have thought of that: post-copulatory sexual selection. J Zool 281:78–93

Birkhead TR, Møller AP (eds) (1998) Sperm competition and sexual selection. Academic, New York

Borgia G (1994) The scandals of San Marco. Q Rev Biol 69:373–375

Bowler PJ (2017) Alternatives to Darwinism in the early twentieth century. In: Delisle RG (ed) The Darwinian tradition in context: research programs in evolutionary biology. Springer, Cham, pp 195–218
Brown JL (1982) The adaptationist program. Science 217:884, 886
Burton-Chellew MN, Dunbar RIM (2015) Hamilton's rule predicts anticipated social support in humans. Behav Ecol 26:30–137
Buss DM (1989) Sex differences in human mate preferences: evolutionary hypothesis testing in 37 cultures. Behav Brain Sci 12:1–49
Buss DM (1995) Evolutionary psychology: a new paradigm for psychological science. Psychol Inq 6:1–30
Buss DM (1998) Sexual strategies theory: historical origins and current status. J Sex Res 35:19–31
Buss DM (2003) The evolution of desire, 4th edn. Basic Books, New York
Buss DM (2016) Evolutionary psychology: the new science of the mind. Routledge, New York
Chapman T, Liddle LF, Kalb JM et al (1995) Cost of mating in *Drosophila melanogaster* females is mediated by male accessory gland products. Nature 373:241–244
Clutton-Brock T (2010) We do not need a sexual selection 2.0-nor a theory of genial selection. Anim Behav 79:e7–e10
Collins FS (2006) The language of god: a scientist presents evidence for belief. Free Press, New York
Coyne J (2009) Why evolution is true. Penguin Publishing Group, New York
Creel S, Creel NM (2015) Opposing effects of group size on reproduction and survival in African wild dogs. Behav Ecol 26:1414–1422
Darwin C (1859) On the origin of species by means of natural selection. John Murray, London
Darwin C (1871) The descent of man, and selection in relation to sex. John Murray, London
Davies NB (1983) Polyandry, cloaca pecking and sperm competition in dunnocks. Nature 302:334–336
Dawkins R (1976) The selfish gene. Oxford University Press, New York
Dawkins R (1981) The blind watchmaker. WW Norton, New York
Dean T, Nakagawa S, Pizzari T (2011) The risk and intensity of sperm ejection in female birds. Am Nat 178:343–354
Delisle RG (2017) From Charles Darwin to the evolutionary synthesis: weak and diffused connections only. In: Delisle RG (ed) The Darwinian tradition in context: research programs in evolutionary biology. Springer, Cham, pp 133–168
Dìaz-Muñoz SL, Duval EH, Krakauer AH et al (2014) Cooperating to compete: altruism, sexual selection and male coalitions. Anim Behav 88:67–78
Driessens T, Huyghe K, Vanhooydonck B et al (2015) Messages conveyed by assorted facets of the dewlap, in both sexes of *Anolis sagrei*. Behav Ecol Sociobiol 69:1251–1264
Dunson DB, Colombo B, Baird DB (2002) Changes with age in the level and duration of fertility in the menstrual cycle. Hum Reprod 17:1399–1403
Eberhard WG (1991) Copulatory courtship and cryptic female choice in insects. Biol Rev 66:1–31
Eberhard WG (1996) Female control: sexual selection and cryptic female choice. Princeton University Press, Princeton
Eberhard WG (2009) Postcopulatory sexual selection: Darwin's omission and its consequences. Proc Nat Acad Sci 106:10025–10032
Edward DA, Chapman T (2011) The evolution and significance of male mate choice. Trends Ecol Evol 26:647–654
Edward DA, Stockley P, Hosken DJ (2015) Sexual conflict and sperm competition. Cold Spring Harb Perspect Biol 7. https://doi.org/10.1101/cshperspect.a017707
Esposito M (2017) The organismal synthesis: holistic science and developmental evolution in the English-speaking world, 1915–1954. In: Delisle RG (ed) The Darwinian tradition in context: research programs in evolutionary biology. Springer, Cham, pp 219–242
Gangestad SW, Thornhill R (2008) Human oestrus. Proc R Soc B 275:991–1000
Gardner A, West SA (2014) Inclusive fitness: 50 years on. Philos Trans R Soc B 369. https://doi.org/10.1098/rstb.2013.0356

Gaulin SJC, McBurney D (2004) Evolutionary psychology. Pearson/Prentice-Hall, Upper Saddle River
Geary DC, Vigil J, Byrd-Craven J (2004) Evolution of human mate choice. J Sex Res 41:27–42
Giffith SC, Owens IPF, Thurman KA (2002) Extra pair paternity in birds: a review of interspecific variation and adaptive function. Mol Ecol 11:2195–2212
Gilby IC, Brent LJN, Wroblewski EE et al (2013) Fitness benefits of coalitionary aggression in male chimpanzees. Behav Ecol Sociobiol 67:373–381
Gottleib A (2012) It ain't necessarily so. New Yorker. http://www.newyorker.com/magazine/2012/09/17/it-aint-necessarily-so
Gould SJ (1976) Biological potential vs. biological determinism. Nat Hist 85:12–16, 18–20, 22
Gould SJ (1981) Hyena myths and realities. Nat Hist 90:16–24
Gould SJ (1982) The guano ring. Nat Hist 91(12–14):17–19
Gould SJ (1984) Only his wings remained. Nat Hist 93:10–18
Gould SJ, Lewontin R (1979) The spandrels of San Marco and the Panglossian paradigm: a critique of the adaptionist programme. Proc R Soc Lond B 205:581–598
Gwynne DT (1984) Sexual selection and sexual differences in Mormon crickets (Orthoptera: Tettigoniidae, *Anabrus simplex*). Evolution 38:011–1022
Gwynne DT (1985) Role reversal in katydids: Habitat influences reproductive behaviour (Orthoptera: Tettigoniidae, *Metaballus* species). Behav Ecol Sociobiol 16:355–361
Gwynne DT, Simmons LW (1990) Experimental reversal of courtship role in an insect. Nature 346:172–174
Hagen EH (2005) Controversial issues in evolutionary psychology. In: Buss DM (ed) The handbook of evolutionary psychology. Wiley, Hoboken, pp 145–173
Hamilton WD (1964) The genetical evolution of social behaviour, I, II. J Theor Biol 7:1–16
Kavanagh EE (2006) Debating sexual selection and mating strategies. Science 312:689–697
Kelly CD (2015) Male-biased sex ratios and plasticity in post-insemination behaviour in the New Zealand stick insect *Micrarchus hystriculeus*. Behaviour 152:653–666
Krams I, Krama T, Igaune K et al (2007) Experimental evidence of reciprocal altruism in the pied flycatcher. Behav Ecol Sociobiol 62:599–605
Kutschera U (2017) Symbiogenesis and cell evolution: an anti-Darwinian research agenda? In: Delisle RG (ed) The Darwinian tradition in context: research programs in evolutionary biology. Springer, Cham, pp 302–332
Levins R, Lewontin R (1985) The dialectical biologist. Harvard University Press, Cambridge
Li NP, Kenrick DT (2006) Sex similarities and differences in preferences for short-term mates: what, whether, and why. J Pers Soc Psychol 90:468–489
Liao X, Rong S, Queller D (2015) Relatedness, conflict and the evolution of eusociality. PLOS One 13. https://doi.org/10.1371/journal.pbio.1002098
Marshall JAR (2011) Group selection and kin selection: formally equivalent approaches. Trends Ecol Evol 26:325–332
Miller SL, Maner JK (2011) Ovulation as a male mating prime: subtle signs of women's fertility influence men's mating cognition and behavior. J Pers Soc Psychol 100:295–308
Miller G, Tybur JM, Jordan BD (2007) Ovulatory cycle effects on tip earnings by lap dancers: economic evidence for human estrus? Evol Hum Behav 28:375–381
Montoya B, Torres R (2015) Male skin color signals direct and indirect benefits in a species with biparental care. Behav Ecol 26:425–436
Moore D, Wigby S, English S et al (2002) Selflessness is sexy: reported helping behaviour increases desirability of men and women as long-term sexual partners. BMC Evol Biol 13. https://doi.org/10.1186/1471-2148-13-182
Nowak MA, Tarnita CE, Wilson EO (2010) The evolution of eusociality. Nature 466:1057–1062
Packer C, Gilbert DA, Pusey AE et al (1991) A molecular genetic-analysis of kinship and cooperation in African lions. Nature 351:562–565
Parker GA (1970) Sperm competition and its evolutionary consequences in insects. Biol Rev 45:525–567

Petrie M, Halliday T (1994) Experimental and natural changes in the peacocks' (Pavo cristatus) train can affect mating success. Behav Ecol Sociobiol 35:213–217
Pigliucci M (2017) Darwinism after the modern synthesis. In: Delisle RG (ed) The Darwinian tradition in context: research programs in evolutionary biology. Springer, Cham, pp 89–104
Queller DC (1995) The spaniels of St. Marx and the Panglossian paradox: a critique of a rhetorical programme. Q Rev Biol 70:485–489
Rachlin H, Jones BA (2008) Altruism among relatives and non-relatives. Behav Proc 79:120–123
Ratnieks FLW, Foster KR, Wenseleers T (2006) Conflict resolution in social insects. Annu Rev Ent 51:581–608
Reeve HK (2000) Review of Unto others: the evolution and psychology of unselfish behavior. Evol Hum Behav 21:65–72
Ridley M (1996) The origins of virtue. Viking, New York
Roberts G (2015) Human cooperation: the race to give. Curr Biol 25:R425–R427
Roberts SC, Havilcek J, Flegr J et al (2004) Female facial attractiveness increases during the fertile phase of the menstrual cycle. Proc R Soc B 271:S270–S272
Roughgarden J (2004) Evolution's rainbow: diversity, gender and sexuality in nature and people. University of California Press, Los Angeles
Roughgarden J (2012) The social selection alternative to sexual selection. Philos Trans R Soc B 367:2294–2303
Roughgarden J, Akçay E (2010) Do we need a sexual selection 2.0? Anim Behav 79:e1–e4
Rumbaugh KP, Trivedi U, Watters C et al (2012) Kin selection, quorum sensing and virulence in pathogenic bacteria. Proc R Soc B 279:3584–3588
Ruse M (1979) Sociobiology – sense or nonsense? D Reidel, Dordrecht
Schultner E, Gardner A, Karhunen M et al (2014) Ant larvae as players in social conflict: relatedness and individual identity mediate cannibalism intensity. Am Nat 184:E161–E174
Shanahan T (2017) Selfish genes and lucky breaks: Richard Dawkins' and Stephen Jay Gould's: divergent Darwinian agendas. In: Delisle RG (ed) The Darwinian tradition in context: research programs in evolutionary biology. Springer, Cham, pp 11–36
Singh M, Boomsma JJ (2015) Policing and punishment across the domains of social evolution. Oikos 124:971–982
Simmons LW (2001) Sperm competition and its evolutionary consequences in the insects. Princeton University Press, Princeton
Slater D (2013) Darwin was wrong about dating. NY Times, Jan 13. http://www.nytimes.com/2013/01/13/opinion/sunday/darwin-was-wrong-about-dating.html?_r=0
Sober E, Wilson DS (1998) Unto others: the evolution and psychology of unselfish behavior. Harvard University Press, Cambridge
Sociobiology Study Group (1976) Sociobiology: another biological determinism. BioScience 26 (182):184–186
Strassmann JE, Page RE Jr, Robinson GE et al (2011) Kin selection and eusociality. Nature 471: E5–E6
Thornhill R (1983) Cryptic female choice in the scorpionfly *Harpobittacus nigriceps* and its implications. Am Nat 122:765–788
Tinbergen N (1963) On the aims and methods of ethology. Z Tierpsychol 20:410–433
Tinbergen N, Broekhuysen GJ, Feekes F et al (1962) Egg shell removal by the black-headed gull, *Larus ridibundus* L.; a behaviour component of camouflage. Behaviour 19:74–117
Trivers RL (2015) Vignettes of famous evolutionary biologists, large and small. Unz Review. http://www.unz.com/article/vignettes-of-famous-evolutionary-biologists-large-and-small/. Accessed 16 Nov 2015
Waage JK (1979) Dual function of the damselfly penis: sperm removal and transfer. Science 203:916–918
Wada T, Takegaki T, Mori T et al (2010) Sperm removal, ejaculation and their behavioural interaction in male cuttlefish in response to female mating history. Anim Behav 79:613–619

West SA, El Moulden C, Gardner A (2011) Sixteen common misconceptions about cooperation in humans. Evol Hum Behav 32:231–262

Westneat DF, Sherman PW, Morton ML (1990) The ecology and evolution of extra-pair copulations in birds. Curr Ornithol 7:331–369

Williams GC (1966) Adaptation and natural selection. Princeton University Press, Princeton, NJ

Wilson DS (1975a) A general theory of group selection. Proc Natl Acad Sci USA 72:143–146

Wilson DS (1977) Structured demes and the evolution of group advantageous traits. Am Nat 111:157–185

Wilson DS, Wilson EO (2007) Rethinking the theoretical foundation of sociobiology. Q Rev Biol 82:327–348

Wilson EO (1975b) Sociobiology, the new synthesis. Harvard University Press, Cambridge

Wilson EO (1976) Academic vigilantism and the political significance of sociobiology. Bioscience 26:183, 187–190

Wu GM, Boivin G, Broideur J et al (2010) Altruistic defence behaviours in aphids. BMC Evol Biol 10:19. https://doi.org/10.1186/1471-2148-10-19

Wynne-Edwards VC (1962) Animal dispersion in relation to social behavior. Oliver and Boyd, London

Darwinism in the Twentieth Century: Productive Encounters with Saltation, Acquired Characteristics, and Development

David J. Depew

Abstract Darwinism is one of several research traditions in evolutionary biology. I identify it, both before and after its unification with genetics, with Darwin's theory of descent by natural selection from a common ancestor. Other traditions include saltationism/mutationism, Lamarckism, and evolutionary developmentalism ("evo-devo"). I argue that Darwinism's continued dominance in evolutionary science reflects its proven ability to interact productively with these other traditions, an ability impressed on it by its founder's example. Evolution by sudden leaps (saltations) is alien to the spirit of Darwinism, but Darwinism advanced its own agenda by incorporating and subverting saltationist themes. Similarly, Lamarckism's belief in the heritability of acquired characteristics has been discredited, but some of the facts to which it seems congenial reappear in genetic Darwinism as phenotypic plasticity and niche construction. These examples help assess challenges to Darwinism's hegemony currently arising from the role of regulatory genes and epigenetic factors in development. Rather than executing already entrenched genetic programs and relying on chance mutation to initiate evolutionary change, the developmental process appears to generate heritable variations that ab initio respond to environmental factors in an adaptive way.

Keywords Baldwin effect • Darwin(ism) • Evo-Devo • Mutation(ism) • Natural selection • Lamarck(ism) • Niche construction • Phenotypic plasticity • Punctuated equilibrium • Saltation(ism)

1 Darwinism's Three Persisting Challenges

In this chapter, "Darwinism" will refer to Darwin's claim that gradual natural selection is the primary (but not the only) cause of evolutionary diversification. It will not refer to Darwin's insistence that all organisms on earth have descended with modification from a common ancestor. In the latter sense, Darwinism was a success from the start.

D.J. Depew (✉)
Bend, OR, USA
e-mail: david-depew@uiowa.edu

R.G. Delisle (ed.), *The Darwinian Tradition in Context*,
https://doi.org/10.1007/978-3-319-69123-7_4

Theories of evolution with multiple starting points disappeared almost overnight. Admittedly, popular culture still identifies Darwinism with denial of species fixism, that is, repudiation of "transformism" as such. "Darwinism" as evolution in general permeates the transcript of the 1925 Scopes Trial in Dayton, Tennessee, and persists to this day in American evangelical circles through the continued rhetorical resonance of that trial (Larson 1997). By the turn of the twentieth century, however, professional biologists were already using "Darwinism" to refer to natural selection's ability (or inability) to explain descent from a common ancestor. After its marriage to population genetics in the 1930s, Darwin's theory of evolution by natural selection acquired a new and improved interpretation that since the end of World War II has organized the space of inquiry in evolutionary biology. This theory or, better, family of related theories is called the Modern Evolutionary Synthesis (Huxley 1942).[1]

From the time of their youthful studies, professional evolutionary biologists have been so deeply immersed in the neo-Darwinian paradigm (as the Modern Synthesis is also called, albeit misleadingly[2]) that they are sometimes surprised to learn that at the turn of the twentieth century Darwinism was widely rumored to be on its "deathbed" as a general theory of evolution (Dennert 1904; Kellogg 1907). Where, we may ask, did these challenges to the power of natural selection come from? Sometimes in combination and sometime as rivals, they came mostly from three generative ideas about evolutionary process that preceded the *Origin of Species*, affected it, and survived it:

Evolutionary developmentalism: the idea that evolution takes place in and by means of inner-driven change in the ontogenetic or embryological process, a historical record of which is preserved in the pattern of phylogeny that ontogeny leaves behind or, as evolutionary developmentalists once believed, recapitulates.

Inheritance of acquired characteristics: the notion that evolution's causal grindstone is the adaptation of competing organisms to resource-scarce environments, but that its driver is not natural selection, as Darwin held, but the adaptive effects of impinging environments on embryos or the direct transmission to offspring of habits acquired postnatally.

Saltationism: whether its causal locus is ontogenetic or takes place at the environment–organism interface, the notion that evolutionary novelties do not result from a gradual process of adaptation at all but from sudden leaps or saltations (from Latin saltus, leap).

[1]Biology, evolutionary biology, evolutionary theory, and philosophy of biology differ but form a continuum. The center of gravity in this chapter is between the second and the third of these forms of inquiry, with glances at the fourth.

[2]"Neo-Darwinism" originally referred to August Weismann's (1834–1914) belief that natural selection working exclusively on germ-line heritability is the sole cause of evolution. "Hard heredity" is a necessary condition for the Darwinism of the Modern Evolutionary Synthesis, but the Synthesis rejects the "all-sufficiency (*Allmacht*) of natural selection" and takes a population-level view of evolutionary processes.

Darwin himself took these ideas seriously. His theory was persuasive in part because the "one long argument" of the *Origin* respectfully reported their attractions, provided evidence against their prowess, and sometimes reformulated them in ways that assigned them auxiliary roles (Darwin 1859: 459). Later editions of the *Origin* eventually swelled to almost twice the size of the first because Darwin persisted in this simultaneously accommodative and refutative process (Depew 2009).

In this chapter, I will protract Darwin's discursive strategy into the history of twentieth-century Darwinism by arguing that:

1. The same *leitmotivs* that presented themselves as challenges to Darwin's Darwinism also presented themselves to neo-Darwinians, but this time coming mostly from the new field of genetics. For example, the recovery and reanalysis of Mendel's experiments on peas in 1900 turned saltationism into genetic mutationism.
2. The Modern Synthesis acquired its hegemony by responding to these three sorts of challenges in ways that were as persuasive, if not always as polite, as Darwin's. Since its consolidation in the 1940s, the Synthesis has maintained its dominance by simultaneously refuting and co-opting new forms of saltationism, Lamarckism, and developmentalism.
3. In attaining and retaining its primacy, the Synthesis has from the start fused empirical discoveries with conceptual adjustments, a process more common in scientific inquiry than is sometimes recognized, and more legitimate. Since the end of the 1960s, conceptual adjustments have played an especially prominent role in containing threats carrying traces of other approaches. Examples of potential threats include supposedly directional, and so far forth Lamarckian, mutation in bacteria; the "punctuated" pattern of macro-evolutionary diversification, in which the production and retention of variation is clustered around sudden, and in an attenuated sense saltational, speciation events; and the increasingly widespread perception that the regulatory sectors of metazoan genomes are highly conserved across taxa and respond to environmental variables in ways that appear to revive the prospects of evolutionary developmentalism.

Some of these threats have waned, by either empirical reanalysis or conceptual accommodation or both. In recent years, however, evolutionary developmentalism ("evo-devo") has been issuing especially strong challenges to the Modern Synthesis. The responsiveness of regulatory genes to environments seems to be the proximate cause of variations that exhibit adaptive characteristics as soon as they appear (Pigliucci 2017). One implication is that heritable factors resemble less and less the rigid genetic programs envisioned by early molecular biologists. What gives this perception an even greater *frisson* or air of danger, however, is that it comes close to contravening a rock-solid conceptual principle of the Modern Synthesis in all of its versions: that organisms and their traits can be said to be adapted only if they evolve from a multigenerational, and so far forth gradual, process in which natural selection, working on nondirected (chance) variation in conjunction with other evolutionary factors, amplifies the frequencies of some genotypes in a population and diminishes or eliminates the frequency of others

(Dobzhansky 1937). It comes as no surprise that recent work on gene expression and related topics has precipitated a debate among respected evolutionary biologists about whether "all is well" with the Modern Synthesis or whether it needs a "rethink" (Wray et al. 2014; Laland et al. 2014). By the end of the chapter, my view will be clear: All is not well, but if it can be made well it will by conceptual adjustments no more or less radical than those the Darwinian tradition has already undergone in its discursively polymorphic and polytypic career (Weber and Depew 2003; Depew and Weber 2017).

To many Darwinians the successes I chart will seem more continuous and cumulative than I portray them. In taking the part of other traditions in evolutionary thought I am not throwing cold water on Darwinism's past triumphs or doubting its ability to rise to new occasions. Rather, I am drawing attention to the contingency of this process. It could have failed if conceptual innovations such as the probabilistic revolution, and not just empirical discoveries and methodological innovations, had not pointed the way forward. Hence, in considering current challenges to Darwinism, it is both sobering and salutary to consider, with the help of many of the essays in this volume, just how tough things looked for Darwinism at various points in its career.

2 Saltation, Old and New

The idea of "biology" construed as a unified science of life that would integrate the natural-historical fields of systematics, comparative anatomy, paleontology, and biogeography with functional studies of morphology, inheritance, physiology, embryology, and biochemistry is no older than the turn of the nineteenth century. The term "biologie" was first used in 1800 in an unpublished manuscript by Lamarck.[3] Göttingen's Gottfried Reinhold Triviranus wrote a multi-volume work called *Biologie* in 1802. No more important site for the flourishing of biology in this sense or its cautious embrace of transformism existed than the Museum of Natural History and Jardin des Plantes in Paris. Its three titans—George Cuvier, Étienne Geoffroy Saint-Hilaire, and Lamarck—lived, worked, and argued there for over 40 tumultuous years during which France changed from an absolute monarchy—the Jardin des Plantes was originally the Jardin du Roi—to a democratic republic, a military dictatorship, and in 1830 a constitutional monarchy in whose shadow their most famous debate was conducted (Appel 1987).

Cuvier, a comparative anatomist, presumed that the parts of organisms are functionally adapted to each other in ways that fit species for life in particular environments. [His phrase "conditions of existence" refers not to these environments, as it does in Darwin, but to the tight internal organization that equips

[3]Lamarck used the term publicly for the first time in 1802. In Latin, "biologia" occurred in the 1760s in the philosophical works of Christian Wolff and his disciples but with a different meaning.

organisms to live in and from them (Grene and Depew 2004; Reiss 2009).] So axiomatic did Cuvier regard this heuristic principle that he bragged that from a single bone he could reconstruct the entire anatomy and way of life of the extinct species he and his colleague Alexandre Brogniart unearthed in the Paris chalk. Geoffroy challenged Cuvier's conviction that animal kinds fall into four distinct, primordial phylum-level body plans: *vertebrata*, *mollusca*, *articulata*, and *radiata*. Cuvier's functionalism, Geoffroy argued, obscures structural elements whose transformations run across all body plans. The same elements recur in the most unlikely places, sizes, and combinations. The gill cover in fish, for example, shows up as the tiny bones in the mammalian inner ear. Still, even if Geoffroy was not a functionalist like Cuvier or an evolutionist like Lamarck, he was enough of a holist to maintain that all the parts of an organism must be in balance with each other. If a new species arises, accordingly, it must result from a sudden fracture followed by a more or less simultaneous rebalancing of parts. In this way, "Monstrosities could become the founding fathers (or mothers) of new species by instantaneous transition from one form to the next" (Hallgrímsson and Hall 2011: 18). This is classical saltation.

When Darwin's *Origin* was translated (poorly) into French in 1860 it entered a conceptual field prestructured by Cuvier, Geoffroy, and Lamarck's disputations. In 1864, Albert von Kölliker embraced Darwin's case for unity of descent but, seeing it through Geoffroyian eyes, discounted his gradualist and selectionist etiology. Something like this happened closer to home as well. In embracing descent with modification from a common ancestor, the comparative anatomist Thomas Henry Huxley, Darwin's polemical champion, warned him, "You have loaded yourself with an unnecessary difficulty in assuming [the ancient principle] that *natura non facit salta* [nature does not make leaps]" (Huxley to Darwin November 23, 1859, #2544).[4] Huxley's opinion reflects an anatomist's appreciation of organic unity. This led him to discount Darwin's declaration that, "My theory [of evolution by natural selection] would break down if it could be demonstrated that any complex organ existed which could not possibly have been formed by numerous, successive, slight modifications" (Darwin 1859: 189).[5]

Huxley was not alone in failing to accept, or perhaps even see, what Darwin called "the paramount power" of natural selection (Darwin 1868; Gayon 1997). Like many others he implicitly construed natural selection more as eliminating the antecedently unfit than as gradually evolving the adapted, as Darwin believed. Puzzles about how eliminative selection could possibly drive evolutionary progress were important sources of what Peter Bowler has called "the non-Darwinian

[4]Texts from letters to and from Darwin are cited by their identifying numbers in the Darwin Correspondence Project, https://www.darwinproject.ac.uk/letters

[5]For Darwin, adaptations evolve by natural selection in order to and because they perform biological functions. He included the functionalist but anti-evolutionist Cuvier on his short list of heroes but not the incipiently evolutionist but structuralist Geoffroy (Darwin to Ogle January 17, 1882, #13622). His other heroes, Aristotle and Linnaeus, were also non-evolutionary functionalists.

revolution" that took place after Darwin's death and persisted in some biological (and social scientific) fields into the 1930s (Bowler 1988, 2013, 2017; Gayon 1995, 1998; Stocking 1968). During this period, evolutionary developmentalism was ascendant. Descent with modification was construed as ontogeny writ large and ontogeny as phylogeny writ small. Like the passage from embryo to adult, phylogenic diversification was thought to clamber up a ladder of progress by inner-driven ("orthogenetic"), not contingent environmental, causes. Human races, unfortunately, were graded the same way.

To be sure, toward the end of the developmentalist interregnum faithful Darwinian gradualists in Great Britain pioneered the use of statistics to show that character gradients correlate with identifiable environmental changes in effecting subspecific evolution by directional natural selection (Provine 1971). It wasn't until the middle decades of the twentieth century, however, that the makers of the Modern Synthesis extended this "biometrical" approach to traits to the evolution of species and higher taxa in an empirically plausible way (Dobzhansky 1937; Mayr 1942; Simpson 1944). This temporary failure, together with the prestige germ-line-only heredity had recently acquired through August Weismann's polemics, made it almost inevitable that when Mendelism appeared in 1900 orthogenetic developmentalism would be displaced not by natural selection but by genetic mutationism and hence by a new form of saltationism.[6] Henry de Vries in the Netherlands, Gregory Bateson in Great Britain, and later the German-trained geneticist Richard Goldschmidt in America all championed the anti-Darwinian idea that genetic mutations of large effect—macromutations, in later parlance—are the creative factor in evolutionary change.[7] If Goldschmidt was mocked and pilloried for this view, it was because he had the misfortune of defending it after the Modern Synthesis had gained a foothold (Goldschmidt 1940; Mayr 1980). From the perspective of the recent turn to evolutionary developmentalism, Goldschmidt appears more as a prophet before his time than as a saltationist born too late (Gould 1980b, c).

It would be wrong to call Thomas Hunt Morgan, discoverer of the chromosomal locus of genes and father of transmission genetics, a saltationist in the sense(s) that Geoffroy, Bateson, and Goldschmidt were. Still, until the end of his life Morgan retained enough of their presuppositions to believe that mutation is the creative factor in evolution (Morgan 1935; Beatty 2016). By the 1930s, he had come to see that populations contain a great deal of standing genetic variation. Assuming, however, that only "the more extreme individuals of the population" can effect a novel redistribution of genotypes, Morgan concluded that no matter how much variation may collect as recessive alleles in heterozygotes or how widely distributed

[6]"New ideas about [germ-line or hard] heredity emerged in part out of an enthusiasm for the concept of evolution by jumps or saltations, reflecting an anti-adaptationist position" (Bowler 2013: 195, 2017).

[7]The phase "factors of [organic] evolution" was first used in Herbert Spencer (1887); the list of candidates is still growing. "Creative factor" was probably due to the influence of Henri Bergson's *Creative Evolution* (1907), even if Morgan, Dobzhansky, and others ascribed evolutionary innovation and direction to factors other than Bergson's inner-driven, intuitively apprehended source of change (*élan vital*) (Loison and Herring 2017).

through a population this variation may be, evolutionary advance will not resume until a variant powerful enough to trigger a new spasm of differential reproduction happens to arise (Morgan 1935: 130; Beatty 2016). A variant may take a while to spread, but there is at least a whiff of saltation in the discontinuity with which the evolutionary process as Morgan saw it starts and stops.

The Modern Synthesis undermined Morgan's model of evolution by capitalizing on the statistically ingenious population geneticist R. A. Fisher's refutation of de Vries and Bateson. Fisher demonstrated mathematically that the probability of a macro-mutation spreading through a population is very low, even under selection, but surprisingly high for mutations with small but continuously additive effects on comparative reproductive output (Fisher 1930). For this reason, Fisher cast natural selection, not the mutations that are its necessary condition, as the creative factor in evolution. Over trans-generational time it gradually evolves adaptations, evolution's leading edge.

In the late 1930s, Theodosius Dobzhansky, who during the decade he spent as a guest researcher in Morgan's lab apprised his host of the extent of variation his Russian mentors had uncovered in natural populations, developed a theory of raciation and speciation that made empirical use of population-genetic theorems derived by Fisher's American rival, Sewall Wright (Adams 1994a, b on Dobzhansky's Russian background; Provine 1986 on his use of Wright). Wright and Dobzhansky argued that genetic drift—the mathematically predictable tendency of genetic variation to spread indiscriminately, and hence by chance, to new generations in small populations—allows useful variations to get a toehold that natural selection can ramp up to the point of species-defining genetic isolation (Wright 1932; Dobzhansky 1937). The idea bore fruit. Mayr's evidence that speciation takes place at the periphery of a species' range is its biogeographical expression (Mayr 1942, 1963). Simpson argued that the process is extrapolable to the evolutionary genesis of higher taxa (Simpson 1944; Pigliucci 2017).

The paleontologist Stephen Jay Gould maintained that by its heyday in the 1950s and 1960s, the Mayr–Dobzhansky–Simpson version of the Modern Synthesis had become, regrettably in his view, less chancy and more selectionist, gradualist, adaptationist, and so far forth Darwinian than its Wright-inflected prototype (Gould 1983). Although the original theory was never as non-selectionist as Gould made out, the claim remains true for at least two reasons. First, Fisher's followers at Oxford proved that many species-marking traits hitherto taken by naturalists to be selectively neutral (because they stay constant enough to serve the classificatory purposes of museum taxonomists) are finely adapted to specific environments (Lack 1947; Kettlewell 1955, 1956). This led to an expectation, soon called "adaptation*ism*," that traits should be presumed to have selectionist etiologies until proven otherwise (Gould and Lewontin 1979).

A second factor in what Gould called "the [adaptationist] hardening of the Modern Synthesis" was that in 1947, a decade after the publication of his Wright-influenced account of speciation, Dobzhansky began finding evidence that natural selection evolves not just adapted traits tied to specific environments but mechanisms for adapting that make the evolution of new races and species more probable in the face of persistent but variable environmental change (Depew 2011). Diploid (and polyploid) chromosomal structures, for example, favor what was later called "evolvability" because they retain variation in heterozygotes that may prove adaptive under new conditions. Indeed, Dobzhansky maintained that heterozygotes are often selectively favored because in rapidly changing environments they are inherently adaptive.[8] Later editions of *Genetics and the Origin of Species* expand on what he told his Columbia University colleague L. C. Dunn in 1947: In natural populations, the higher reproductive rate of variation-preserving heterozygotic chromosomal loci diminishes when not yet fully reproductively isolated local races of fruit flies are crossed, but goes up again when they have become genetically isolated species (Dobzhansky 1951, 1970). This result, Dobzhansky informed Dunn, offers "one of the most elegant proofs of natural selection known," since it shows that natural selection does not depend on mutation, haphazard migration, or genetic drift but by itself can induce hybrid sterility between races-turning-into-species even as it maintains hybrid vigor within them once they are genetically closed (Dobzhansky to Dunn, April 26, 1947, Dunn Papers, American Philosophical Society).

Gould lamented the eclipse of Wright's "pluralism" about evolutionary factors because he was on the lookout for non-selective processes to account for the "punctuated" pattern he and his collaborator Niles Eldredge had spotted in phylogenetic history, according to which gene frequency changes are concentrated around nodes of speciation instead of being smoothly distributed across an ongoing process of adaptive improvement ("phyletic gradualism"), as Simpson postulated (Eldredge and Gould 1972). Gould took the punctuated character of phylogenesis as evidence that organisms are internally integrated in ways that constrain the work of natural selection. Stasis is the norm (Turner 2017). When change occurs it comes in sudden bursts of speciation. This echo of Geoffroy-like saltation and hint of regression to Morgan's start-and-stop theory of evolution led protectors of orthodoxy to point out that gradual does not mean constant. Simpson himself had allowed, indeed required, natural selection to work at different rates (Simpson 1944). Moreover, a paleontologist's "instantaneous" is consistent with a population geneticist's "gradual." "One hundred thousand years," wrote Mayr,

[8]Fisher proved mathematically that under certain conditions natural selection can favor heterozygotes, but, unlike Dobzhansky, he did not assign an evolutionary function to this scenario (Fisher 1930). Dobzhansky's encounter with French evolutionists may have been a source of the distinction he drew between adaptations to specific environments (which can be traps) and heterotic adaptations for adapting (Loison and Herring 2017). As a young man, Dobzhansky read Bergson. He devoted his career to showing that natural selection can explain tendencies that Bergson's followers ascribed to an inner drive that can be philosophically intuited but not experimentally proven. When he argued that, "Nothing in biology makes sense except in the light of evolution," Dobzhansky's overt target was creationism, but his point has wider significance (Dobzhansky 1973). In contrast to traits that evolutionary and non-evolutionary observers alike can agree are adaptive—the differently shaped beaks of the finches Darwin found on neighboring islands in the Galapagos, for example—evolutionary history's most important adaptations cannot even be seen by pre- and anti-evolutionary biologists, let alone be explained by them.

> would be instantaneous for a species experiencing a 10-million-year stasis. The semantic problem is evident when we consider that all population evolution ... is gradual. It is obvious from the recent controversy that the chronology of speciation events cannot be established by paleontological analysis. Rather, it will have to be inferred from an analysis of currently living speciating species ... In freshwater fishes it may take less than four thousand years (Mayr 1992).

This was not telling Gould anything he did not know or had not already said, albeit without Mayr's implication that no revision in the general direction of saltation is required, that "all is well." Gould's aim was not to refute the Modern Synthesis, but to free it from constraining assumptions, most harking back to the context in which Darwin was arguing, by "expanding" the ways in which its full explanatory toolkit can be put to work. Evolutionary factors—mutation, migration, drift, selection (of various sorts on various objects), and sheer stochasticity—are allowed to combine differently at different levels of a hierarchy of biological structuration ranging from proteins below to higher taxa above (Gould 1980a). Gould proposed to accommodate his own brand of paleontology to his "expanded Synthesis" by liberating the Synthesis from the Procrustean box of Darwinian gradualism and adaptationism in which it had confined itself. *Pace* its founders, he maintained that macroevolution is sufficiently discontinuous with evolution at and below the species level to exhibit a kind of selection process—species selection, which ranges over differences in the fecundity and longevity of clades—that is more open than evolution at the organismic level to survival and extinction by sheer accident (Gould 1980a, 1989; Delisle 2017; Shanahan 2017; Pigliucci 2017).

Among the attractions of Gould's expanded Synthesis is that it was consonant with, and may have helped catalyze, the mass conversion of systematists to Willi Henning's "phylogenetic systematics," which began in the 1970s (Eldredge and Tattersal 1975; Delisle 2001). More commonly called "cladism" (after the Greek word for branch), phylogenetic systematics stipulates that only points where lineages split should count in classifying (Hennig 1950). By contrast, Mayr, Simpson, and especially Huxley insisted that taxonomic practice must take into account gradual phyletic evolution to higher grades as well as the splitting off of new clades. From the cladist perspective, this view seemed to harbor a prejudice that Linnaeus, for one, inherited from the medieval Great Chain of Being. Science, Darwinism, and a fortiori what passes as common sense are still having trouble freeing themselves from this "higher-lower" metaphysical picture.

The idea of expanding the Synthesis also helped slow an anti-Darwinian tendency in molecular biology. By the late 1960s, molecular geneticists, whose intellectual roots more often lay in biochemistry than natural history, had shown that the genetic code for amino acids, and hence for the proteins they compose, substitutes mutated nucleotides at a constant rate without affecting biological function. This theorem is called "neutral mutationism" and sometimes "non-Darwinian evolution" (Kimura 1968; King and Jukes 1969; see Pigliucci 2017). Even a protein as functional and deeply entrenched in evolutionary history as cytochrome c, which in its long career has presumably been subjected to a wide range of environmental pressures, shows a mutation rate constant enough to reveal the ticking of a "molecular clock" that tracks evolutionary time's branching pathways (Zuckerkandl and Pauling 1965). Gould's proposal was simple. This might be true of protein evolution, but only philosophical reductionism implies that what goes for proteins goes for everything else. In admitting a variety of evolutionary factors, multiple levels and units of selection, and rates of change, Gould was defending the Darwinism of the Modern Synthesis by expanding its conceptual framework to include semi-saltationist scenarios and developmental constraints.

This is not, however, the end of the story. It has since been shown that molecular clocks tick at different speeds in different lineages and that the occurrence of neutral or nearly neutral mutations is not as random as defenders of "non-Darwinian evolution" have assumed, since the third position of each codon, on which functioning depends least, does most of the mutating (Lee and Ho 2016; Moorjani et al. 2016).[9] Discoveries of this sort have encouraged adaptationists to believe that, far from saving the Modern Synthesis, downplaying Darwin's gradualist and adaptationist axioms undermines it. The paleontologist Simon Conwy Morris is among those who see natural selection, adaptation, and gradualism permeating the biological hierarchy. He has resisted Gould's effort to enlist him as a supporter by appealing to convergent adaptation by natural selection to restore the continuity between micro- and macroevolution postulated by Dobzhansky, Mayr, Huxley, and Simpson (Gould 1989; Morris 2003).

To be sure, many evolutionary biologists remain as convinced as Gould that important facts do not fit the adaptationist paradigm. If they also happen to believe that the Modern Synthesis is inseparable from that paradigm, their resentment in being asked to choose between up-to-date evolutionary biology and the pretensions of the adaptationist worldview to be the premier *defensor scientiae* in our time can take the form of rejecting not just the Modern Synthesis but the Darwinian tradition generally. Part of the difficulty arises from identifying Darwinism with a rigid paradigm rather than seeing it as a historically evolving research tradition with a core set of principles that is often confused with a shifting periphery of replaceable working assumptions (Lakatos 1970; Laudan 1977). Levit and Hossfeld (2017) as well as Delisle (2017) also invoke this historiographical heuristic. Interpreting Darwinism's gradualism as constancy of rate may be one of those confusions.

[9]Intelligent design creationists have glommed on to neutral mutation and evenly ticking molecular clocks as reasons for disputing not just natural selection but evolution itself. Not surprisingly, they have been loath to take note of complications suggesting the workings of natural selection after all (Hofmann 2017).

3 Lamarckism and Darwinism: A Dialectical Relationship

Lamarck, an invertebrate biologist, was an evolutionist who did not embrace unity of descent. On the contrary, he held that evolution from the simple sorts of organisms he studied began from several points. From spontaneously generated origins, it moves toward complexity on its own steam but also relies on the direct effect of environments and the heritability of habits acquired by repeated use to adapt organisms and their offspring to particular environments.

Lamarck's theory of evolution is called *transformisme* for a good reason. His evolutionary process implies that species do not (always) go extinct but instead are slowly transformed into descendant species. Darwin's youthful voyage on the *Beagle* solidified his belief in extinction, and lots of it, and set him searching for an evolutionary mechanism far more under the sometimes lethal influence of external circumstances than the blend of Lamarckism and neo-Geoffroyean unity of descent retailed by his foil, the [then anonymous] "author of the *Vestiges of Creation*," Robert Chambers (Darwin 1859, Introduction; Chambers 1844; Secord 2000). Darwin retained use inheritance as a secondary cause of adaptation, but it played second fiddle because Darwin was sure that only his theory of natural selection could explain both the extent of wreckage in life's history and the beautiful "co-adaptations of organic beings to each other and to their physical conditions of life" (Darwin 1859, Introduction). It is no accident that Darwin did not use "transformism" in the *Origin* and used "evolving" only once.

The late nineteenth-century "non-Darwinian revolution" studied by Bowler made inner-driven complexification evolution's primary driver, reversed the primacy Darwin assigned to natural selection over the heritability of acquired characteristics, and treated the direct effect of environments on embryos and neonates as a more important form of Lamarckian adaptation than use inheritance. Natural selection was accepted as a factor, but it was largely relegated to the task of weeding out organisms that fail to adapt to the exigencies of the Malthusian scarcity under which all living beings were thought to labor (Spencer 1887). Accordingly, August Weismann was flying in the face of a widespread consensus when he insisted that acquired characteristics of every stripe come too late in the developmental process to be heritable and that only gradual natural selection working on chance variation in the germ line (not identified as genetic mutations for another decade) could cause evolution (Weismann 1889).

It is a myth that Weismann's neo-Darwinism—so called because it did away with Darwin's pluralism about inheritance and embraced the "all-sufficiency" of natural selection—was universally accepted as soon as he reported that chopping off the tails of a few generations of mice failed to show the heritability of taillessness. On the contrary, the Paris Municipal Council funded France's first chair of evolutionary biology, at the Sorbonne, with a view to supporting Lamarck. The University of Paris's interest in so arcane a topic sprang in part from the association of neo-Darwinism with eugenics at a time when France was preoccupied with raising its birth rate so that its military would be in a position to take revenge on the Germans for their occupation of France in 1870 (Cook 1999; Burian and Gayon 1999). The chair's first occupant, the zoologist Alfred Giard, urged funding an "*institut transformiste*" whose reliance on the high standards of French experimentalism established by Claude Bernard would demonstrate Weismann's errors. When Mendelism bolstered the fortunes of Weismann's hard inheritance, a graduate student in a French university was charged with proving that Morgan's work on Mendelian ratios in fruit flies was compromised by improper laboratory methods and safeguards. When he proved instead that Morgan was right, his Ph.D. supervisor disowned him (Burian and Gayon 1999: 317). After the heritability of acquired characteristics proved resistant to experimental validation, French Lamarckians turned to its inner-driven side (Loison and Herring 2017). It was not until 1946 that France established a university chair of genetics.[10]

The French were not the only latter-day Lamarckians. At the turn of the twentieth century, American biologists, especially paleontologists and botanists, had an even more pronounced neo-Lamarckian bent.[11] The biological research stations at Woods Hole, Massachusetts, and Cold Spring Harbor, New York, were initially funded with the aim of opposing Weismann. Fearing that his pan-selectionism would undermine the very possibility of evolutionary and social progress—after all, genes adapt organisms only to local, evanescent environments—and regarding as inconclusive a recent high-profile debate on this subject between Weismann and Spencer, Henry Fairfield Osborn, paleontological curator at the American Museum of Natural History, challenged the Marine Biological Laboratory at Woods Hole to conduct experiments in which:

> ... [A]n organism with an environment or habit A is transferred to environment or habit B, and after one or more generations exhibits variation B. This organism is then re-transferred to environment or habit A. If it still exhibits, even for a single generation or transitorily, any of the variations B, the experiment is a demonstration of the inheritance of ontogenetic variations (Osborn 1895: 97).

---

[10] A caveat. In the interwar period, a talented circle of French geneticists, many with training in and funding from other countries, began experiments in physiological genetics. This work, conducted in research institutes, positioned Boris Ephrussi, André Lwoff, Francois Jacob, and Jacques Monod (who with Jacob discovered the *lac* operon, the first regulatory genetic mechanism to be understood) to take the lead in studies of gene regulation in the 1960s, in the process restoring the decisive importance of experiment (Burian and Gayon 1999; Loison and Herring 2017). At that time, molecular geneticists in America were still preoccupied with nailing down the genetic code and finding the mechanisms and pathways of protein production.

[11] American neo-Lamarckism, prominent in the nineteenth century, did not last far into the twentieth. Unlike their French counterparts, American evolutionary biologists in the interwar period embraced genetic determinism and negative eugenics—eugenics aimed at preventing the supposed unfit from reproducing in contrast to the positive eugenics that flourished in the United Kingdom, which aimed at breeding a fitter governing class—in ways that tended to support the racism with which the United States still struggles (Kevles 1985). Dobzhansky worked with American anthropologists to develop a version of the Modern Synthesis that opposed all three: genetic determinism, eugenics, and racism (Jackson and Depew 2017).

Two years later, in 1896, the Welsh ethologist Conwy Lloyd Morgan and the American child psychologist James Mark Baldwin appealed to Osborn's experimental design to propose a "new factor in evolution" that Baldwin called "organic selection" (Baldwin 1896; Lloyd Morgan 1896). Habits acquired through instruction—in songbirds as well as humans, Lloyd Morgan pointed out—can be protracted by Lamarckian inheritance across an indefinite number of generations until genetic variations independently arise to support them, as Darwin and Weismann required. The reasoning was that, while variation in any and all directions is constantly arising, variants other than adaptive will either be selected against or have no effect at all (Depew 2003). Osborn jumped on the bandwagon. Why not? Organic selection saved evolutionary progress and made social progress almost axiomatic (Osborn 1896; Rainger 1991). In ascribing the concept of adaptedness to social inheritance, Obsorn predictably put the accent on the Lamarckian side. By contrast, in publications after 1896, Baldwin predicated organic selection and adaptation of the genetic changes that on his hypothesis eventually support initially learned and socially transmitted behaviors.

There the issue lay until the early 1950s, when the developmental geneticist C. H. Waddington likened his experiments in producing genetic change by heat-shocking fruit flies to Baldwin's scenario (Waddington 1953). This comparison suggested that genetic variation is both environmentally inducible and incipiently oriented in a favorable direction. Accordingly, what Waddington called genetic assimilation imposed a burden on supporters of "the Baldwin effect," as Simpson dubbed it, to show that it was not Lamarckian (Simpson 1953). The burden proved increasingly hard to meet, especially after the "central dogma of molecular biology" set germ-line inheritance and Mendel's "laws" in stone, thereby further hardening the Modern Synthesis by conceiving of DNA sequences as "coding for" particular traits and seeming to require that random mutations in the genetic code must initiate evolutionary change. The three musketeers of the Modern Synthesis, Dobzhansky, Mayr, and Simpson, had from the start predicated adaptation of gene frequency changes but had not stipulated that mutation or recombination must come first. Because they were worrying about Lysenko's Lamarckism in the Soviet Union, however, and did not want to put the Synthesis at odds with Crick and Watson's discovery of the ultimate (albeit not proximate) source of genetic variation, they distanced themselves from the phenomena to which Waddington and before him Baldwin were pointing. They regarded them as expressions of previously fixed genes with "wide norms of reaction" and "phenotypic plasticity" (Dobzhansky 1970; Levins and Lewontin 1985: 94–95, with specific reference to Waddington; Gilbert 1994: 153). That the environment in Baldwin's effect is cultural and Waddington's genetic assimilation is induced by violence contributed to the sense that even if they do occur these phenomena are unnatural and infrequent.

This is not an unreasonable interpretation. Dobzhansky's theme had always been that wherever it can natural selection fixes genotypes in populations that express themselves differently, and adaptively, in a wide range of environments, thereby enhancing the ability of lineages to ride over environmental contingencies. In a sense, the phenotypic plasticity that informs the mid-century Synthesis reconstructed Lamarckian themes by turning what Lamarck regarded as the *explanans* of a transformational process into an *explanandum* well explained by Darwin's model of chance variation and selective retention (Dobzhansky 1951; Levins and Lewontin 1985: 95). "When a trait becomes plastic," Dobzhansky wrote, "it exhibits more and more 'Lamarckian' modifications without thereby altering the [inherited] trait in the offspring" (Dobzhansky to Montagu, July 20, 1947, Montagu Papers, American Philosophical Society).

Richard Lewontin, Dobzhansky's former student and Gould's sometime coauthor, was even more forthright. In the 1960s, Dobzhansky began fancying that trans-generational natural selection uses genetic variation to solve "problems" posed to populations by environmental "challenges" (Dobzhansky 1962: 17). Lewontin saw this trope as complicit with the adaptationist and deterministic hardening of the Synthesis that Dobzhansky himself had scorned when he began viewing natural selection as favoring genotypes with wide norms of reaction. He proposed to liberate his mentor's earlier thought from his challenge-response conceit by throwing cold water on the very concept of adaptation. He portrayed organisms as making their own niches, not passive products of, because reacting to, genetic and environmental pressures. They are agents in their lived worlds because, far from merely obeying internal and external forces, they tailor developmental resources to their needs. "The environment of an organism," Lewontin writes,

> is not an independent, preexistent set of problems to which an organism must find solutions, for organisms not only solve problems, they create them in the first place. There is no organism without an environment, and no environment without an organism. 'Adaptation' is the wrong metaphor and needs to be replaced by 'construction' (Lewontin and Levins 2007: 231; see also Levins and Lewontin 1985: 99–104; Lewontin 1982).

Niche construction, as it has come to be called, has made the quasi-Lamarckian theme of agency in the unhardened Synthesis, including the Baldwin effect, more salient (Odling-Smee et al. 2003). Its advocates have been prominent among those calling for a "rethink" of the Modern Synthesis (Laland et al. 2014). Their case has profited from a sea change in genetics. The image of a genetic program running recursively, irreversibly, and autonomically through a fixed sequence from DNA to RNA to protein until a potentially useful mutation in the genetic code happens to pop up emerged in the immediate wake of Crick and Watson's great discovery in 1953. In the schools and popular press, this picture still passes as the sum and substance of genetics and as the cornerstone of trait-adaptationist, genetic-determinist versions of the Modern Synthesis, in which we are told that there is a gene for this and a gene for that. But progress in understanding gene regulation since the pioneering work of Jacques Monod, Francois Jacob, and André Lwoff has rendered this picture not just incomplete but misleadingly upside down. In turning genes on and off in the ontogenetic process, regulatory sectors of the genome slice and dice RNA as the dynamic interaction between an organism's development and its environmental circumstances requires—so much so that it is difficult to identify a particular chunk of DNA as "the gene for x," or even as a gene at all (Burian and Kampourakis 2013).[12]

Nor is DNA a splendidly isolated molecule. It often comes wrapped, sometimes loosely, sometimes tightly, around proteins and encrusted with chemical side chains, such as methyl groups. These affect whether DNA is transcribed into RNA and RNA is translated into protein. At least in single-celled organisms, these effects are "epigenetically" heritable enough to affect evolution. What is in question is whether they do so merely by providing pluralistic Darwinism with another source of naturally selectable variation or, in their openness to external influence, circumvent this process by forging a more immediate link between organism and environment. DNA's advantages as an information bank are inseparable from its chemical stolidity, to be sure, but it can be modified by tricks microbes learned long ago and that genetic biotechnicians are becoming increasingly skilled at mimicking and exploiting. When combined with revitalizing phenotypic plasticity as developmental plasticity (Schlichting and Pigliucci 1993; Pigliucci 2007; Schlichting 2008), the environmentally sensitive "epigenome"—genes in combination with other factors involved in transcription, translation, and protein folding—now coming into view has made life easier for niche constructionists, Baldwin boosters, and latter-day Lamarckians. Mary Jane West Eberhard refers to her extensively documented claim that genes are following not leading indicators of evolutionary change as Baldwinian (West-Eberhard 2003). In discussing epigenetics, Eva Jablonka and Marion Lamb speak of a "Lamarckian dimension" (Jablonka and Lamb 1995). Even genetic assimilation has been getting another look (Pocheville and Danchin 2017).

In framing one-way information flow from DNA to RNA to protein as molecular biology's "central dogma," Watson helped turn evolutionary inquiry into a well-institutionalized techno-scientific discipline, in part by creating an impression of straight-line progress in genetics from Mendel to Weismann to Morgan to his and Crick's decoding of DNA to the Human Genome project to gene therapy.[13] A strongly trait-adaptationist view of evolutionary dynamics, and a corresponding view of organisms as decomposable assemblies of parts and their functions, lurks in the background. Ironically, however, this program has rained on its own parade by turning up facts suggesting that neo-Darwinians have falsely generalized from the metazoa that served them as model organisms. Neo-Darwinian principles apply well enough to lineages whose modes of generating variation, differentially retaining it, and patterns of phylogenetic diversification depend on rigid command and control of somatic cells by sequestered genes (Keller 2000; Newman and Müller 2000). Things are more fluid, however, in other biological kingdoms.

[12]"It takes an enormous amount of biological machinery for genes to be expressed; exactly which parts of the genome are processed depends on specific settings and structure of that machinery" (Burian and Kampourakis 2013: 613).

[13]Crick pointed out in 1970 that his 1956 version of the central dogma was not as dogmatic as the (unnamed) Watson's insistence that information must flow unidirectionally in temporal order from DNA to RNA to protein. See Burian and Kampourakis (2013: 616, n. 27).

Evolutionary bacteriologists and botanists recognize in the systems they study phenomena such as endosymbiosis, horizontal gene transfer, and strategies that enhance evolvability by biasing variation in directions from which organisms or colonies will benefit and may even anticipate (Margulis 1992; Kutschera 2017; Doolittle 1999; Cairns 1988; Foster 2007; Goldenfeld and Woese 2007; Woese and Goldenfeld 2009). It is doubtful whether the very notion of species applies to microbial systems, let alone the "biological species concept" favored by the founders of the Modern Synthesis.

It is common for Darwinians of the Strict Observance to dismiss these notions as "Lamarckian." By using it as a term of abuse, the enforcers of the central dogma broadened "Lamarckian" to include anything even hinting of violations of the formula "random genetic variation plus [and before] natural selection." However, the growing realization that there are more living things on earth than Weismann, Morgan, or Watson dreamed of has prompted Jablonka and Lamb to suggest that Lamarckism construed as broadly as it has come to be has a better claim to theoretical generalizability than the Darwinian tradition (Jablonka and Lamb 1995, 2005). The suggestion is that the explanatory range of the Modern Synthesis is biased toward organisms to which our own size and senses have adapted us. Its command over the range of living systems is no wider than the command of classical (Newtonian) physics over the full range of physical systems.

Jablonka and Lamb's suggestion can be tempting when the Modern Synthesis is viewed in terms of "selfish" chunks of DNA that "create" organisms as assemblies of optimally adapted traits (Dawkins 1989). It is less tempting, however, when "Lamarckism" is liberated from its status as a junk category and the Synthesis is made even more pluralist than Gould proposed. In adaptationism, the conflict between Lamarckism and Darwinism is categorical. In Darwinian pluralism, they are dialectically intertwined in ways I have tried to sketch in this section.

4 Evolutionary Developmentalism, Lost and Found

The fact that organisms are beings that develop was central to Aristotle's (384–322 BCE) seminal studies of biology. The creationists and materialists of his day thought of living beings as assemblies of parts. They differed on whether these parts aggregate by design or chance (Sedley 2007). Aristotle simultaneously refuted both by adducing facts showing that the commingling of male and female seminal elements triggers off a self-propelled, end-oriented (*teleios*) process in which an originally indeterminate matter forms itself into a progressively more differentiated and individuated whole of hierarchically organized parts. These parts carry out the morphological, physiological, behavioral, and cognitive functions that in concert allow members of each species to flourish in the environment that affords it appropriate resources (*Generation of Animals* 2.1.735al0-25; 2.4.740al-24). For Aristotle, the process of ontogeny (normally) culminates in the act of reproduction, thereby setting off a new round of a cycle that for each lineage shows no sign of beginning or ending, creation or extinction (*On the Soul* 2.4.415a26-30).

Aristotle's "epigenetic" view, as it came to be called, was preserved in medical schools (with accretion of creationist and corpuscular elements) from the time of Galen (129–216 CE), who was physician to the Roman emperors Marcus Aurelius and his heir Commodus, to that of William Harvey (1578–1657), who treated James I of England. In the eighteenth century, Caspar Friedrich Wolff (1733–1794) called on epigenesis to help him refute preformationism, which accommodated the end-oriented and functional aspects of organisms to Enlightenment mechanism by sleight of hand. On the sixth day of creation, according to preformationists, God placed in the egg or sperm—there was a dispute about which—a nested series of miniature little humans (*homunculi*) which thereafter come mechanically rolling out (evolving in the original sense of the term) generation after generation (Richards 1992).[14]

In the transformism stirring within the new science of *Biologia* at the end of the eighteenth century, there lurked a notion that the complexification of individual development mirrors and recapitulates phylogenetic history. Darwin accepted Karl Ernst von Baer's weak recapitulationism, which conceded that the embryos of advanced species resemble the adult forms of more primitive kinds (Darwin 1859, Chap. XIII; Nyhart 2009). For this reason, many early readers interpreted the *Origin* as restating "the development hypothesis." The strong recapitulationism of Darwin's self-proclaimed disciple Ernst Haeckel, in which organisms in the course of their development are said to run though the *adult* stages of earlier forms, gave them precious little reason to think otherwise. For Haeckel, there was a time when each human passed through a fish-like stage. This background helps us see why it was easy for Darwinism, or rather Haeckel's *Darwinismus*, to give way to the orthogenetic ontogeny–phylogeny parallelism of the later nineteenth century (Gould 1977).

The geneticist Morgan was trained as an embryologist. Aware as he was, however, that efforts to find inherited factors supporting recapitulation had so far been in vain, he set development aside to focus experimentally on the chromosomal mechanism of Mendelian inheritance. He always meant to get back to embryology armed with new insights (Allen 1979; Maienschein 2016 doubts he ever left it). Instead, his "transmission genetics" was taken up into population genetics and the

[14]Classical epigenesis and contemporary epigenetics do not refer to the same thing but do have historical connections. Neo-Darwinism reduced the scope of nongenetic forms of heritability, such as cytoplasmic inheritance, almost to zero. Those who defended the latter, notably C. H. Waddington, referred to all aspects of inheritance as "epigenetic." The epigenome includes genes but goes beyond them. By stressing the ontogenetic locus in which an array of reproductive factors interact as "developmental resources," current advocates of the evolutionary significance of epigenetic modifications contest whether DNA is the sole carrier of biological heritability. Seeing residues of preformationism lurking in the notion of molecules that carry and transmit "information," they sometime call for a new form of preformationism's ancient antagonist, epigenesis (Oyama et al. 2001).

Modern Synthesis. No one associated with these research programs denied that organisms develop, but they did sense that viewing the evolutionary process from the perspective of "population thinking" yields insights that focusing on the development of individual organisms casts into shadow (Mayr 1980). Organisms develop but don't evolve. Spatiotemporally bounded populations don't develop, but they do evolve—by gradual shifts in the distribution of genotypes in interbreeding populations over trans-generational time. It is true that life cycle strategies are naturally selected in accord with ecological circumstances. It is also true that, following his Russian colleagues I. I. Schmalhausen and I. M. Lerner, Dobzhansky incorporated development into the Modern Synthesis by distinguishing modes of natural selection—diversifying, stabilizing, balancing—that reflect ontogeny (Dobzhansky 1970; Gilbert 1994; Depew 2011). But in both unhardened and hardened versions of the Synthesis, these phenomena bear only on how variation is differentially distributed in populations, not on how it becomes available for selection in the first place. It is the latter process that after a century of separation has brought evolutionary developmentalism into dialogue with Darwinism again, minus the red herring of recapitulation. By making ontogeny the cause of phylogeny rather than the other way around, developmental biology revealed its experimental prowess until well into the twentieth century (Esposito 2017). The maturation of developmental genetics in our own time is helping evolutionary developmental biology emerge from its eclipse by population genetics—and helping the latter come to terms at last with development.

Since the 1980s, the debate about "evo-devo" has rotated around three successive ways of envisioning "the return of the organism." In protesting the adaptationist fractioning of organisms into aggregates of separate traits, the "process structuralist" Brian Goodwin insisted that species are self-organizing natural kinds, not (just) historical lineages (Goodwin 1994). A second approach was no less opposed to adaptationism but more open to Darwinism. In Gould's "punctuated equilibrium," genetic change clusters around speciation events for the same reason one cannot presume that natural selection can or will always come up with optimally adapted traits. The many-layered structuration organisms acquire in the developmental process constrains both the availability of genetic variation and the scope and path of adaptive natural selection, thereby showing why lineages are as vulnerable to fortune as Gould took them to be (Maynard Smith et al. 1985).

A third, more recent approach has caused Darwinians to pay more attention to evo-devo, in part because it highlights empirical discoveries more than conceptual revisions or metaphysical theorizing about what organisms or species (really) are. The fundamental insight is that in controlling the developmental process, regulatory sectors of the genome turn structural, protein-specifying genes on and off only with a great deal of help from other "developmental resources." In doing so, regulatory genes do not constrain genetic variation, as Gould had it. On the contrary, they make it available in the form of shifts in the timing, placement, and rate of gene expression which almost immediately affect adaptedness to particular environmental conditions (Alberch and Alberch 1981; Newman and Müller 2000; Carroll 2005; Gilbert and Epel 2009; Pigliucci and Müller 2010; Pigliucci 2017).

Two linked observations support this claim. The first is that regulatory genes are highly conserved across lineages. HOX genes, for example, which control bilateral symmetry, are as ancient as the first metazoa. The second observation concerns the source of ontogenetic differences if genes aren't. DNA works in conjunction with epigenetic mechanisms such as the methylation of DNA, RNA in its multiple roles, enzymatic transcriptases, and hormones, to open the developmental process to organismic and environmental signals. These processes are not so much violations as evasions of the central dogma of molecular biology. Still, in conjunction with the fact that it can be spliced in many alternate ways, they suggest that DNA does not contain a code that encrypts the information for making traits in a quasi-preformationist way, but is one of a number of developmental resources that interact dynamically in the reproductive cycle (Moss 2003; Oyama et al. 2001). The resulting view is more than a return of the organism in Aristotle's substantialist sense. It is a re-conception of organisms as developmental processes that are more deeply embedded in their ecologies than even niche constructionists typically envision.[15]

From the perspective of evo-devo, setting development aside as population-genetic Darwinism does, if only for tactical reasons, forecloses the very possibility of understanding evolutionary dynamics. That is why the degree of continuity and discontinuity between the Modern Synthesis and evo-devo, especially in the third sense I have discriminated, currently divides advocates of an "extended Synthesis" from those who would replace it, either with a post-Synthesis form of Darwinism or with a resolutely post-Darwinian form of evolutionary developmentalism. Having surveyed the impressive ability of the Modern Synthesis to maintain its continuity by incorporating insights that initially seemed to threaten it, I would be the last to discount its ability to adjust. Still, I hazard to guess that whatever emerges just inside or just beyond the conceptual boundaries of the Modern Synthesis will affect how adaptation, speciation, and the origins of higher taxa are interpreted.

To be sure, the Synthesis has done better on the topic of speciation than Darwin, who merely hoped that adaptive natural selection combined with geographical isolation would eventually be shown to lead to lineage splitting. The Synthesis furthered his cause by using its characteristic "population thinking" to identify genetic, not just environmental, isolating mechanisms as marking off the boundaries of species (Dobzhansky 1937, 1951; Mayr 1963). The molecular-genetic revolution of the mid-twentieth century gave the Synthesis new tools for tracking gene frequencies but didn't affect its fundamental approach to this

---

[15]Embedding organisms in ecological systems brings into view the lawful thermodynamic imperatives to which ecological systems must conform. In thermodynamically open, far from equilibrium systems, variation and selection of efficient dissipative pathways is inevitable. These physical and chemical imperatives permit, or even encourage, the emergence of developmental systems in which variation and selection take specifically biological forms. A lesson favorable to integrating evo-devo and Darwinism is that adaptive natural selection properly so called can take place only in developmental systems, which in turn are entrained with the environments by which they are co-defined.

topic.[16] Point mutations in protein production have little to do with the speciation process and so throw little light in. By contrast, switching regulatory genes on and off is so closely linked to phylogenetic branching that it validates the sweeping homology Darwin saw in the history of life by making visible the architecture of diversification (Carroll 2005).

Evo-devo highlights the unexpected speed with which speciation can occur. Earlier, I quoted the gradualist Mayr as conceding that, "In freshwater fishes [speciation] may take less than four thousand years." In fact, even without benefit of gene flow from a "founder" population, it has been reported that speciation occurred in ten generations in certain freshwater sticklebacks, a genus of sub-arctic fish (McKinnon and Rundle 2002). It is a good guess that the same or similar genetic shifts were in play in the transition of sticklebacks from their ancestral ocean-going habitat to bays, streams, and lakes as glaciers receded. These include but are not restricted to reduction or full elimination of the protruding dorsal spines that give these fish their name by deletion of specific base pairs in the HOX gene *Pitx1* and a range of associated shifts in the frequency of gene sequences coding for proteins (McKinnon and Rundle 2002; Colosimo et al. 2004, 2005; Chan et al. 2010; Jones et al. 2012). Experiments show that dorsal spine expression is affected almost as soon as the diet of saltwater sticklebacks has been changed (Wund 2008; Pfennig et al. 2010). One can readily see in the parallel evolution of species in isolated environments the plasticity of genotypes previously fixed in ancestral populations. Recasting phenotypic plasticity as developmental plasticity may allow extended versions of the Modern Synthesis to accommodate stickleback speciation without undue strain (Schlichting and Pigliucci 1993; Schlichting 2008; Pigliucci 2017). Still, this case study, if properly attested, makes it temping to cast gene regulation in the developmental process, which recruits supportive genetic changes in dynamic interaction with environment, as the creative factor in evolution, not gradual natural selection (Gilbert and Epel 2009). If so, continuity with the Modern Synthesis may demand amending, and not just extending, what counts as its core and what is peripheral. If this cannot be done in a way that preserves and further illuminates the vast amount of knowledge that has been accumulated under the aegis of the Modern Synthesis, even as new knowledge about the role of developmental genetics is added, evolutionary theory may escape the gravitational field of the Modern Synthesis, and perhaps of the Darwinian tradition more generally. I wouldn't bet on it. But it is important to acknowledge the current situation of debate in evolutionary biology without prejudging or foreclosing it.

Evo-devo's effect on genetic Darwinism's conception of adaptation may be more challenging than its effect on speciation. Earlier, we noted the Modern Synthesis's insistence that gradual natural selection is the proper cause of

[16]Coyne and Orr (2004) summarize the methods and results of speciation research. That Coyne is an opponent of expanded, extended, or new syntheses is not unconnected with his understandable desire to defend real achievements of which he is a direct heir and contributor (Coyne and Orr 1989; Coyne 2009).

adaptation. The first occurrence of a useful genetic variant is by definition not an adaptation. On this view, natural selection cannot be restricted to eliminating the already unfit or to merely retaining mutations that happen from day one to have beneficial effects. The Modern Synthesis makes it a matter of principle that the adaptedness of populations and the adaptations acquired by members of these populations evolve over a number of generations by a cumulative process in which genotypes having a positive effect on a population's rate of reproduction are differentially propagated, and in which in consequence the mean distribution of chance variation bends in the direction of the adaptive process (Beatty 2016).[17] This is why Dobzhansky said, "Selection is ... much more than a sieve retaining lucky and losing the unlucky mutations" and why Julian Huxley remarked, "The statement that selection is a destructive agency is not true if it is meant as merely destructive ... It has a share in evolutionary creation. Neither mutation nor selection alone is creative of anything important in evolution; but the two in conjunction are creative" (Dobzhansky 1962: 430–431; Huxley 1942: 28; see also Mayr 1980: 2, 18).

Here, too, issues of timing come to the fore. Earlier, we reviewed the question of how many generations it takes to evolve an adaptation. This issue arises for and within the orthodox framework of population genetic Darwinism, not in opposition to it. It concerns the minimal conditions necessary for abiding by Darwin's gradualist axiom. Directed mutation, organisms rather than genes as initiators and beneficiaries of adaptive change, environments eliciting evolutionary novelties by modifying genetic material, and other controversial claims that have agitated the scene of inquiry since the 1970s pose different, potentially more heterodox questions. Recent suggestions "that variation is not random, that there is more to inheritance than genes, and that there are multiple routes to the fit between organisms and environments" have challenged not just adaptationism but the idea of adaptation as construed by the Modern Synthesis (Laland et al. 2014).

Lewontin has proposed disarming the question by deconstructing adaptation as an ideologically suspect residue of natural theology allowed to live an undeservedly beatific afterlife in many versions of modern Darwinism. Once this illusion has been dispelled, Lewontin implies that the slow steady dialectic between experimentation, observation, and theorizing in a truly scientific evolutionary science may resume (Levins and Lewontin 1985). Advocates of evo-devo have a different solution. The concept and reality of adaptation are to be retained but transferred from gene frequency changes that meet certain criteria back to organisms considered as developmental processes (Nicholson 2014; Walsh 2015). It is not populations that adapt to environmental contingencies but organisms that adapt environmental resources to their needs. Tracking gene frequencies is useful, even indispensable, in bringing evolutionary biology's *explananda* into view, but it cannot identify evolutionary causes, which are to be found in the dynamic interrelation between ontogeny and ecology (Ariew and Matthen 2002; see Millstein et al. 2009; Hodge 2016 to the

[17]Lamarkism co-opted by Darwinism yet again!

contrary). This perspective threatens the conceptual coherence of the Modern Synthesis. Waddington seems prophetic when, following the lead of the embryologist Wilhelm Roux, he proclaimed that the problems of evolutionary biology would never be solved until biologists come back to development (Needham 1984; Gilbert 1994). What this entails remains to be seen.

5 Conclusion: Intertwined Traditions in Evolutionary Biology

Bowler has provocatively contributed to the current debate by using the genre of counter-factual history to make an informed guess that nineteenth-century evolutionary developmental biology would probably have arrived somewhere close to where it is now even if Darwin had never lived (Bowler 2013). The very existence of Bowler's thought experiment testifies to the seriousness with which evo-devo is currently being received. Bowler does not mean that natural selection would not have been discovered along the way. Weldon and other biometricians, he says, would have found their way to it in the course of applying advanced statistical-probabilistic analysis to shifts in trait distributions in species under identifiable environmental pressures (Bowler 2013 199). But in their hands natural selection would have taken its place in a continuous research tradition with Lamarckian, saltationist, and developmentalist features. In fact, Bowler argues that the current state of knowledge would have been reached earlier because it would not have been delayed by the ideological fireworks set off by Darwin's *Origin*. Descent with modification from a common ancestor and natural selection as an evolutionary agent would long since have acquired secure places in the common sense of modernity because, never having had to endure the rise of literalism about *Genesis* I, neither would we have had to endure Dawkins's atheistic provocations or Daniel Dennett's testimony that in Darwin's hands natural selection is the "best idea that anyone ever had" because it "eats like a universal acid" through religious illusions (Dawkins 2006; Dennett 1995). This is theology more than science.

In Bowler's book what counts as Darwinism is not far removed from what Gould called "Darwinian fundamentalism" (Gould 1997). It is true that in recent decades gene-by-gene, trait-by-trait adaptationism, especially applied to animal and human behavior, passes as Darwinism's highest achievement, final justification, and hence defining mark. In this chapter, I have tried to suggest that Darwinism is better identified as a research tradition whose unifying thread is not just natural selection but natural selection viewed as its founder was the first to view it: as the creative factor in evolutionary change. I have argued that its continuity has been achieved by following Darwin's lead in attempting to meet challenges arising from competing orientations in evolutionary biology. In contrast to Dawkins' and Dennett's, my Darwinism is far from triumphalist. I have acknowledged that whether it can retain its conceptual integrity and dominance in evolutionary studies in the face of challenges from evo-devo is at least as problematic as whether it could survive mutationism or revitalized forms of Lamarckism once were.

My account raises the question of whether the rival traditions that Bowler pictures as making continuous progress under the sober impulse of ideology-free inquiry may actually have reached their present condition by interacting with each other and with Darwinism in much the same way Darwinism interacted with them. These interactions would have found their way toward reliable facts and explanations by working through, not ignoring, ideological pressures that affect them all. Historians of biology who wish to pursue research along these lines will find a clue through the maze in the lively disputes about the "creative factor" in evolution that began in the 1880s and continue to this day.[18]

References

Adams M (ed) (1994a) The evolution of Theodosius Dobzhansky. Princeton University Press, Princeton

Adams M (1994b) Introduction: Theodosius Dobzhansky in Russia and America. In: Adams M (ed) The evolution of Theodosius Dobzhansky. Princeton University Press, Princeton, pp 3–11

Alberch P, Alberch J (1981) Heterochronic mechanisms of morphological diversification and evolutionary change in the neo-tropical salamander, Bolitoglossa occidental. J Morphol 167:249–264

Allen G (1979) Thomas Hunt Morgan: the man and his science. Princeton University Press, Princeton

Appel T (1987) The Cuvier-Geoffrey debate: French biology in the decades before Darwin. Oxford University Press, Oxford

Ariew A, Matthen M (2002) Two ways of thinking about fitness and natural selection. J Philos 49:55–83

Baldwin JM (1896) A new factor in evolution. Am Nat 30:441–451; 536–553

Beatty J (2016) The creativity of natural selection in the modern evolutionary synthesis Part I: Darwin, Darwinism, and the mutationists. J Hist Biol 49:659. https://doi.org/10.1007/s10739-016-9456-5

Bowler P (1988) The non-Darwinian revolution. John Hopkins University Press, Baltimore

Bowler P (2013) Darwin deleted: imagining a world without Darwin. University of Chicago Press, Chicago

Bowler PJ (2017) Alternatives to Darwinism in the early twentieth century. In: Delisle RG (ed) The Darwinian tradition in context: research programs in evolutionary biology. Springer, Cham, pp 195–218

Burian R, Gayon J (1999) The French school of genetics: from physiological and population genetics to regulatory molecular genetics. Annu Rev Genet 33:313–349

Burian R, Kampourakis K (2013) Against 'genes for': could an inclusive concept of genetic material replace gene concepts? In: Kampourakis K (ed) The philosophy of biology: a companion for educators. Springer, Dordrecht, pp 597–628

Cairns J, Overbaugh J, Miller S (1988). The origin of mutants. Nature 335:142–145

[18]My thanks to Richard Delisle, Jean-Baptiste Grodwohl, Jim Hofmann, and Bruce Weber for helping me improve this chapter.

Carroll S (2005) Endless forms most beautiful: the new science of evo-devo. W.W. Norton, New York
Chambers R (1844) Vestiges of the natural history of creation. Churchill, London
Chan Y, Marks M, Jones F, Villarreal G, Shapiro M, Shannon M, Brady D, Southwick A, Absher D, Grimwood J, Schmutz J, Myers R, Petrov D, Jónsson B, Schluter D, Bell M, Kingsley D (2010) Deletion of a *Pitx1* enhancer. Science 327:302–305
Colosimo P, Peichel C, Nereng K, Blackman B, Shapiro M, Schluter D, Kingsley D (2004) The genetic architecture of parallel armor plate reduction in threespine sticklebacks. PLoS Biol 2 (5):E109
Colosimo P, Hosemann K, Balabhardra S, Villareal G, Dickson M, Grimwood J, Schmutz J, Myers R, Schluter D, Kingsley D (2005) Widespread parallel evolution in sticklebacks by repeated fixation of ectodysplasin alleles. Science 307:1928–1933
Cook G (1999) Neo-Lamarckian experimentalism in America: origins and consequences. Q Rev Biol 74:417–437
Coyne J (2009) Are we ready for an "extended evolutionary synthesis"? http://whyevolutionistrue.wordpress.com/2009/02/16/are-we-ready-for-an-extended-evolutionary-synthesis. Accessed 4 Aug 2015
Coyne J, Orr A (1989) Patterns of speciation in Drosophila. Evolution 43:362–381
Coyne J, Orr A (2004) Speciation. Sinauer, Sunderland
Crick F (1956) On protein synthesis. Symp Soc Exp Biol 12:139–163
Crick F (1970) The central dogma of molecular biology. Nature 227(5258):561–563
Darwin C (1859) On the origin of species. John Murray, London (Harvard University Press, Cambridge, Facsimile of 1st ed)
Darwin C (1868) The variation of animals and plants under domestication. John Murray, London, 2 vol
Dawkins R (1989) The selfish gene. Oxford University Press, Oxford (2nd edn, 1st edn 1976)
Dawkins R (2006) The god delusion. Houghton Mifflin, Boston
Delisle R (2001) Adaptationism versus cladism in human evolution studies. In: Corbey R, Roebroeks L (eds) Studying human origins: disciplinary history and epistemology. University of Amsterdam Press, Amsterdam, pp 107–121
Delisle RG (2017) From Charles Darwin to the evolutionary synthesis: weak and diffused connections only. In: Delisle RG (ed) The Darwinian tradition in context: research programs in evolutionary biology. Springer, Cham, pp 133–168
Dennert E (1904) At the deathbed of Darwinism (trans: O'Harra E, Peschges H). German Literary Board, Burlington (First German edition, 1903)
Dennett D (1995) Darwin's dangerous idea. Simon & Schuster, New York
Depew D (2003) Baldwin and his many effects. In: Weber B, Depew D (eds) Evolution and learning. MIT Press, Cambridge, pp 3–31
Depew D (2009) The rhetoric of Darwin's origin of species. In: Ruse M, Richards R (eds) The Cambridge companion to the origin of species. Cambridge University Press, Cambridge, pp 237–255
Depew D (2011) Adaptation as process: the future of Darwinism and the legacy of Theodosius Dobzhansky. Stud Hist Philos Bio Biomed Sci 42:89–98
Depew D, Weber B (2017) Developmental biology, natural selection, and the conceptual boundaries of the modern evolutionary synthesis. Zygon 52:468–490
Dobzhansky T (1937) Genetics and the origin of species, 1st edn. Columbia University Press, New York
Dobzhansky T (1951) Genetics and the origin of species, 3rd edn. Columbia University Press, New York
Dobzhansky T (1962) Mankind evolving. Yale University Press, New Haven
Dobzhansky T (1970) Genetics of the evolutionary process. Columbia University Press, New York
Dobzhansky T (1973) Nothing in biology makes sense except in the light of evolution. Am Biol Teach 35:125–129

Doolittle WF (1999) Phylogenetic classification and the universal tree. Science 284 (5423):2124–2129
Eldredge N, Gould SJ (1972) Punctuated equilibria: an alternative to phyletic gradualism. In: Schopf T (ed) Models in paleobiology. Freeman, San Francisco, pp 82–115
Eldredge N, Tattersal I (1975) Evolutionary models, phylogenetic reconstruction, and another look at hominid phylogeny. Contrib Primatol 5:218–242
Esposito M (2017) The organismal synthesis: holistic science and developmental evolution in the English-speaking world, 1915–1954. In: Delisle RG (ed) The Darwinian tradition in context: research programs in evolutionary biology. Springer, Cham, pp 219–242
Felicity C. Jones, Manfred G. Grabherr, Yingguang Frank Chan, Pamela Russell, Evan Mauceli, Jeremy Johnson, Ross Swofford, Mono Pirun, Michael C. Zody, Simon White, Ewan Birney, Stephen Searle, Jeremy Schmutz, Jane Grimwood, Mark C. Dickson, Richard M. Myers, Craig T. Miller, Brian R. Summers, Anne K. Knecht, Shannon D. Brady, Haili Zhang, Alex A. Pollen, Timothy Howes, Chris Amemiya, Jen Baldwin, Toby Bloom, David B. Jaffe, Robert Nicol, Jane Wilkinson, Eric S. Lander, Federica Di Palma, Kerstin Lindblad-Toh, David M. Kingsley (2012) The genomic basis of adaptive evolution in threespine sticklebacks. Nature 484 (7392):55–61
Fisher RA (1930) The genetical theory of natural selection. Oxford University Press, Oxford
Foster PL (2007) Stress-induced mutagenesis in bacteria. Crit Rev Biochem Mol Biol 42:373–397
Gayon J (1995) Sélection naturelle ou survie des plus aptes? Éléments pour une histoire du concept de fitness dans la théorie évolutionniste. In: Blanckaert C, Fischer J, Rey R (eds) Nature, histoire, société: essais en hommage à Jacques Roger. Klincksieck, Paris, pp 263–287
Gayon J (1997) The 'paramount power' of selection: from Darwin to Kauffman. In: Della Chiara J (ed) Structure and norms in science, Synthese. Kluwer Academic, Dordrecht, pp 265–282
Gayon J (1998) Darwinism's struggle for survival. Cambridge University Press, Cambridge (1st French edition 1992)
Gilbert S (1994) Dobzhansky, Waddington and Schmalhausen: embryology and the modern synthesis. In: Adams M (ed) The evolution of Theodosius Dobzhansky. Princeton University Press, Princeton, pp 143–154
Gilbert S, Epel D (2009) Ecological developmental biology. Sinauer, Sunderland
Goldenfeld N, Woese C (2007) Biology's next revolution. Nature 445:369
Goldschmidt R (1940) The material basis of evolution. Yale University Press, New Haven, CT
Goodwin B (1994) How the leopard changed its spots: the evolution of complexity. Princeton University Press, Princeton
Gould SJ (1977) Ontogeny and phylogeny. Harvard University Press, Cambridge
Gould SJ (1980a) Is a new and general theory of evolution emerging? Paleobiology 6:119–130
Gould SJ (1980b) The return of hopeful monsters. Nat Hist 86:22–30 (Reprint: Gould, SJ (1980) The panda's thumb. Norton, New York, pp 186–193)
Gould SJ (1980c) The uses of heresy: introduction to reissue of Goldschmidt 1940. Yale University Press, New Haven, CT
Gould SJ (1983) The hardening of the modern synthesis. In: Grene M (ed) Dimensions of Darwinism. Cambridge University Press, Cambridge, pp 71–93
Gould SJ (1989) Wonderful life: the Burgess shale and the nature of history. Norton, New York
Gould SJ (1997) Darwinian fundamentalism. NY Rev Books 44:34–37. http://www.nybooks.com/issues/1997/06/12/
Gould SJ, Lewontin RC (1979) The spandrels of San Marco and the panglossian paradigm: a critique of the adaptationist programme. Proc R Soc Lond Ser B 205(1161):581–598
Grene M, Depew D (2004) The philosophy of biology: an episodic history. Cambridge University Press, Cambridge
Hallgrímsson B, Hall B (2011) Variation: a central concept in biology. Academic, New York
Hennig W (1950) Grundzüge einer Theorie der phylogenetischen Systematik. Deutscher Zentralverlag, Berlin

Hodge MJS (2016) Chance and chances in Darwin's early theorizing and in Darwinian theory today. In: Ramsey G, Pence C (eds) Chance and evolution. University of Chicago Press, Chicago, pp 41–75
Hofmann J (2017) Rate variation during molecular evolution: creationism and the cytochrome c molecular clock. Evol Educ Outreach 10:1. https://doi.org/10.1186/s12052-017-0064-4
Huxley J (1942) Evolution: the modern synthesis. Allyn and Unwin, London
Jablonka E, Lamb M (1995) Epigenetic inheritance and evolution: the Lamarckian dimension. Oxford University Press, Oxford
Jablonka E, Lamb M (2005) Evolution in four dimensions: genetic, epigenetic, behavioral, and symbolic variation in the history of life. MIT Press, Cambridge
Jackson J, Depew D (2017) Darwinism, democracy, and race: American anthropology and evolutionary biology in the twentieth century. Routledge, London
Keller EF (2000) The century of the gene. Harvard University Press, Cambridge
Kellogg V (1907) Darwinism today. Holt, New York
Kettlewell B (1955) Selection experiments on industrial melanism in the Lepidoptera. Heredity 9:323–342
Kettlewell B (1956) Further selection experiments on industrial melanism in the Lepidoptera. Heredity 10:287–301
Kevles D (1985) In the name of eugenics. Harvard University Press, Cambridge
Kimura M (1968) Evolutionary rate at the molecular level. Nature 217(5129):624–626
King J, Jukes T (1969) Non-Darwinian evolution. Science 164(3881):788–798
Kutschera U (2017) Symbiogenesis and cell evolution: an anti-Darwinian research agenda? In: Delisle RG (ed) The Darwinian tradition in context: research programs in evolutionary biology. Springer, Cham, pp 302–332
Lack D (1947) Darwin's finches. Cambridge University Press, Cambridge
Lakatos I (1970) Falsification and the methodology of scientific research programmes. In: Lakatos I, Musgrove (eds) Criticism and the growth of knowledge. Cambridge University Press, Cambridge, pp 91–195
Laland K, Uller T, Feldman M, Sterelny K, Müller G, Moczek A, Jablonka E, Odling Smee J (2014) Does evolutionary theory need a rethink? yes, urgently. Nature 514:161–162
Larson E (1997) Summer for the gods. Basic Books, New York
Laudan L (1977) Progress and its problems. University of California Press, Berkeley
Lee M, Ho S (2016) Molecular clocks. Curr Biol 26:R387–R407
Levins R, Lewontin R (1985) The dialectical biologist. Columbia University Press, New York
Levit GS, Hossfeld U (2017) Major research traditions in 20th century evolutionary biology: the relations of Germany's Darwinism with them. In: Delisle RG (ed) The Darwinian tradition in context: research programs in evolutionary biology. Springer, Cham, pp 169–194
Lewontin R (1982) Organism and environment. In: Plotkin H (ed) Learning, development and culture: essays in evolutionary epistemology. Wiley, New York, pp 151–170
Lewontin R, Levins R (2007) Biology under the influence: dialectical essays on ecology, agriculture, and health. Monthly Review Press, New York
Lloyd Morgan C (1896) On modification and variation. Science 4:733–740
Loison L, Herring E (2017) Lamarckian research programs in French biology (1900–1970). In: Delisle RG (ed) The Darwinian tradition in context: research programs in evolutionary biology. Springer, Cham, pp 243–270
Maienschein J (2016) Garland Allen, Thomas Hunt Morgan, and development. J His Biol 49 (4):587–601
Margulis L (1992) Symbiosis in cell evolution: microbial communities in the Archean and proterozoic eons. Freeman, San Francisco
Maynard Smith J, Burian R, Kauffman S, Alberch P, Campbell J, Goodwin B, Lande R, Raup D, Wolpert L (1985) Developmental constraints and evolution. Q Rev Biol 60:265–287
Mayr E (1942) Systematics and the origin of species. Columbia University Press, New York
Mayr E (1963) Animal species and evolution. Harvard University Press, Cambridge

Mayr E (1980) Prologue. In: Mayr E, Provine W (eds) The evolutionary synthesis. Harvard University Press, Cambridge, pp 1–48

Mayr E (1992) Speciational evolution or punctuated equilibria. In: Somit A, Peterson S (eds) The dynamics of evolution. Cornell University Press, Ithaca, pp 21–48

McKinnon J, Rundle H (2002) Speciation in nature: the threespine stickleback model system. Trends Ecol Evol 7:470–488

Millstein R, Skipper R, Dietrich M (2009) (Mis)interpreting mathematical models: drift as a physical process. Philos Theor Biol 1:e002. http://quod.lib.umich.edu/p/ptb/6959004.0001.002?view=text;rgn=main

Moorjani P, Amorin C, Arndt P, Przeworski M (2016) Variation in the molecular clock of primates. PNAS. www.pnas.org/cgi/doi/10/1073/pnas.1600374113d

Morgan TH (1935). The scientific basis of evolution, 2nd edn. Norton, New York (1st edn, 1932)

Morris SM (2003) Life's solution: inevitable humans in a lonely universe. Cambridge University Press, Cambridge

Moss L (2003) What genes can't do. MIT Press, Cambridge

Needham J (1984) Forward. In: Ho M, Saunders P (eds) Beyond neo-Darwinism: an introduction to the new evolutionary paradigm. Academic, London, pp vii–viii

Newman S, Müller G (2000) Epigenetic mechanisms of character formation. J Exp Zool 288:304–2317

Nicholson D (2014) The return of the organism as a fundamental explanatory concept in biology. Philos Compass 9:347–359. https://doi.org/10.1111/phc3.12128

Nyhart L (2009) Embryology and morphology. In: Ruse M, Richards R (eds) The Cambridge companion to the origin of species. Cambridge University Press, Cambridge, pp 237–255

Odling-Smee J, Laland K, Feldman M (2003) Niche construction: the neglected process in evolution. Princeton University Press, Princeton

Osborn HF (1895) The hereditary mechanism and the search for the unknown factors of evolution. In: Biological lectures delivered at the marine biological laboratory of Wood's Holl [sic] in the summer of 1894. Ginn, Boston, pp 79–100

Osborn HF (1896) A mode of evolution requiring neither natural selection nor the inheritance of acquired characteristics. Trans NY Acad Sci 15:141–142; 148

Oyama S, Griffiths P, Gray R (eds) (2001) Cycles of contingency: developmental systems and evolution. MIT Press, Cambridge

Pfennig D, Wund M, Snell-Root E, Cruickshank T, Schlichting C, Moczek A (2010) Phenotypic plasticity's impacts on diversification and speciation. Trends Evol Ecol 25:459–467

Pigliucci M (2007) Do we need an extended evolutionary synthesis? Evolution 61:2743–2749

Pigliucci M (2017) Darwinism after the modern synthesis. In: Delisle RG (ed) The Darwinian tradition in context: research programs in evolutionary biology. Springer, Cham, pp 89–104

Pigliucci M, Müller G (eds) (2010) Evolution: the extended synthesis. MIT Press, Cambridge

Pocheville A, Danchin E (2017) Genetic assimilation and the paradox of blind variation. In: Huneman P, Walsh D (eds) Challenging the modern synthesis. Oxford University Press, Oxford

Provine W (1971) The origins of theoretical population genetics. University of Chicago Press, Chicago

Provine W (1986) Sewall Wright and evolutionary biology. University of Chicago Press, Chicago

Rainger R (1991) An agenda for antiquity: Henry Fairfield Osborn and vertebrate paleontology at the American Museum of Natural History, 1890–1935. University of Alabama Press, Tuscalousa

Reiss J (2009) Not by design: retiring Darwin's watchmaker. University of California Press, Berkeley

Richards R (1992) The meaning of evolution. University of Chicago Press, Chicago

Schlichting C (2008) Hidden reaction norms, cryptic variation, and evolvability. Annu Rev Acad Sci 1133:187–203

Schlichting C, Pigliucci M (1993) Control of phenotypic plasticity via regulatory genes. Am Nat 142:366–370
Secord JA (2000) Victorian sensation: the extraordinary publication, reception, and secret authorship of Vestiges of the natural history of creation. University of Chicago Press, Chicago
Sedley D (2007) Creationism and its critics in antiquity. University of California Press, Berkeley
Shanahan T (2017) Selfish genes and lucky breaks: Richard Dawkins' and Stephen Jay Gould's: divergent Darwinian agendas. In: Delisle RG (ed) The Darwinian tradition in context: research programs in evolutionary biology. Springer, Cham, pp 11–36
Simpson GG (1944) Tempo and mode in evolution. Columbia University Press, New York
Simpson GG (1953) The Baldwin effect. Evolution 7:110–117
Spencer H (1887) The factors of organic evolution. Appleton, New York
Stocking G (1968) Race, culture, and evolution: essays in the history of anthropology. University of Chicago Press, Chicago
Turner DD (2017) Paleobiology's uneasy relationship with the Darwinian tradition: stasis as data. In: Delisle RG (ed) The Darwinian tradition in context: research programs in evolutionary biology. Springer, Cham, pp 333–352
Waddington C (1953) Genetic assimilation of an acquired character. Evolution 7(2):118–126
Walsh D (2015) Organisms, agency, and evolution. University of Cambridge Press, Cambridge
Weber B, Depew D (2003) Evolution and learning: the Baldwin effect reconsidered. MIT Press, Cambridge
Weismann A (1889) Essays upon heredity. Clarendon Press, Oxford
West-Eberhard MJ (2003) Developmental plasticity and evolution. Oxford University Press, Oxford
Woese K, Goldenfeld N (2009) How the microbial world saved evolution from the Scylla of molecular biology and the Charybdis of the modern synthesis. Microbiol Mol Biol Rev 73:14–21
Wray G, Hoekstra H, Futuyma D, Lenski R, Mackay T, Schluter D, Strassma J (2014) Does evolutionary theory need a rethink? no, all is well. Nature 514(161):163–164
Wright, S (1932). The roles of mutation, inbreeding, crossbreeding, and selection in evolution. Proc. 6th Int. Cong. Genet. 1: 356–366.
Wund M (2008) A test of the 'flexible stem' model of evolution: ancestral plasticity, genetic cccomodation, and morphological divergence in the threespine stickleback radiation. Am Nat 172:449–462
Zuckerkandl E, Pauling L (1965) Evolutionary divergence and convergence in proteins. In: Bryson V, Vogel H (eds) Evolving genes and proteins. Academic, New York, pp 97–166

Darwinism After the Modern Synthesis

Massimo Pigliucci

Abstract The Modern Synthesis in evolutionary biology took place in two major phases from the 1920s through the 1940s. It achieved a reconciliation between the original Darwinism and the newly discovered Mendelism, which gave birth to the field of population genetics, the mathematical core of evolutionary theory. Since that time, the field has further evolved, encompassing entirely new areas of inquiry, leading to new empirical findings, and developing new conceptual tools. In the 1960s, for instance, the invention of gel electrophoresis made it possible for the first time to directly estimate gene frequencies in natural populations. While developmental biology was initially still excluded from the synthesis, in the 1980s new techniques led to the emergence of the field of evolutionary developmental biology (evo-devo). The parallel molecular revolution unfolded initially independently of evolutionary studies, until the turn of the twenty-first century, when genomics became "comparative" in nature and began importing techniques from phylogenetic analysis. The reevaluation of phenomena like phenotypic plasticity and epigenetic inheritance led to a rebirth of interest in the complexities of the genotype–phenotype map, triggering the formulation of concepts like robustness, modularity, and evolvability. Paleontology came roaring back with the theory of punctuated equilibria in the 1970s and 1980s. These many strands brought about the need for a further expansion of evolutionary theory, catalyzing discussions concerning a possible Extended Synthesis. "Darwinism" has never been more alive and pregnant with opportunities for both theoretical advancements and philosophical reflection.

Keywords Neo-Darwinism • Modern synthesis • Extended evolutionary synthesis • Alternatives to Darwinism • Structure of evolutionary theory • Population genetics • Paradigm shifts

---

M. Pigliucci (✉)
Department of Philosophy, City College of New York: North Academic Center, 5/144, 160 Convent Avenue, New York, NY 10031, USA
e-mail: massimo@platofootnote.org

R.G. Delisle (ed.), *The Darwinian Tradition in Context*,
https://doi.org/10.1007/978-3-319-69123-7_5

1 Introduction

Scientific theories are always provisional accounts of how the world works (Winther 2015), intrinsically incomplete, and expected to be replaced by better accounts as science progresses (Pigliucci 2016). The theory of evolution, colloquially referred to as "Darwinism," is, of course, no exception. It began in 1858 with joint papers presented to the Linnaean Society by Darwin and Wallace and was formalized shortly thereafter in *On the Origin of Species* (Darwin 1859). The original theory featured two conceptual pillars: the idea of common descent (which was accepted by a number of scholars even before Darwin), and that of natural selection as the chief mechanism of evolution, and the only one capable of generating adaptation.

The first bit of tinkering took place shortly thereafter, when Wallace himself, together with Weismann (1893), proposed to drop any reference to Lamarckian theories of heredity[1] because of the newly proposed notion of the separation between germ and somatic lines, thus generating what is properly known as neo-Darwinism (such neat separation between cell lineages, incidentally, turned out to have plenty of exceptions, especially outside the animal world). After undergoing a temporary crisis, as a result of increasing skepticism from paleontologists (Turner 2017) and developmental biologists (Bowler 1983, 2017), we enter the well-documented two phases of the Modern Synthesis (Delisle 2017, though I disagree with him that the connections between developments are "weak and diffuse"): first the reconciliation between Mendelism and Darwinism, leading to the birth of population genetics (Fisher 1930; Wright 1931; Haldane 1932), and later the expansion to include fields like natural history, population biology, paleontology, and botany (Dobzhansky 1937; Huxley 1942; Mayr 1942; Simpson 1944; Stebbins 1950).

What happened to "Darwinism" after 1950? The Modern Synthesis (MS) reigned as the dominant paradigm in the field, rather unchallenged until the late 1980s and early 1990s. At which point a number of authors, coming from a variety of disciplines, began to question not so much the foundations but the accepted structure of the MS. By the very late twentieth-century and early twenty-first-century, calls to replace the MS with an Extended Evolutionary Synthesis (EES) had begun to grow loud, and to be countered by equally loud voices raised in defense of the MS (West-Eberhard 2003; Jablonski 2005; Love 2006; Newman et al. 2006; Pigliucci 2007; Pigliucci and Muller 2010; Laland et al. 2014). How did this happen, and what does it mean for the current status and future of evolutionary theory? To understand this we need to step back for a moment and take a broad view of conceptual developments in the biological sciences during the second half of the twentieth century.

[1] See Loison and Herring (2017) about the surprising persistence of neo-Lamarckian ideas in France.

2 Biology in the Second Half of the Twentieth Century: A Bird's Eye View

The second half of the twentieth century has been an incredibly exciting time for biology, a period that has put the discipline on the map at least at the same level of interest as physics, the alleged queen of sciences, and arguably even more so. Let me remind the reader of some of the major developments that have made this possible, because they all—directly or indirectly—eventually fed into the current discussion over the MS versus the EES as dominant conceptual frameworks in evolutionary biology.

A major breakthrough in one of the foundational fields of the Modern Synthesis, population genetics, came with the invention of gel electrophoresis, which for the first time made it possible to directly assess first protein and then gene frequencies in large samples drawn from natural populations. While research on electrophoresis began as early as the 1930s with Arne Tiselius (1937), it was the breakthrough work of Lewontin and Hubby (1966) on gel electrophoresis that set population genetics on fire. The unexpected discovery—which actually went against the general assumptions of the MS—was, as the authors put it, that "there is a considerable amount of genic variation segregating in all of the populations studied . . . [it is not] clear what balance of forces is responsible for the genetic variation observed, but [it is] clear the kind and amount of variation at the genic level that we need to explain" (Lewontin and Hubby 1966: 595). This problem posed by a much larger degree of heterozygosity than expected in natural populations eventually led to a revolution in population genetics, and also directly to the origination of the impactful neutral theory of molecular evolution (Kimura 1983).

The neutral theory was a landmark conceptual development because for the first time since Darwin it challenged the primacy of natural selection as an agent of evolutionary change. To be sure, Kimura and colleagues didn't think that *phenotypic* evolution occurred in a largely neutral fashion, but if it turned out that much of what goes on at the molecular level is independent of selective processes, then this would force biologists to abandon simplistic models of the so-called Genotype → Phenotype map (Alberch 1991; Wagner and Zhang 2011), opening the question of how largely neutral molecular variation can give rise to non-neutral phenotypic outcomes. Eventually, the debate about the neutral theory—which raged on intensely for a number of years—was settled with a sensible and empirically consistent compromise: a lot of molecular variation is "near-neutral" (Otha 2002), which means that the role of stochastic processes such as genetic drift at the molecular level is significantly higher than might have been expected on the basis of a face-value reading of the tenets of the Modern Synthesis (Lynch 2007).

What could possibly connect the near-neutral molecular level with the obviously functional and therefore likely selected phenotypic level? The obvious answer was: development. The only problem was that developmental biology had famously been left out of the Modern Synthesis, with Mayr sometimes blaming the developmental biologists themselves, who were allegedly not interested in evolution. (I will leave it to historian of science to determine if, and to what extent, the charge may be substantiated; see also Depew 2017.) The fact of the matter is that developmental biology was glaringly missing from the synthesis of the 1930s–1950s, despite well-known (but later abandoned or deeply revised) early notions, such as Ernst Haeckel's "ontogeny recapitulates phylogeny" (Haeckel 1866; Gould 1985).

Things began to change as an offshoot of yet another revolution in biology: the rapid advances made in molecular biology after the discovery of the structure of DNA (Watson and Crick 1953). While molecular biology kept accelerating its pace independently of organismal biology for several decades—until their confluence in the era of evolutionary genomics (Caetano-Anollés 2010)—in the late 1970s, Lewis (1978) discovered the existence of homeotic genes regulating embryonic patterns of development in *Drosophila*. It soon turned out that this and similar classes of regulatory genes are both widespread and evolutionarily conserved, so that they are one of the major keys to the understanding of the complex interplay among genotype, development, and phenotype.

This approach flourished into the field of evolutionary developmental biology, or evo-devo (Müller 2007), and one of its major contributions so far has been a marked shift of emphasis in the study of morphology and development, from the sort of classical population genetic studies focused on structural genes to an emphasis on regulatory genes and their potential to help us build a credible theory of the origin of evolutionary novelties (Pigliucci 2008a). As Prud'homme et al. (2007: 8605) put it:

> Because most animals share a conserved repertoire of body-building and -patterning genes, morphological diversity appears to evolve primarily through changes in the deployment of these genes during development. … Morphological evolution relies predominantly on changes in the architecture of gene regulatory networks and in particular on functional changes within CREs [cis-regulatory elements]. … Regulatory evolution: (i) uses available genetic components in the form of preexisting and active transcription factors and CREs to generate novelty; (ii) minimizes the penalty to overall fitness by introducing discrete changes in gene expression; and (iii) allows interactions to arise among any transcription factor and downstream CRE.

The picture that emerges from this and many other studies is not incompatible with the simple models that were incorporated into the Modern Synthesis, but it does present us with a much more complex and nuanced understanding of genetic, developmental, and phenotypic evolution, so much so that it is little wonder that people have been increasingly referring to the current, very much in flux, version of evolutionary theory as the Extended Synthesis.

I have already mentioned the molecular biology revolution initiated in the 1950s, which eventually led to the genomic revolution. Both these radical developments initially affected evolutionary biology only indirectly, by providing increasingly powerful new analytical tools, such as gel electrophoresis, and later on gene sequencing. But inevitably genomics itself became an evolutionary science, once technical developments made it possible to sequence entire genomes more quickly and cheaply, and molecular biologists fully internalized, as Dobzhansky (1973) famously put it, that nothing in biology makes sense except in the light of evolution. The structure and function, as well as the sheer diversity, of genomes are themselves not understandable if not through evolutionary lenses, so that genomics and evolutionary biology currently represent a rare example of synergism between scientific disciplines: the first provides tools for the latter to advance, while the second one allows for a theoretical understanding of the data that the first one accumulates at such a heady pace.

While of course other disciplines within biology have made progress during the second part of the twentieth century—ecology, for instance (Real and Brown 1991)—the last bit of this panoramic view I wish to briefly comment on concerns yet another area of inquiry that had played only a secondary role during the Modern Synthesis: paleontology. The field had always been a thorn in the side of Darwinism, since many paleontologists early on had rejected the Darwinian insight, proposing instead the idea that macro-evolutionary change was qualitatively distinct from the sort of micro-evolution that Darwin famously modeled on the basis of plant and animal breeding. Indeed, it was this very rejection, together with the apparent incompatibility of Mendelism and Darwinism, that led to the abovementioned period of "eclipse" of the Darwinian theory at the turn of the twentieth century (Bowler 1983).

Paleontology's early alternative to Darwinism took the shape of orthogenetic theory (Coulter 1915; Jordan 1920; Lipman 1922), which in turn was essentially a scaled-up version of Lamarckism (Bowler 2003; Loison and Herring 2017), since it postulated an inner vital force responsible for long-term evolutionary trends, which many paleontologists saw as otherwise inexplicable within the Darwinian framework. It was George Gaylor Simpson's (1944) magistral role within the Modern Synthesis to clear away any remnants of orthogenesis from paleontology, doing for that field what Fisher, Haldane and Sewall Wright had done for Mendelian genetics: he convincingly argued that the sort of so-called "micro"-evolutionary processes accounted for by Darwinism could be extrapolated to geological timescales, thus yielding the *appearance* of macro-evolutionary changes of a qualitatively different nature. In reality, Simpson argued, the second is simply a scaled up version of the former.

Simpson, however, was arguably *too* successful, essentially making paleontology a second-rate handmaiden to population genetics while overlooking the potential for its original contributions—theoretical as well as empirical—to the overall structure of evolutionary theory. Eventually, Simpson's "conservatism," so to speak, led to a backlash: Niles Eldredge and Stephen Jay Gould (1972), the *enfants terribles* of modern paleontology, published their landmark paper proposing the theory of punctuated equilibria (Turner 2017; Depew 2017), according to which evolution, when seen at the macroscopic scale, works by fits and starts: long periods of stasis during which not much appears to be happening in a given lineage, interrupted by sudden "bursts" of phenotypic change. The theory was immediately misunderstood by many population geneticists, who thought that Eldredge and Gould were attempting to revive Goldschmidt's (1940) old idea of "hopeful monsters," i.e., of instantaneous evolutionary change resulting from genome-wide restructuring. To be fair, at some point Gould's own anti-establishment rhetoric, and the fact that creationists often mentioned him in their support, contributed to the confusion. But in fact, the sort of punctuations that Eldredge and Gould saw in the fossil record takes place over tens of thousands of generations, thus leaving plenty of time for standard Darwinian processes to do their work. As they pointed out later on in the debate (Gould and Eldredge 1977), the real novel issue is that of prolonged stasis, over millions of years, not the allegedly (but not really) "instantaneous" change. A major class of explanation proposed especially by Gould (1985) had to do with developmental processes and "constraints," which nicely connects the new paleontology (Jablonski 2005; Jablonski and Shubin 2015) with the emerging field of evo-devo, making both of them into pillars of the ensuing Extended Synthesis in evolutionary biology.

3 Shifting Theoretical Ground: S. J. Gould's *Magnus Opus*

The many conceptual and empirical advances in evolutionary biology during the second half of the twentieth century that I have briefly sketched above naturally led to a broader theoretical turmoil. More and more people felt like the Modern Synthesis was increasingly becoming too restrictive a view of evolution to keep playing the role of biology's "standard model." This group included Rollo (1995), Schlichting and Pigliucci (1998), West-Eberhard (2003), Jablonka and Lamb (2005), among others. But arguably none made a more concerted, if partial and highly redundant, effort than Stephen Jay Gould in his *magnum opus*, *The Structure of Evolutionary Theory* (2002).

The Structure is comprised of two parts, one tracing the history of evolutionary ideas, both pre- and post-Darwin, and the second one presenting Gould's view of current theoretical debates within the field. While the constructive part of the book focuses too much on paleontology and multilevel selection, Gould correctly identified three conceptual pillars of Darwinism that got imported wholesale into the Modern Synthesis:

1. *Agency*: the locus of action of natural selection. For Darwin, this was the individual organism, while within the MS the focus expanded to the gene, thus leading to an overall increase of agency. Gould advocated further expansion, to include multiple levels of selection, from the gene to the individual to kin groups to species. This suggestion is perfectly in line with that of other authors advocating an Extended Evolutionary Synthesis (Okasha 2006; Wilson and Wilson 2008).
2. *Efficacy*: the causal power of natural selection relative to other evolutionary mechanisms. According to Darwin, natural selection is the chief mechanism of evolutionary change, and certainly the only one capable of producing adaptation. The MS formally described—via population genetic theory—four additional mechanisms: mutation, recombination, migration, and genetic drift. Gould adds a positive role for developmental constraints to the picture, and advocates of the EES further expand on this theme, including concepts such as those of evolvability (Pigliucci 2008b), facilitated variation (Kirschner and Gerhart 2005), and niche construction (Laland and Sterelny 2006), among others.

3. *Scope*: the degree to which natural selection can be extrapolated from micro- to macro-evolutionary outcomes. As we have seen, this has been controversial early on, with the MS settling for the same basic picture proposed by Darwin: so-called macro-evolutionary processes are simply micro-evolutionary ones writ large. Gould, of course, questions this, on the basis of the already discussed theory of punctuated equilibria. Proponents of the EES also doubt the received view, suggesting that species selection and group-level ecological characteristics may partially, though not entirely, decouple micro- from macro-evolution (Jablonski 2000, 2008).

If Gould's general take is right, then, evolutionary theory has changed over time and the process can best be tracked conceptually by keeping tabs on changes in the agency, efficacy, and scope of natural selection within the theory. This, incidentally, makes natural selection *the* fundamental idea in biological evolution, and rightly so. No other concept, not even that of common descent, has had such a complex and convoluted history within the field. Moreover, what the EES is attempting to do can also be understood within Gould's framework, as sketched above.

4 What Is the Extended Evolutionary Synthesis?

As we have seen, the latter part of the twentieth century and the beginning of the twenty-first century have seen a renewed debate about the status of contemporary evolutionary theory, with a number of calls for an expansion of the Modern Synthesis into an Extended Evolutionary Synthesis. But what does the latter look like, at the current state of the discussion?

I provided an early sketch of it in a paper published in *Evolution* back in 2007 (Pigliucci 2007), and an updated and expanded version of that sketch has recently been put out by Laland et al. (2015).[2] My early analysis began by noting that philosopher Karl Popper famously interpreted the MS as a theory of genes, lacking a comparable theory of forms (i.e., phenotypes). The field got started, however, as a theory of forms in Darwin's days, with genetics taking on a fundamental role only after the rediscovery of Mendel's work at the turn of the twentieth century. Consequently, I suggested, a major goal that an EES aims for is an improvement and unification of our theories of genes and of forms. This, seems to me, may best be achieved through an organic grafting of novel concepts onto the foundational structure of the MS, particularly evolvability (Pigliucci 2008b), phenotypic plasticity (Pigliucci 2001; West-Eberhard 2003), epigenetic inheritance (Jablonka and Lamb 2005), complexity theory (Hoelzer et al. 2006), and the theory of evolution in highly dimensional adaptive landscapes (Gavrilets 1999).

[2]For those interested in a closer look, a new website has been launched to track the development of research and to foster outreach activities concerning the EES (http://extendedevolutionarysynthesis.com/).

Contra Delisle's (2017) provocative analysis, I don't think one needs to look for an "essence" of Darwinism in order to recognize the multifarious research traditions within either the MS or the EES. I also disagree with Delisle that biologists have been using Darwinian notions in a (highly) "inconsistent" way. While no scientific theory, especially as complex and pliable over time as evolutionary theory, is characterized by perfect coherence, it takes a lot of inconsistency to conclude that the MS and the original Darwinism were linked by "weak and diffused connections only" and that remains true also for the connections between the MS and the EES that interest me here.

As a case in point, Laland et al.'s (2015) paper is both more focused[3] and more systematic than previous attempts, explicitly attempting to clear away inconsistencies in previous works. They begin with a comparison of core assumptions of the MS versus the EES. To give you an idea of what they are getting at, here are the entries for inheritance:

Genetic inheritance (MS): "Genes constitute the only general inheritance system. Acquired characters are not inherited."

Inclusive inheritance (EES): "Inheritance extends beyond genes to encompass (transgenerational) epigenetic inheritance, physiological inheritance, ecological inheritance, social (behavioural) transmission and cultural inheritance. Acquired characters can play evolutionary roles by biasing phenotypic variants subject to selection, modifying environments and contributing to heritability."

They then run through a series of alternative interpretations of important evolutionary phenomena according to the two frameworks. For instance, in the case of developmental plasticity:

MS: "conceptualized as a genetically specified feature of individuals (i.e., a reaction norm) that can evolve under selection and drift. Focus is on the conditions that promote adaptive evolution of plastic versus non-plastic phenotypes. The primary evolutionary role of plasticity is to adjust phenotypes adaptively to variable environments. Plastic responses regarded as pre-filtered by past selection."

EES: "considers reducing plasticity to a genetic feature to be explanatorily insufficient. Retains an interest in adaptive evolution of plasticity, but also focuses on how plasticity contributes to the origin of functional variation under genetic or environmental change, and how the mechanisms of plasticity limit or enhance evolvability, and initiate evolutionary responses. Many plastic responses viewed as reliant on open-ended (e.g., exploratory) developmental processes, and hence capable of introducing phenotypic novelty."

[3]Their paper is focused on four specific areas: evo-devo, developmental plasticity, what they call inclusive inheritance, and niche construction.

Moreover, Laland et al. (2015) provide readers with a comparison of different predictions originating from the competing frameworks. For instance, in the case of the relationship between genetic and phenotypic change:

MS: "genetic change causes, and logically precedes, phenotypic change, in adaptive evolution."
EES: "phenotypic accommodation can precede, rather than follow, genetic change, in adaptive evolution."

Laland et al. (2015) also present a graphical outline of the structure of the Extended Evolutionary Synthesis, as they see it (their Fig. 2, p. 8). It is instructive to comment on a number of features of their model. Phenotypic evolution—the target of explanation of the entire framework, just as it was for Darwin—is assumed to be affected by three classes of processes: those that generate novel variation, those that bias selection, and those that modify the frequency of heritable variation.

Beginning with the first class, these processes include classical ones like mutation, recombination, gene expression, and developmental regulatory processes. But also EES-specific ones like environmental induction (of developmental processes), niche construction, phenotypic accommodation, and facilitated variation. The second class (processes that bias selection) include only EES-related entries: developmental bias and niche construction, while the third class (processes that affect heritable variation) are all classical (mutation pressure, selection, drift, and gene flow) but are in turn affected by the previous class.

The resulting picture is one of complete and, seems to me, highly coherent, meshing of the MS and the EES perspectives, where the latter *adds* to but does not really replace any of the previously recognized mechanisms. Which brings me to the next question I wish to address concerning the most recent developments of the now more than 150-year-old Darwinian tradition: is the proposed shift from the MS to the EES akin to a Kunhian paradigm shift?

5 Is Evolutionary Biology Going Through a Paradigm Shift?

One of the most controversial aspects of the discussion surrounding the MS versus EES debate is the extent to which the new framework is claimed to be distinct from the old one. At one extreme, there are scientists who simply reject the idea that the EES presents much that is new, claiming that whatever new concepts are being advanced were in fact already part of the MS, either implicitly or explicitly (see Laland et al. 2014, part 1). At the opposite extreme, some supporters of the EES have been making statements to the effect that the new framework somehow amounts to a rejection of fundamental aspects of Darwinism (Jablonka and Lamb 1999; Depew and Weber 2011), akin to what philosopher Thomas Kuhn (1962) termed a "paradigm shift" within the discipline, thus aligning themselves with a tradition that can be fairly characterized as anti-Darwinian (Kutschera 2017). My own position has always been that the truth lies somewhere in the middle: the EES is significantly different from the MS, and yet the change does not reflect any kind of scientific revolution within modern biology, but rather more of the same process that has led us from the original Darwinism to neo-Darwinism to the MS itself.

Kuhn famously argued—on the basis, crucially, of examples drawn exclusively from physics—that science goes through an alternation of two phases: during "normal" or "puzzle solving" science, practitioners are focused on addressing specific issues from within a given theoretical framework and set of methods (the "paradigm"), which itself is not the target of empirical testing or conceptual revision. From time to time, however, a sufficient number of "anomalies," or unresolved puzzles, accumulate and precipitate a crisis within the field. At that point scientists look for a new paradigm, better suited to take into account the insofar unresolved issues. If they find it, the new framework is quickly adopted and deployed in turn to guide a new phase of normal science.

Kuhn suggested a number of approaches to tell whether a paradigm shift has occurred (or, in our case, is in the process of occurring). These include five criteria for theory comparison, as well as three classes of potential incommensurability between theories. Let's begin by examining the five criteria: (1) accuracy, (2) consistency (internal and with other theories), (3) explanatory scope, (4) simplicity, and (5) fruitfulness of the accompanying research program. Here is how the MS and EES compare, in my mind, according to the Kuhnian criteria:

Accuracy, MS: building on the original Darwinism, it has produced quantitative accounts of the change over time of the genetic makeup of natural populations.

Accuracy, EES: incorporates the same methods and results of both the original Darwinism and the MS, adding the explanation of developmental and other self-organizing biological phenomena.

Consistency, MS: as internally consistent as any major scientific theory, features explicit external links to genetics, molecular biology, and ecology.

Consistency, EES: same degree of internal and external consistency as the MS, with the addition of external links to developmental biology, genomics, and complexity theory, among others.

Scope, MS: new facts about the biological world that are explained have been consistently uncovered for the past several decades.

Scope, EES: further expands the scope of the MS by explicitly including questions about the origin of evolutionary novelties, the generation of biological form, and the problem of genotype–phenotype mapping.

Simplicity, MS: uses a limited number of mechanisms (natural selection, genetic drift, mutation, migration, assortative mating) to account for evolutionary change over time.

Simplicity, EES: makes use of all the mechanisms of the MS, adding a number of others such as epigenetic inheritance, evolvability, facilitated (i.e., self-emergent) variation, etc.

Fruitfulness, MS: has a history of more than 70 years of vigorous research programs, building on the previous fruits of the original Darwinism.

Fruitfulness, EES: builds on the ongoing research program of the MS but has also already led to empirical (e.g., emergent properties of gene networks and of cell assemblages) and conceptual (e.g., evolvability, evolutionary capacitors) discoveries, though of course it is very much a work in progress as of the moment of this writing.

Even this brief survey ought to make it clear that the MS → EES is not a paradigm shift, but rather an organic expansion. Then there is the second test proposed by Kuhn to consider, a test in a sense more stringent, that of incommensurability. If two theories are incommensurable in even one of the three classes, a good argument can be made that a paradigm shift is occurring. The classes in question are methodological, observational, and semantic.

Methodological incommensurability refers to the notion that different paradigms lead scientists to pick different "puzzles" as objects of research, as well as to the idea that scientists then develop distinct approaches to the solution of those puzzles. The EES takes on board the same puzzles, and the same set of approaches, of the MS, but it also adds new puzzles (such as the appearance of so-called evolutionary novelties, like eyes, feathers, spines, and so forth), which were largely untouched, or dealt with only superficially, by the MS. It further adds new approaches, like interpretations of evolutionary changes in terms of niche construction, developmental plasticity, or epigenetic inheritance.

Observational incommensurability is tightly linked to the idea that observations are theory dependent: what is considered a "fact" within one theoretical context may not be such in a different theoretical context. For instance, in pre-relativity physics there was a (supposed) fact of the matter that some kind of substance, referred to as ether, had to be present in space in order for light to travel through it. After the famous Michelson–Morley experiment demonstrating that there was no such thing as ether, the relevant fact became the constancy of the speed of light and therefore the relativity of frames of reference. Nothing like that seems to be happening in evolutionary biology at the moment: the very same facts that have been catalogued and explained by the MS enter into the empirical corpus of the EES, to be further expanded with new facts that come to the forefront because of the additional conceptual advancements (e.g., the fact of geological stasis of certain lineages, noted as such only after Eldredge and Gould (1972)'s proposal of the theory of punctuated equilibria).

Semantic incommensurability has to do with shifts in the meaning of terms used by scientists, one of Kuhn's examples being that of "mass," which is a conserved, static quantity in Newtonian mechanics, but becomes interchangeable with energy within the framework of Einstein's relativity. Again, I do not discern any analogous shift in the terminology used by proponents of the MS versus EES. Key biological concepts, such as species, genes, phenotypes, niche, and so forth, retain similar and perfectly commensurable meanings, even though our understanding of their referents becomes increasingly sharp.

It seems, therefore, that Darwinism after the Modern Synthesis has proceeded along similar lines to those followed by Darwinism before the MS: a continuous expansion of both empirical knowledge and conceptual understanding, an expansion that is likely to continue for the remainder of the current century and beyond.

6 Conclusion: A Bit of House Cleaning

This volume on Darwinian research traditions is also an opportunity to call for a bit of house cleaning, so to speak, on the part of evolutionary biologists and philosophers of science. For instance, it is truly astounding that in France the Modern Synthesis, and in particular population genetics, was not included in standardized university curricula, or addressed within main research programs until the 1970s. According to Loison and Herring (2017), and against the Darwinian picture that was developing abroad, French life scientists opposed various forms of Lamarckism throughout the twentieth century, and some of that attitude still lingers. There is no good scientific reason for that, and it is hard not to pin such an attitude on sheer nationalism and the cultural worship of Lamarck. Needless to say, that sort of thing has no place in a mature science. The French are not the only culprits here, and the fact that there are "German," "Russian," and other "traditions" within evolutionary biology (Levit and Hossfeld 2017) is a little bizarre.

It's also somewhat surprising that behavioral biologists are allegedly (Alcock 2017) still clinging to simplistic notions from sociobiology and evolutionary biology, which have long since been debunked (Kaplan 2006). It's not the basic idea that behaviors, and especially human behaviors, evolve by natural selection and other means that is problematic. The problem, rather, lies with some of the specific claims made, and methods used, by evolutionary psychologists. Contra Alcock (2017), it seems to me crucially beneficial for behavioral biology to take on board notions like multilevel selection (Okasha 2006).

It is also both surprising and problematic that some researchers are still pursuing non-"mechanistic" or non-"physicalist" research programs (Esposito 2017), whatever that means. Indeed, a major point of the EES is to help bring the focus back on the organism and even the ecosystem, and yet—as I just argued above—this does not require a wholly alternative synthesis at all. I agree, however, with Kutschera (2017) on the suggestion that there is no reason to treat notions such as evolution by symbiosis as somehow anti-Darwinian, *pace* the notorious attempts by Lynn Margulis to push that interpretation.

In fact, Depew (2017) is correct when he states that Darwinism has advanced its own agenda by incorporating (not without internal resistance, to be fair) a variety of themes proposed by its critics, including "saltationism" (punctuated equilibrium) and "Lamarckism" (epigenetic inheritance, phenotypic plasticity, and niche construction). This is fine, so long as we understand that the terms within scare quotes above are to be understood in a modern, radically updated sense, and not along the lines of what, respectively, Richard Goldschmidt (1940) and Lamarck were thinking. As Shanahan (2017) points out, it is this inherent flexibility of Darwinism that has allowed people with views as divergent as Stephen Jay Gould and Richard Dawkins to (rightly) claim the Darwinian mantle. But this is neither just a rhetorical move nor somehow indicative of serious problems inherent in the Darwinian approach. In the end, the Darwinian tradition(s) in evolutionary biology are best understood as a wide ranging family of conceptual and research approaches, always in dialectic dialogue with each other, always in a constructive tension that transcends the agendas and (sometimes strong) personalities of the many individual scientists that recognize themselves as intellectual descendants of Charles Darwin.

References

Alberch P (1991) From genes to phenotypes: dynamical systems and evolvability. Genetics 84:5–11

Alcock J (2017) The behavioral sciences and sociobiology: a Darwinian approach. In: Delisle RG (ed) The Darwinian tradition in context: research programs in evolutionary biology. Springer, Cham, pp 37–60

Bowler PJ (1983) The eclipse of Darwinism: anti-Darwinian evolution theories in the decades around 1900. Johns Hopkins University Press, Baltimore

Bowler PJ (2003) Evolution: the history of an idea. University of California Press, Oakland

Bowler PJ (2017) Alternatives to Darwinism in the early twentieth century. In: Delisle RG (ed) The Darwinian tradition in context: research programs in evolutionary biology. Springer, Cham, pp 195–218

Caetano-Anollés G (ed) (2010) Evolutionary genomics and systems biology. Wiley-Blackwell, Hoboken

Coulter JM (1915) A suggested explanation of orthogenesis in plants. Science 42:859–863

Darwin C (1859) The origin of species by means of natural selection; or the preservation of favoured Racesn in the struggle for life. http://tinyurl.com/oyadykh. Accessed 26 Oct 2016

Delisle RG (2017) From Charles Darwin to the evolutionary synthesis: weak and diffused connections only. In: Delisle RG (ed) The Darwinian tradition in context: research programs in evolutionary biology. Springer, Cham, pp 133–168

Depew DJ (2017) Darwinism in the 20th century: productive encounters with saltation, acquired characteristics, and development. In: Delisle RG (ed) The Darwinian tradition in context: research programs in evolutionary biology. Springer, Cham, pp 61–88

Depew DJ, Weber BH (2011) The fate of Darwinism: evolution after the modern synthesis. Biol Theor 6:89–102

Dobzhansky T (1937) Genetics and the origin of species. Columbia University Press, New York

Dobzhansky T (1973) Nothing in biology makes sense except in the light of evolution. Am Biol Teach 35:125–129

Eldredge N, Gould SJ (1972) Punctuated equilibria: an alternative to phyletic gradualism. In: Schopf TJM (ed) Models in paleobiology. Freeman Cooper, San Francisco, pp 82–115

Esposito M (2017) The organismal synthesis: holistic science and developmental evolution in the english-speaking world, 1915–1954. In: Delisle RG (ed) The Darwinian tradition in context: research programs in evolutionary biology. Springer, Cham, pp 219–242

Fisher RA (1930) The genetically theory of natural selection. Oxford University Press, Oxford

Gavrilets S (1999) A dynamical theory of speciation on holey adaptive landscapes. Am Nat 154:1–22

Goldschmidt R (1940) The material basis of evolution. Yale University Press, Yale

Gould SJ (1985) Ontogeny and phylogeny. Harvard University Press, Cambridge
Gould SJ (2002) The structure of evolutionary theory. Harvard University Press, Cambridge
Gould SJ, Eldredge N (1977) Punctuated equilibria: the tempo and mode of evolution reconsidered. Paleobiology 3:115–151
Haeckel EHP (1866) Generelle Morphologie der Organismen. Allgemeine Grundzüge der Organischen Formen-wissenschaft, Mechanisch Begründet Durch die von Charles Darwin Reformirte Descendenztheorie. G. Reimer. https://archive.org/details/generellemorphol01haec. Accessed 4 Nov 2016
Haldane JBS (1932) The causes of evolution. Princeton University Press, Princeton
Hoelzer GA, Smith E, Pepper JW (2006) On the logical relationship between natural selection and self-organization. J Evol Biol 19:1785–1794
Huxley JS (1942) Evolution: the modern synthesis. MIT Press, Cambridge
Jablonka E, Lamb MJ (1999) Epigenetic inheritance and evolution: the Lamarckian dimension. Oxford University Press, Oxford
Jablonka E, Lamb MJ (2005) Evolution in four dimension: genetic, epigenetic, behavioral, and symbolic variation in the history of life. MIT Press, Cambridge
Jablonski D (2000) Micro- and macroevolution: scale and hierarchy in evolutionary biology and paleobiology. Paleobiology 26:15–52
Jablonski D (2005) Evolutionary innovations in the fossil record: the intersection of ecology, development, and macroevolution. J Exp Zool 304B:504–519
Jablonski D (2008) Species selection: theory and data. Annu Rev Ecol Evol Syst 39:501–524
Jablonski D, Shubin NH (2015) The future of the fossil record: paleontology in the 21st century. Proc Natl Acad Sci USA 112:4852–4858
Jordan DS (1920) Orthogenesis among fishes. Science 52:13–14
Kaplan JM (2006) More misuses of evolutionary psychology. Metascience 15:177–181
Kimura M (1983) The neutral theory of molecular evolution. Cambridge University Press, Cambridge
Kirschner MW, Gerhart JC (2005) The plausibility of life: resolving Darwin's Dilemma. Yale University Press, Yale
Kuhn T (1962) The structure of scientific revolutions. University of Chicago Press, Chicago
Kutschera U (2017) Symbiogenesis and cell evolution: an anti-Darwinian research agenda? In: Delisle RG (ed) The Darwinian tradition in context: research programs in evolutionary biology. Springer, Cham, pp 302–332
Laland K, Sterelny K (2006) Seven reasons (not) to neglect niche construction. Evolution 60:1751–1762
Laland K, Uller T, Feldman M, Sterelny K, Muller GB et al (2014) Does evolutionary theory need a rethink? Nature 514:161–164. http://tinyurl.com/myohvp2. Accessed 26 Oct 2016
Laland K, Uller T, Feldman M, Sterelny K, Muller GB et al (2015) The extended evolutionary synthesis: its structure, assumptions, and predictions. Proc R Soc B 282. https://doi.org/10.1098/rspb.2015.1019
Levit GS, Hossfeld U (2017) Major research traditions in 20th century evolutionary biology: the relations of Germany's Darwinism with them. In: Delisle RG (ed) The Darwinian tradition in context: research programs in evolutionary biology. Springer, Cham, pp 169–194
Lewis EB (1978) A gene complex controlling segmentation in *Drosophila*. Nature 276:565–570
Lewontin RC, Hubby JL (1966) A molecular approach to the study of genic heterozygosity in natural populations. II Amount of variation and degree of heterozygosity in natural populations of Drosophila pseudoobscura. Genetics 54:595–609
Lipman CB (1922) Orthogenesis in bacteria. Am Nat 56:105–115
Loison L, Herring E (2017) Lamarckian research programs in French biology (1900–1970). In: Delisle RG (ed) The Darwinian tradition in context: research programs in evolutionary biology. Springer, Cham, pp 243–270
Love A (2006) Evolutionary morphology and evo-devo: hierarchy and novelty. Theor Biosci 124:317–333

Lynch M (2007) The origins of genome architecture. Sinauer, Sunderland
Mayr E (1942) Systematics and the origin of species. Columbia University Press, New York
Müller GB (2007) Evo-devo: extending the evolutionary synthesis. Nat Rev Genet 8:943–949
Newman SA, Gabor F, Muller GB (2006) Before programs: the physical origination of multicellular forms. Int J Dev Biol 50:289–299
Okasha S (2006) Evolution and the levels of selection. Cambridge University Press, Cambridge
Otha T (2002) Near-neutrality in evolution of genes and gene regulation. Proc Natl Acad Sci USA 99:16134–16137
Pigliucci M (2001) Phenotypic plasticity: beyond nature and nurture. Johns Hopkins University Press, Baltimore
Pigliucci M (2007) Do we need an extended evolutionary synthesis? Evolution 61:2743–2749
Pigliucci M (2008a) What, if anything, is an evolutionary novelty? Philos Sci 75:887–898
Pigliucci M (2008b) Is evolvability evolvable? Nat Rev Genet 9:75–82
Pigliucci M (2016) Progress in science. Plato's Footnote. http://tinyurl.com/PlatoFootnote1, http://tinyurl.com/PlatoFootnote2, http://tinyurl.com/PlatoFootnote3. Accessed 26 Oct 2016
Pigliucci M, Muller GB (eds) (2010) Evolution – the extended synthesis. MIT Press, Cambridge
Prud'homme B, Gompel N, Carroll SB (2007) Emerging principles of regulatory evolution. Proc Natl Acad Sci USA 104:8605–8612
Real LA, Brown JH (1991) Foundations of ecology. Classic papers with commentaries. University of Chicago Press, Chicago
Rollo CD (1995) Phenotypes: their epigenetics, ecology and evolution. Chapman & Hall, London
Schlichting CD, Pigliucci M (1998) Phenotype evolution, a reaction norm perspective. Sinauer, Sunderland
Shanahan T (2017) Selfish genes and lucky breaks: Richard Dawkins' and Stephen Jay Gould's: divergent Darwinian agendas. In: Delisle RG (ed) The Darwinian tradition in context: research programs in evolutionary biology. Springer, Cham, pp 11–36
Simpson GG (1944) Tempo and mode in evolution. Columbia University Press, New York
Stebbins GL (1950) Variation and evolution in plants. Columbia University Press, New York
Tiselius A (1937) A new apparatus for electrophoretic analysis of colloidal mixtures. Trans Faraday Soc 33:524–531
Turner DD (2017) Paleobiology's uneasy relationship with the Darwinian tradition: stasis as data. In: Delisle RG (ed) The Darwinian tradition in context: research programs in evolutionary biology. Springer, Cham, pp 333–352
Wagner GP, Zhang J (2011) The pleiotropic structure of the genotype-phenotype map: the evolvability of complex organisms. Nat Rev Genet 12:204–213
Watson JD, Crick FHC (1953) Molecular structure of nucleic acids. Nature 171:737–738
Weissman A (1893) The germ-plasm: a theory of heredity. Charles Scribner's Sons. http://tinyurl.com/gq9m765. Accessed 26 Oct 2016
West-Eberhard MJ (2003) Developmental plasticity and evolution. Oxford University Press, Cambridge
Wilson DS, Wilson EO (2008) Evolution 'for the good of the group'. Am Sci 96:380–389
Winther RG (2015) The structure of scientific theories. Stanford Encyclopedia of Philosophy. http://plato.stanford.edu/entries/structure-scientific-theories/. Accessed 26 Oct 2016
Wright S (1931) Evolution in Mendelian populations. Genetics 16.97–159

Human Evolution as a Theoretical Model for an Extended Evolutionary Synthesis

Adam Van Arsdale

Abstract Humans have occupied a paradoxical position within the history of evolutionary studies. On one end, humans have been central to both the academic motivation of the field and the public tensions surrounding evolution. Simultaneously, humans have been cast aside as a poor model organism for understanding the processes that underlie evolutionary theory. As a result, anthropologists who work within an evolutionary context, often chided as being two decades behind mainstream biology, have come to occupy a unique position with respect to the understanding of how evolution operates on humans. Incorporating theoretical developments from a diverse set of related evolutionary fields, biological anthropologists have begun to gather empirical data on the unique evolutionary processes that have shaped our own evolutionary path. Some of the important components that have emerged in human evolutionary studies—biocultural feedback systems, culturally mediated niche construction, and technological ratchet effects—have shed new light not only on how human evolution has proceeded but also on the range of capabilities of evolution more broadly. While not rejecting traditional neo-Darwinian theory and the importance of genetic inheritance, these new developments have highlighted the tremendous complexity afforded by the cumulative action of both selective and neutral evolutionary forces across a range of inheritance modes. Rather than a poor evolutionary model, many of these evolutionary processes are best, or perhaps only, observable in humans. The traits which have structured critical transitions in our hominin past—encephalization, expanded childhood development, and generative language—open up new windows into thinking about an Extended Evolutionary Synthesis.

Keywords Hominin evolution • *Homo erectus* • Darwinism • Modern synthesis • Encephalization

---

A. Van Arsdale (✉)
Department of Anthropology, Wellesley College, 106 Central Street, Wellesley, MA 02481, USA
e-mail: avanarsd@wellesley.edu

R.G. Delisle (ed.), *The Darwinian Tradition in Context*,
https://doi.org/10.1007/978-3-319-69123-7_6

1 Background

In Charles Darwin's *On the Origin of Species* (1859: 488), he famously limits his references to human evolution to a single line near the end of the volume, writing that through the study of evolution, "light will be thrown on the origin of man and his history." Twelve years later, with the publication of *The Descent of Man* (1871), Darwin expands on human evolution in much greater detail, using it as an opportunity to develop the concept of sexual selection. However, even in this work, Darwin begins on a cautionary note:

> When we confine our attention to any one form, we are deprived of the weighty arguments derived from the nature of the affinities which connect together whole groups of organisms—their geographical distribution in past and present times, and their geological succession (Darwin 1871: 1).

Darwin's awareness of the significance for humanity in understanding the evolution of humans, coupled with his reticence to focus exclusively on humans as a model for evolution, provides an interesting framing for the question considered in this essay. There exists a great deal of interest and curiosity in the knowledge of our shared human evolutionary past but considerable apprehension in reversing the equation and considering what humans can help us understand about how evolution operates. We use evolution to understand our origins as a species, but what does the story of our own evolution reveal about the forces of evolutionary change?

Today, it is unusual to find a researcher whose specialization is the behavioral and morphological evolution of humanity in a biology department (human genetics as a focus in molecular biology being a notable exception). Instead, researchers who focus on human evolution are more often found in departments of anthropology, anatomy divisions of medical schools, or more recent incarnations such as departments of human evolutionary biology.

This contemporary dynamic is the result of an evolution within scientific fields of evolutionary study and was not always the case. As Strier (2016) notes, Section H (Anthropology) of the American Association for the Advancement of Science (AAAS) dates back to 1882. The American Anthropological Association (AAA), the largest association of anthropologists in the world, grew out of AAAS in 1902. The American Association of Physical Anthropologists (AAPA), the largest association of anthropologists whose study is focused on humans and grounded in evolutionary theory, also took origin from the AAAS. First proposed in 1924, a newly chartered AAPA held joint meetings with the AAAS in 1930, and met jointly with both AAAS and the AAA in 1932 (Strier 2016). These facts illustrate the evolutionary origin of the scientific study of human evolution within the broader natural science and evolutionary studies academic realm.

Indeed, several evolutionary biologists intimately connected with the Modern Synthesis were directly involved in the major conversations on human evolution as late as the mid-twentieth century (Smocovitis 2012). Among his endlessly prolific works, Ernst Mayr wrote on the "*Taxonomic categories in fossil hominids*" (1950).

When anthropologist Frank Livingstone provided the first formal critique of the biological race concept in 1962, the respondent on the paper was Theodosius Dobzhansky (Livingstone and Dobzhansky 1962). Again, outside genetics, it is rare to find those points of overlap not just in theory, but in subject, within biology and anthropology today. Stephen Jay Gould's voluminous *The Structure of Evolutionary Theory* (2002), as one example of this trend, only makes passing reference to hominid evolution in its more than 1300 pages.

The drift of human evolutionary studies away from mainstream evolutionary research, or vice versa, in the period after World War II is understandable (Barkan 1996). The revelations of the atrocities of science engineered under the National Socialist regime of Germany, especially those focused on human subjects, made public by the Nuremberg Trials were a watershed moment in twentieth-century human biology (Marks 2008). Given the historical focus of anthropology on race, and the prominence of race-based perspectives on human evolution within anthropology prior to WWII, it is easy to understand the movement away from studies focused on humans in mainstream biology (Washburn 1951; Caspari 2009).

The resistance to E. O. Wilson's *Sociobiology: The New Synthesis* (1975) from areas of the social sciences critical of any hint of biological determinism only furthered this trend (Sahlins 1976; Lieberman 1989; see also Alcock 2017). Humans are too complex to distinguish between genetic and environmental ("cultural") effects. We live too long to look at trans-generational changes in allele frequencies. The data needed to study evolution for humans is too messy. Humans, quite simply, are not a good model organism for the study of evolution. Or so the logic went.

Despite the sidelining of humans within evolutionary studies, humans remain a major focus of the public facing side of evolutionary studies (Plutzer and Berkman 2008). Major fossils relevant for human evolution are disproportionately represented on the covers of *Science* and *Nature*. Documentaries on evolution rarely bypass, and more often than not highlight as a central topic, human evolution. While often devoting entire spacious halls to narratives of human evolution, organisms like *Drosophila* or *Caenorhabditis elegans* (or even *Mus musculus*) rarely get the public coverage warranted by their importance within the scientific process itself.

On the more controversial side of things, it is the evolution of humans, rather than evolutionary theory more broadly, which often raises legal and political challenges to the teaching of evolution or public acceptance of evolution (Lynn et al. 2017). Likewise, the acceptance of scientific knowledge itself, regardless of its evolutionary content, often is strongly correlated with one's understanding of the application of evolution to humans (Pobiner 2016).

In the time period that human evolution has drifted away from the center of evolutionary studies, traditional biological sciences have also been involved in critical self-examination of its foundational framework (Laland et al. 2014). These debates encompass a broad range of topics and developments within the fields of evolutionary studies but can be summarized as discontentment (or a lack of discontentment) with the traditional gene-centric view of the Modern Synthesis (Blute 2017; Laland 2017). For some researchers, the major developments within evolutionary theory over the past 80 years—neutral theory, renewed engagement with evolutionary perspectives on development, epigenetics and complex genomic structure, and hierarchically structured plasticity—have shifted the main focus of evolutionary causation away from natural selection and raised questions as to whether the traditional neo-Darwinian framework remains the best approach to understanding and presenting the action of evolution (Jablonka and Lamb 2014). These are not, it should be pointed out, arguments that "evolution is wrong" but instead are arguments about where the focus is placed on the processes of evolutionary change through time.

Conversations around this issue are equally fascinating and frustrating (Pigliucci 2007, 2017; Szathmáry 2015). On the one hand, it is possible that these conversations can highlight the extraordinary capacity of evolution to create complex patterns of variation in a seemingly endless and changing number of ways (Pigliucci 2008; Jablonka and Lamb 2014). On the other hand, some see these efforts as undermining public understanding of evolution, unnecessarily focusing on ephemeral "exceptions" rather than central tendencies in evolution, and potentially prioritizing novel theoretical perspectives over interpretation of basic observational data (Wray and Hoekstra 2014; Futuyma 2015).

Mindful of the cautionary note sounded by Darwin at the beginning of this essay, in what will follow I will nevertheless argue that this is a unique and important moment to reconsider the role that human evolution plays in the mainstream of evolutionary biology (Fuentes 2016). The unique trajectory of our own evolutionary past highlights both the possibilities and the limitations of an Extended Evolutionary Synthesis. There is compelling evidence that specific changes in our evolutionary past, beginning with the emergence of the genus *Homo*, initiated a fairly unique manifestation of neo-Darwinian processes involving multiple modes of trans-generational inheritance with significant evolutionary consequences. And yet, the specific nature of the changes that have occurred in humans highlights the exceptional status of the circumstances under which humans evolved, perhaps thereby serving as the exception that proves the rule when it comes to the need for new theoretical understandings of evolution, as opposed to simply new ways of understanding existing theory and data.

2 Humans as a Model Organism in Evolutionary Studies

At the heart of many of the discussions around a "new" Evolutionary Synthesis are varying views about how well integrated are emerging bodies of theory in ecology and evolution (Laland et al. 2015), with a healthy smattering of discussion of demography (Lowe et al. 2017) and development added to the mix (Gilbert et al. 2015). As it happens, multiple lines of evidence associated with recent human evolution highlight the critical role played by changing patterns of demography, development, and ecology in shaping our own evolutionary trajectory. As such, an examination of recent human evolution provides a valuable lens into ongoing and active debates about the nature of evolutionary theory.

The particular importance of humans in this discussion can be observed by focusing on the nature of the evolutionary transition at two major points in our past. Traditionally, narratives of human evolution (scientific or public) tend to focus on two transitions: the origin of hominins and the last common ancestor with extant apes (e.g., Gibbons 2007) and the African origin of "modern" *Homo sapiens* toward the end of the Pleistocene (Stringer 2012). These two evolutionary transitions are certainly important, but for the purposes here, a more telling perspective comes from examining the emergence of the genus *Homo*, sometime around 2 million years ago (and *Homo erectus sensu lato*, in particular), and the widespread shift to agricultural subsistence systems over the past 12,000 years. These transitions encompass a significant shift in the pattern of human evolution in the case of the origin of *Homo* and a major ecological change with the origin of agricultural food systems that highlights the magnitude of the evolutionary frame shift that has occurred in our lineage.

2.1 The Origin of Homo

The transition from *Australopithecus* to *Homo* is one marked by fairly subtle changes in morphology but changes that have a profound effect on the ecology of subsequent hominins and the ensuing pattern of evolution seen in humans and our ancestors (McHenry and Coffing 2000). In particular, the transition from *Australopithecus* to *Homo* involves the evolutionary investment in mechanisms that enhance plasticity and enable long-term changes to the basic pattern of human evolution.

Most anthropologists place the origin of *Homo* somewhere in East Africa near the Pliocene–Pleistocene transition (e.g., Villmoare et al. 2015; Schrenk et al. 2015). However, current consensus is lacking on the exact nature, both taxonomic and biological, of these earliest specimens, sometimes referred to simply as early *Homo* (Wood and Collard 1999; Wolpoff 1999; Van Arsdale and Wolpoff 2013; Berger 2013; Antón et al. 2014) As such, it is more useful to focus on the less controversial initial appearance of *Homo erectus*, also likely in East Africa (Antón 2003; Spoor et al. 2007), though perhaps most clearly evident in its earliest appearance with fossil material from the site of Dmanisi, Georgia, at approximately 1.8 MA (Lordkipanidze et al. 2007, 2013; Van Arsdale and Lordkipanidze 2012). *Homo erectus* presents the convergence of several morphological and behavioral traits seen heterogeneously in earlier hominin fossils, including the diverse assemblage of potential early *Homo* fossils.

Morphologically, an evolutionary shift from Australopithecines to *Homo erectus* can be seen across several broad features. One of the clearest distinguishing features of *Homo erectus* is an expanded body size, particularly involving elongated hind limbs (Rose 1984; Walker and Leakey 1993; Antón 2003; Lordkipanidze et al. 2007). The Nariokotome partial skeleton, dating to roughly 1.6 MA, provides a near complete glimpse of an extremely long-legged individual that would have been in the vicinity of six feet in height as an adult (Walker and Leakey 1993; Dean and Smith 2009). The longer limbs of *Homo erectus* have been associated with greater long-distance locomotor efficiency (Steudel-Numbers 2006; Pontzer et al. 2010; Pontzer 2012). This attribute appears to have been associated with the utilization of a broader range of food resources, including higher quality food resources relative to earlier hominins (Walker 1981; Organ et al. 2011).

Further evidence of a dietary shift toward higher quality food resources comes from the reduced masticatory apparatus in *Homo erectus* relative to earlier hominins. This reduction includes significantly reduced post-canine dentition (Wood 1992; Gabunia and Vekua 1995; Kaifu 2006; Suwa et al. 2007; Zaim et al. 2011), an increase in the gracility of the mandibular corpus (Rightmire 1981; Antón 2003; Van Arsdale and Lordkipanidze 2012), reduced temporal fossa size corresponding to a reduced temporalis muscle (Wolpoff 1975; Demes and Creel 1988), and a possible shift in the biomechanics of hominin chewing (Teaford et al. 2002). There is at least some evidence for a genetic basis to this suite of masticatory changes in the form of a nonsense mutation to *MYH16* gene in the human lineage dating to the Plio-Pleistocene boundary, though additional research needs to be done to confirm this link (Stedman et al. 2004).

Importantly, the reduction in masticatory apparatus, particularly in the cranial musculature and post-canine dentition, is not limited to *Homo erectus* at this time. *Australopithecus sediba* from South Africa, possibly concurrent with the earliest *Homo erectus* in East Africa, also shows a reduced masticatory apparatus, but absent many of the other changes seen in *Homo erectus*, most notably a lack of brain expansion (Berger et al. 2010). The presence of coexisting hominin lineages in the Lower Pleistocene, but featuring a different combination of traits, provides a natural test for the significance of the combination of traits present in *Homo erectus* and distinguishes the eventual evolutionary success of *Homo erectus* relative to these other lineages.

The focus on higher quality food resources was conditioned on the utilization of cultural technology, including intentionally flaked stone tools. Recent findings have pushed the earliest appearance of stone tools to at least 3.3 million years (Harmand et al. 2015), but by the time of *Homo erectus*, tool technologies have taken on aspects of complexity involving form, acquisition, specialization, and usage, which highlight a dynamically changing role within *Homo erectus* ecology (Shipman and Walker 1989; Braun et al. 2009; Ferring et al. 2011; Hovers and Braun 2009). This ecological shift included a broadening, or at least increased variability, in the diet (Ungar et al. 2006), as well as an increase in energetic quality (Leonard and Robertson 1992). The reduction in masticatory capabilities seen in *Homo erectus*, particularly when compared to contemporary and possibly sympatric robust Australopithecine lineages, highlights the important role of stone tool technology in separating the ecological niche of these two lineages.

The convergence of larger body size, reduced masticatory apparatus, and improved dietary quality (aided by the application of cultural technology) all relate to the most significant feature of *Homo erectus*, an increase in the size and encephalization of the brain. Beginning just after 2 million years ago, some members of the hominin lineage begin displaying an absolute and relatively larger brain (Ruff et al. 1997; Antón 2003; Lee and Wolpoff 2003; Rightmire 2004; Baab 2008). In addition to an expansion in absolute and relative brain size, fossil endocasts reveal evidence of a shift toward a more human-like brain morphology (Falk 1987; Bruner and Holloway 2010; Berger et al. 2015).

Changes to the brain observed in *Homo erectus* also reflect related changes to development, life history, and ecology of this species. The brain is an energetically expensive tissue that expands in *Homo erectus* without an associated change in basal metabolic rate (Aiello and Wheeler 1995; Aiello 2007). The ecological transition to higher quality food resources thus appears to be a necessary precursor for adequately fueling an expanded brain, consistent with the emerging evidence of complex tool use in earlier Pliocene hominins (McPherron et al. 2010; Harmand et al. 2015) and, in a less durable fashion, nonhuman primates (Van Schaik et al. 1999; Whiten et al. 1999; Matsuzawa 2008). Again, the contrast between the dietary ecology of *Homo erectus* and the contemporaneous robust Australopithecines is telling. *Homo erectus* appears to have fully committed, with both an encephalized brain and a reduced gut, to a more intensified, high-quality, diet. The contemporaneous Australopithecines, no doubt also engaged in diverse dietary strategies, retained the ability, likely fully realized in species like *Australopithecus boisei*, of targeting low-quality, high-volume food resources (Cerling et al. 2011). This contrast is helpful not only in highlighting the different niche in *Homo* but also in identifying the points of departure in the evolutionary trajectory in *Homo*. It is difficult to imagine a later member of the genus *Homo* biologically specializing in low-quality food resources without the aid of the kinds of cultural technology we see in contemporary farming populations (see below).

The transition to a relatively larger brain required not only an ecological shift but also a life history change. Relative to nonhuman primates or earlier hominins, *Homo erectus* had a more rapid pattern of brain growth after birth in addition to an overall extended period of brain development (Leigh 2006; DeSilva and Lesnik 2008; Snodgrass et al. 2009; DeSilva 2011; Grabowski 2016). Developing larger brains in infants and toddlers more rapidly and for a longer period of time requires a subsequent change in parental, and particularly maternal, investment (Aiello and Key 2002; Aiello and Wells 2002; Ellison 2009; Dunsworth et al. 2012). The expansion and increasing energetic investment in early development in *Homo erectus*, coupled with the increased energetic demands and higher quality diet, all attest to an overall energetic/ecological intensification within *Homo erectus* relative to earlier hominins and nonhuman primates.

The above changes in *Homo erectus* are all supported by morphological evidence from the existing hominin fossil record or material evidence in the Plio-Pleistocene archaeological record. However, they also gesture toward a set of behavioral changes that are harder to directly examine in the fossil and archaeological record but also have at least circumstantial support. These behavioral changes—population expansion and dispersal, decreasing mortality, increasing social complexity, and increasing behavioral plasticity—all have tremendous importance for how the pattern of evolution and the underlying processes are inferred for our shared, recent evolutionary past.

The least contentious consequence of the above changes in *Homo erectus* is an overall expansion in the geographic range of *Homo erectus* relative to earlier hominins. In a relatively narrow window of time between 1.8 and 1.6 million years ago, hominins (and generally considered to be *Homo erectus*) expand from an exclusively African fossil clade to one found in the Caucasus (Dmanisi), China (Nihewan Basin), and Indonesia (Java) (Swisher et al. 1994; Sémah et al. 2000; Gabunia et al. 2001; Zhu et al. 2004). After evolving in Africa for at least 3 million years, within a short window of time, *Homo erectus* becomes the first hominin to expand outside of Africa, occupying a range of habitats that dramatically expands the existing range of seasonality (both temperature and humidity) and ecosystems occupied by earlier hominins (Tappen 2009; Potts and Teague 2010).

Another subtle shift that occurs with *Homo erectus* is evidence of a reduction in the rate of adult mortality. Analysis of hominin dental remains throughout the Plio-Pleistocene identifies a small but significant increase in the fraction of adult remains that might be considered "older adults" at the time of death as opposed to young adults (Caspari and Lee 2004). Even a slight change in patterns of adult mortality may have led to significantly different dynamics structured around life history, including the potential for an associated increase in longevity (O'Connell et al. 1999). The effects of an increase in adult survivorship could be compounded by the geographic expansion observed in *Homo erectus* and the potential increase in the number of accessible and suitable environments for occupation. More populations with more individuals living longer create a greater number of natural laboratories for processes of evolution to operate.

There is also direct evidence at the site of Dmanisi for increased longevity in *Homo erectus* with the presence of a completely edentulous specimen, showing extensive resorption of both the mandibular and maxillary alveolar regions, indicating sustained survival in the absence of a functioning masticatory apparatus (Lordkipanidze et al. 2006; Van Arsdale and Lordkipanidze 2012). This specimen is not sufficient to argue directly for social care in early *Homo erectus*, but it does provide the earliest evidence in humans of the potential to survive with some combination of cultural technology. This view is consistent with the evidence associated with the energetic demands of pregnancy and childhood development, as well as the acquisition of large-sized prey (Bunn 2001), of an overall uptick in the intensity and significance of social networks of hominins.

2.2 *Pleistocene Hominin Evolution*

The evolutionary changes associated with *Homo erectus* described above, viewed individually, may be considered unremarkable. However, when viewed in conjunction with the evidence drawn from the archaeological record indicating changing behavioral patterns, they provide compelling evidence of a significant change in the properties that define the pattern of human evolution. More importantly, this pattern can be observed by looking at the Pleistocene fossil and archaeological record that

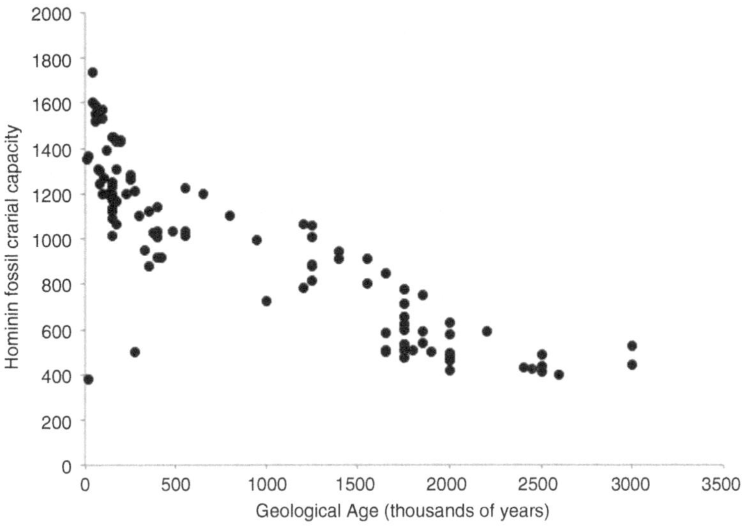

Fig. 1 Pleistocene hominin cranial capacity of major fossil specimens across the Pleistocene. The initial evidence of encephalization begins shortly after 2 million years ago, concurrent with the origin of *Homo erectus*, marking the onset of a trend that continues until about 30,000 years ago

follows the origin of *Homo erectus*, suggesting that the evolutionary changes associated with humans are durable across a geological timeframe.

What makes the transition to *Homo* noteworthy is the extent to which the brain becomes both a focus of evolutionary pressure and a mechanism for establishing the evolutionary landscape governing hominins in the Pleistocene. The behavioral, technological, and dietary changes that predate the origin of *Homo erectus* indicate the beginning of a shift in at least some Australopithecines to a higher quality diet, involving more intensive extraction of food resources (presumably relatively scarce fat components, in particular). At a certain point, however, evident by at least 1.8 million years ago in *Homo erectus*, this led to the onset of a clear pattern of brain expansion seen throughout Pleistocene hominins (see Fig. 1).

The gradual expansion in hominin brain size throughout the Pleistocene is paralleled by other transitions that reflect the impact and scope of this change. The most significant trends are those indicating increased demographic potential (Caspari and Lee 2004), increasing cultural complexity (McBrearty and Brooks 2000), and increasing habitat range and diversity (Dennell and Roebroeks 2005; Finlayson 2005).

As the brain became a more important part of the hominin ecological adaptation, it simultaneously became a more efficacious mechanism for the transmission of evolutionarily relevant information across generations and between individuals. The brain takes on a dual importance in Pleistocene hominins as both a shaper of phenotype, and therefore target for selective processes, and at the same time a nongenetic alternative hereditary system. Increasing social complexity, coupled with extended periods of childhood development and elevated cognitive abilities, establishes an increasingly durable pathway for the transmission of behavioral, and potentially symbolic, forms of evolutionary information. As others have noted, in this scenario, the brain becomes positioned as the fulcrum in an evolutionary "ratchet" model of human cognition and culture (Tomasello 1999; Tennie et al. 2009).

What exactly is being "ratcheted" up in the above model of human evolution is up for interpretation, particularly when the amorphous term of "culture" is invoked? An examination of the fossil evidence is helpful. Brain size is, of course, one thing that is steadily increasing, along with the aforementioned demographic increases, range expansion, and technological complexity. But another way of viewing this scenario is that it is the breadth of the hominin niche itself that is being expanded. As cognitively mediated technology/culture becomes a more important aspect of phenotype in hominins, the range of conditions under which hominins might persist also expands. Importantly, this expansion in niche and phenotypic plasticity is at least partially mediated by nongenetic modes of transmission. The development of new tool technologies allows for procurement of different kinds of food resources (Wilkins et al. 2012). The application of fire to food processing allows for enhanced nutrient extraction from the environment (Attwell et al. 2015). The long-range exchange of material goods helps facilitate cultural and technological practices across a wider range of environments (Langley and Street 2013). These kinds of traditions develop, propagate, and change on the basis of cultural factors, not genetic ones.

It is reasonable to ask whether or not these patterns are supported by an analysis of the cranial morphology in Pleistocene hominins. Does the pattern of niche expansion in Pleistocene hominins fit the predictions associated with a dramatic shift in the pattern of human evolution and is such a change visible in the hominin fossil record?

In order to address this question, the below analysis looks at the pattern of cranial variation observed across 143 Plio-Pleistocene hominin fossils, a sample representing the majority of the well-preserved cranial specimens available for study at this time (Table 1). A diverse array of views exist as to the appropriate taxonomic classification of these specimens individually and the number and identity of the taxonomic groups to which they belong. As a starting point, this study builds off recent genetic evidence that suggests anatomically modern *Homo sapiens* admixed with multiple populations of archaic hominins, including at a minimum Neandertals and Denisovans (Sankararaman et al. 2016; Wall and Brandt 2016; Nielsen et al. 2017). Assuming that the admixture between *Homo sapiens* and these archaic populations reflects a theoretical maximum amount of genetic divergence, patterned across time and space, through which an evolutionarily continuous lineage can be maintained, the combined *Homo sapiens*–Neandertal sample is used as a lens to evaluate the pattern of cranial variation observed throughout the remainder of the Pleistocene.

The 143 cranial specimens used in this study are variable in their degree of preservation. In order to maximize the comparability of pairs of specimens, up to 188 cranial measurements were used for comparison, with a minimum threshold for inclusion in the study of 20 measurements. For each specimen pair, an average

Table 1 List of cranial specimens

Amud 1	La Chaise 2	Skhul 5
Arago 21	Laetoli 18	Skhul 6
Atapuerca 5	Lantian	Skhul 9
Biache	Lazaret 1	Spy 1
Bodo	LB 1	Spy 2
Ceprano	Maba	Solo 1
Chao Hu	MH 1	Solo 3
Chesowanja 1	MK 1	Solo 4
Daka	MK 9	Solo 5
Dali	MK 37	Solo 6
Dmanisi 2280	Narmada	Solo 8
Dmanisi 2282	Ndutu	Solo 9
Dmanisi 2700	OH 5	Solo 10
Dmanisi 3444	OH 9	Solo 11
Ehringsdorf	OH 12	SK 12
Eliye Springs	OH 13	SK 13
Florisbad	OH 16	SK 46
Fontechevade	OH 24	SK 47
Herto 16/1	Omo 1	SK 48
Herto 16/5	Omo 2	SK 49
Hexian	Petralona	SK 52
Jebel Irhoud 1	Qafzeh 3	SK 79
Jebel Irhoud 2	Qafzeh 5	SK 80
Jinniushan	Qafzeh 6	SK 83STS 5
Kabwe	Qafzeh 7	STS 13
Kanjera 1	Qafzeh 9	STS 17
KNM-WT 15000	Qafzeh 11	STS 19
KNM-WT 17000	Saccopastore 1	STS 22
KNM-WT 17400	Saccopastore 2	STS 52
KNM-ER 406	Saldanha	STS 53
KNM-ER 732	Sale	STS 71
KNM-ER 1470	Saltzgitter	STS 1511
KNM-ER 1590	Sambungmachan 1	STW 52
KNM-ER 1805	Sambungmachan 3	STW 252
KNM-ER 1813	Sambungmachan 4	STW 505
KNM-ER 3732	Sangiran 2	Steinheim
KNM-ER 3733	Sangiran 3	Swanscombe
KNM-ER 3883	Sangiran 4	Reilingen
KNM-ER 13750	Sangiran 10	Tabun 1
Krapina A	Sangiran 12	Trinil 1
Krapina B	Sangiran 17	Vértesszolos
Krapina C	Shanidar 1	ZKD D1
Krapina D	Shanidar 2	ZKD E1

(continued)

Table 1 (continued)

Krapina E	Shanidar 4	ZKD H3
Krapina 16	Shanidar 5	ZKD L1
Kromdraai	Skhul 2	ZKD L2
La Chaise 0	Skhul 4	ZKD L3
		Zuttiyeh

normalized pairwise distance was calculated on the basis of mutually preserved homologous measurements:

$$\frac{1}{n}\left[\sum_{1}^{n}\left(\text{abs}(x_a - x_b)*\sum_{1}^{z}\frac{\text{abs}(x_1 - x_z)}{\mu_z}\right]$$

A conservative taxonomic scheme was employed, consisting of a mixed *Homo sapiens*–Neandertal sample, specimens assigned to *Homo heidelbergensis*, *Homo erectus*, and five species of Australopithecines (*A. africanus, A. sediba, A. aethiopicus, A. boisei, A. robustus*).

Figure 2 displays the results of a pairwise comparison of cranial variation of Pliocene and Pleistocene hominin crania.

A number of interesting results emerge from this analysis. First, while there is considerable noise within the pairwise comparisons for any given group (more on that below), the mean for those groups assigned to Pleistocene *Homo* fit a linear model through time. In contrast, the samples of pairwise Australopithecine comparisons, in aggregate, are consistently elevated from this trend in their degree of morphological divergence.

In this analysis, *Homo sapiens*, Neandertals, *Homo heidelbergensis*, and *Homo erectus* do not show major gaps or discontinuities suggestive of a significant change in the overall pattern of evolution. In contrast, the division between *Homo* and *Australopithecus* is marked by such a discontinuity, as shown by a secondary analysis of these results. If the 95th percentile of variation observed in the *Homo sapiens*–Neandertal sample is used as a guide for the expected level of morphological variation within an evolving lineage, it is possible to explore the percentage of pairs in each group of pairwise comparisons that exceeds that level (Table 2). The significance of these values is assessed via a resampling approach that generates an expected level of variation for a randomly resampled set of pairwise comparisons equivalent in sample size to that observed in this study. None of the pairwise samples within *Homo* show a significant fraction of highly divergent pairwise comparisons (though they do show an increasingly amount of variation as the temporal gap increases). In contrast, all of the available Australopithecine comparisons show statistically significant fractions of pairwise comparisons that are more divergent than the observed 95th percentile of *Homo sapiens*–Neandertal comparisons.

The apparent "noise" in Fig. 2 might also be indicative of the strong effects of genetic drift on shaping aspects of Pleistocene hominin crania morphology. This result has been suggested by previous research (Ackermann and Cheverud 2004; Roseman 2016) and likely reflects the strong evolutionary constraint imposed on Pleistocene hominins given their demographic limitations. Parsing the data from

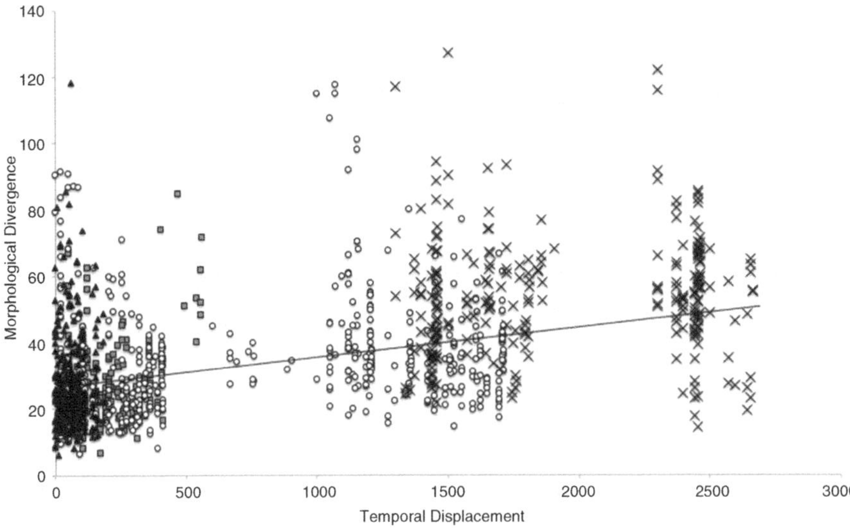

Fig. 2 Morphological divergence in a sample of 1201 pairwise comparisons of Plio-Pleistocene hominin crania. The *y*-axis displays the degree to which any pair of crania differ on the basis of homologous linear measurements. The *x*-axis depicts the best estimate of the temporal distance separating the pair of fossils. Solid triangles indicate pairwise comparisons from a mixed *Homo sapiens*–Neandertal sample. Gray squares include pairwise comparisons between *Homo sapiens*–Neandertals and *Homo heidelbergensis* specimens. Open circles include pairwise comparisons between *Homo sapiens*–Neandertals and *Homo erectus* specimens. The trendline depicts a model of the expected level of divergence given an estimated temporal displacement between any two specimens. The trendline intersects nearly exactly with the mean morphological and temporal divergence in the samples including pairwise comparisons with *Homo heidelbergensis* and *Homo erectus*. Pairwise comparisons involving specimens assigned to *Australopithecus* (*A. africanus*, *A. robustus*, *A. boisei*) are displayed with a gray X. Relative to comparisons with *Homo*, these pairings show substantially elevated levels of morphological divergence

above more closely, within pairwise groupings there exists a significant relationship between morphological variation and geographic separation, in addition to the association with temporal displacement. This observation adds an additional ripple to the evolutionary pattern evident in the morphology of Pleistocene *Homo*, highlighting thc significant constraint imposed by demographic constraint and geographic isolation on hominin populations. Thus, alongside the strong argument for continuous selection acting on the associated pattern of encephalization, genetic drift is likely playing a major and ongoing role in shaping aspects of cranial morphology.

The concordant patterns of change seen in Pleistocene hominins can be understood as part of a novel emergent evolutionary pattern in *Homo*, centered on the brain as both an adaptive and hereditary element. Even while populations of hominins are expanding into diverse and distant environments in the Pleistocene, they show consistent trends in morphological variation, particularly centered around aspects of neurocranial size and structure. This is true, despite the evidence of an expanded role for genetic drift in shaping patterns of population differentiation. That these semi-isolated populations are not fissioning off into distinct

Table 2 Results of pairwise cranial analysis

	A. africanus	*A. robustus*	*A. boisei*	*H. erectus*	*H. heidel.*	Neandertal	*H. sapiens*
A. africanus		65	41	228	78	99	83
A. robustus	0		37	158	57	73	68
A. boisei	21.95	16.22		160	44	65	61
H. erectus	50.88**	47.37**	33.75**		347	525	479
H. heidel.	65.38**	45.61**	47.73**	10.37		132	117
Neandertal	54.55**	28.75**	49.23**	12.38	6.82		169
H. sapiens	56.63**	30.88**	50.26**	12.11	9.40	5.92	

The numbers above the diagonal reflect the number of pairwise comparisons that met the threshold for having at least 20 homologous measurements available for comparison. The numbers below the diagonal are the percentage of pairwise comparisons that exhibited morphological differences greater than the 95th percentile of observed variation in Neandertal–*Homo sapiens* pairings. Asterisks indicate significant differences on the basis of a resampling analysis morphological lineages attests to the expanding hominin niche and its power in reducing the development of long-term reproductive barriers.

While it is difficult to directly identify a mechanism for causation with the encephalization of the Pleistocene hominin brain, it is nearly impossible to avoid its association with increasing social and technological complexity. More advanced cultural and technological capabilities allowed hominins to expand and persist across a greater range of environments. The expansion of hominins into new environments in the Pleistocene does not correlate strongly with significant morphological change; rather, it appears to be driven by increasing plasticity associated with the application of cognitive resources to novel environmental challenges. Hominins in the Pleistocene are able to adapt, increasingly, by creating novel cognitive/cultural/technological variants specifically directed toward environmental challenges, rather than depending on the creation of such novel variation via purely genetic processes.

The Pleistocene archaeological record attests to not just greater cognitive complexity in tool construction but also an increased emphasis on the mechanisms associated with the transmission of that complexity across generations. Tools become more complex throughout the Pleistocene—including a diversification of tool types, materials, and construction technologies—but they also become more consistent in form. Reconstructions of the cognitive processes associated with Pleistocene tool construction attest to the active learning and instruction processes necessary for successful replication of elements of the archaeological record (Nowell and Davidson 2010; Bruner 2014).

The knowledge produced by cognitive archaeology parallels developmental psychology literature contrasting humans and nonhuman primates (Want and Harris 2002; Hare and Tomasello 2005). Humans are extraordinarily good active learners and teachers. The human brain places sensory information from the environment into complex semantic structures in a hypothesis-like manner. In this capacity, the Pleistocene hominin brain acts as an alternative hereditary system, conveying evolutionarily relevant information associated with the construction of phenotype, and thereby subject to evolutionary forces, including natural selection.

2.3 The Origin of Domestication

While the origin of *Homo* appears to include both a morphological and related ecological shift, the transition to domestication-based lifestyles and subsistence strategies in human populations appears to be largely ecological (though with the possibility for substantial lagging morphological changes in response). Importantly, the transition to agriculture is best explained as a direct product, and indeed continuation, of the pattern enabled by the origin of *Homo* (Zeder 2006).

By the end of the Pleistocene, hominin populations, still predicated on subsistence level foraging, had occupied the majority of available environments within Eurasia and Africa. Technological and cultural developments had enabled populations to overcome a wide range of climatic and ecological barriers (Roebroeks and Villa 2011; Boivin et al. 2013; Glantz et al. 2016). Evidence based on faunal exploitation shows that population levels in many of these regions were increasing, a finding consistent with the gradual changes in mortality patterns that began with *Homo erectus* (Stiner et al. 1999). Facing such circumstances, populations in the Late Pleistocene began to show evidence of increased exploitation of marginal plant resources, including the precursors of eventual domesticates (Henry et al. 2011).

Out of this, comes the major transition to domestication-based food resources. Within 10,000 years, less than 1000 generations for humans, the vast majority of the world's people go from a foraging-based food economy to a sedentary, domesticated food economy. The scale of the transition, and its implications for human evolution, is enormous.

It should further be noted that the transition to domestication-based subsistence strategies is not just a major event in human evolutionary history but also a major event in the history of the planet. The emerging utilization of the "Anthropocene" as a concept and time period to understand and focus attention on the consequences of human processes on the world can largely be tied into the transition to agricultural and pastoral food systems (Smith and Zeder 2013).

Like the transition to *Homo*, the establishment of domesticated food systems in human societies is conditioned on both the evolvability of cognitive and genetic systems. However, more so than the transition to *Homo*, this transition is clearly driven by cognitive/technological changes outside of the strict constraints of the human genotype/phenotype complex. Without a doubt, the transition has had and continues to have substantial genetic changes, many of which are implicated in contemporary issues of human health (Cordain et al. 2005). However, the change to a domestication-based food economy was not, itself, predicated on any specifically arising beneficial genetic mutation.

The origin of agriculture is the most significant and widespread example of the utility and importance of an Extended Evolutionary Synthesis for human evolutionary studies. The transition from a subsistence-foraging based ecology to one predicated on domestication is the most significant ecological change in the past 5 million years of hominin evolution. And yet, unlike the origin of *Homo* (or *Homo erectus sensu lato*), it does not correspond with major morphological changes in humans. It also does not evolve in association with a single population but rather occurs independently and in rapid succession across populations in the Near East, Northeast Africa, South Asia, East Asia, and Central and South America. Likewise, it is not associated with a single set of environmental conditions. Instead, what it is associated with is a broadly shared and incredibly flexible human niche, the product of nearly 2 million years of evolution in the genus *Homo*.

Some researchers view the morphological changes associated with the genus *Homo* as a transition in which the emerging hominin brain, and all of its associated cultural technology, comes to replace hominin teeth in their role in energy procurement. In a similar vein, the emergence of agricultural systems represents a wholesale shift of food procurement away from the evolved hominin physiology and onto the hominin environment itself. This hereditary systems associated with the development of agricultural systems are, functionally, cultural (or in Jablonka's nomenclature, symbolic) modes of inheritance.

Interestingly, the focus on domestication as a "cultural phenomenon" is implicit within the historical placement of the study of agricultural origins largely in the archaeological realm, rather than the realm of physical anthropology. In other words, this transition has historically been viewed as a cultural transition in evolution, rather than a biological or genetic transition. This perspective has changed dramatically in recent years given the wealth of ancient genomic data from early agriculturalists, pastoralists, and their predecessors (Haak et al. 2005; Bramanti et al. 2009; Skoglund et al. 2012). A new wave of emerging biological anthropological and genetic research is highlighting the complex demographic and genetic relationship, including genetic components of fitness, between pre-agricultural populations and the agricultural populations that came to replace them almost in their entirety (Skoglund et al. 2014; Racimo et al. 2015).

One way of viewing the evolutionary pattern of humans in the Pleistocene is as a 2-million-year transition period, beginning with the onset of encephalization in *Homo erectus*, and culminating with the ecological transition to agriculture. To be sure, the transition to agriculture does not mark the end of human evolution (Hawks et al. 2007), but rather that permanent establishment of a robust system of evolutionary inheritance running parallel to genetic heredity. The population growth observed in agricultural populations pushed human populations across the threshold of the demographic challenge faced by human and nonhuman primates up to this point (Excoffier and Schneider 1999; Bocquet-Appel 2011). Large brains, even those of nonhuman primates, take a long time to develop with evolutionary costs associated with delayed reproduction and longer interbirth intervals. While hominin evolution has, since its beginning, been marked by steps toward greater reproductive potential in comparison to non-hominin primates (Lovejoy 1981, 2009), population persistence on the order of millennia is likely only a consistent product of the agricultural revolution. The cumulative effects of long-term cultural-technological transmission across generations, critically dependent on long-term population stability and persistence, are therefore only firmly established in the human record over the last 10,000–15,000 years. While the transition to agricultural food production might have had negative effects on overall human population health, it greatly expanded human reproductive potential at the population level (Armelagos et al. 1991). Thus, while the brain is uniquely positioned to be both a mechanism for generating phenotypic novelty and a means of trans-generational transmission, it is biologically constrained in critical ways by processes of development and demography.

3 The Limits and Possibilities of an Extended Evolutionary Synthesis

This essay began with the argument that humans might serve as an important model for examining the need to rethink evolutionary theory given ongoing debates within evolutionary biology. The argument above outlines how the evolution of the human brain over the Pleistocene opens the door to an expanded set of evolutionary processes. The brain's role in shaping patterns of phenotypic plasticity and the subsequent expansion of the human ecological niche, coupled with the role of the brain in constructing a durable mode of transmission via social learning, is a fairly unique evolutionary dynamic. As Pleistocene hominins became more dependent on the brain to facilitate socially and technologically enabled ecological strategies, hominins became better learners, better teachers, and more creative, canalizing a kind of symbolic/behavioral mode of transmission. The human brain is fairly exceptional in its size, degree of complexity, and duration of development by mammalian standards. And yet the properties that gave rise to the human brain are not so unique.

The evolutionary changes that gave rise to *Homo erectus* were predicated on a complex developmental physiology that is broadly shared across mammals. While there appears to be clear evidence of selection favoring encephalization throughout the Pleistocene, many of the identified selective variants in the human genome over the past 50,000 years are associated with regulatory and epigenetic systems (Akey 2009). As Esposito (2017) notes, such interactions are not non-Darwinian so much as they defy the mechanistic interpretations of Darwinian processes often associated with representations in the Modern Evolutionary Synthesis. Nor should they be expected to be limited to any particular lineage given the deep homology of these inherited regulatory and developmental systems. Another key element of the hominin pattern of evolution throughout the Pleistocene is the way in which development is malleable to the traditional forces of Darwinian evolution. Yet, as MacCord and Maienschein (2017) make clear, developmental approaches to understanding biology might have their own historical trajectory but are not inconsistent with the overarching historical trajectory or theoretical groundings of the varied nature of Darwinism (see also Bowler 2017; Levit and Hossfeld 2017; Loison and Herring 2017).

But it is also worth considering the limitations on the human example for understanding of broader patterns of evolution. Many of the critiques of an Extended Evolutionary Synthesis focus on the ephemeral nature of emergent, "non-Darwinian," evolutionary processes in model organisms. In the human case, the power of alternative modes of evolutionary inheritance is fundamentally and significantly limited by demographic constraints. Despite the brain serving as a kind of Swiss army knife of adaptation, hominin populations nearly everywhere they are found, throughout the Pleistocene, were likely always vulnerable to localized extinction. Areas were occupied and then unoccupied for long periods of time. New technological patterns emerged, were lost, and then emerged (seemingly independently) again. Even the emergence of seemingly fundamentally altering technologies, like the controlled use of fire, was not enough to overcome the long-term evolutionary drag of genetic drift on populations extremely limited by their demographic scale (Sandgathe et al. 2011).

And yet, in the human example, the demographic constraints on human populations in the Pleistocene reflect an essential component of the expanded evolutionary processes in hominin evolution. The very features that make human behavioral/symbolic inheritance more durable than the emergent properties of other exceptional organisms simultaneously limited human evolutionary potential. Alternative modes of inheritance predicated on the properties of the hominin brain would not be possible, or at least not have sustained, long-term effects, without long periods of pre-adult development and intense dedication of evolutionary resources to child-rearing. But these are the very traits that limit the demographic potential of Pleistocene hominin populations. Even in the Late Pleistocene, as populations are developing more and more complex forms of cultural technology and occupying a greater range of environments, they remain vulnerable to population extinction and replacement. Thus, the Pleistocene human evolutionary record is marked by the many remainders, both fossil and archaeological, of side branches to the human evolution story, positioned somewhere between the population and species level (Slatkin and Racimo 2016).

The key disruption in human evolution, after the emergence of the genus *Homo*, is the origin of agricultural systems. In many ways, this evolutionary transition, massive in its long-term scope, highlights the complex ways in which we frame discussions of Darwinian or non-Darwinian processes. Historically, this transition has been viewed as "cultural" in the sense that it is not predicated on a genetic change, but this view is incomplete. Genetic change in its most traditional understanding (the Modern Synthesis) is a critical part of this transition, even if it is not the primary driver of the shift toward agricultural food production. Likewise, the "cultural" processes of inheritance associated with the transition do not behave in ways fundamentally dissimilar to genetically inherited and maintained properties, recognizing the complex ways in which genotypic variation becomes expressed. The human transmission to a domesticated mode of subsistence marked the beginning of a new epoch on the planet, the Anthropocene, appropriately reflecting the scope and magnitude of the change our species has and continues to produce. The changes wrought by humans to environments across the globe and the evolutionary trajectory of other organisms are not fully unique in the history of the planet but are certainly exceptional enough to highlight how rare such a transition is likely to be in our planet's evolutionary past.

Looking back at the broader picture, every species on the planet is the product of a unique set of evolutionary events. By taking a comparative approach to understanding the diversity of biological life on the planet, evolutionary scientists have produced reliable and durable theories for the nature of evolutionary change. While the understanding of evolutionary processes, both in the broad context and in specific cases, is constantly being refined by new research, our basic knowledge of the principles of evolutionary change remains grounded in the basic tenets laid out by Darwin himself. Biological variation exists, some of that variation is transferred from one generation to the next, and a variety of natural processes sort that variation through time in regular and predictable ways. Contained within those basic ideas, however, are vast areas open to complexity. Some of this complexity is focused on how variation is shaped throughout the life of an organism via processes of development. Some of this complexity relates to the ways in which variation might be transferred across generations. And still other complexities can be found in how exactly evolutionary processes act, and with what strength, to shape those patterns of variation. As Delisle points out (2017), even the architects of the Modern Synthesis focused not only on distinct areas of this broad paradigm but came to different interpretations about which questions and processes were foundationally important for the understanding of evolution's action. As such, rather than reveal the need for a paradigm shift, the period of self-reflexivity that evolutionary biology is currently engaged in might instead highlight the variable understandings of Darwinian processes (neo- or other) that exist and, in many ways, have always been a part of the field of evolutionary biology.

In this context, the study of a single lineage, such as humans and our evolutionary predecessors, affords the opportunity to explore the boundary conditions of existing theory more than offer a paradigmatic shift in thinking. The above essay argues that now is an appropriate time to more broadly engage humans as a model organism within evolutionary studies. For reasons both scientific and political, humans are often positioned at the center of descriptive conversations about evolution but rarely the subjects of research on evolutionary processes and theory. This is a missed opportunity to improve basic literacy around issues of evolution and science, as well as a missed opportunity to incorporate one of the more remarkable and well-documented organisms into broader evolutionary thinking.

The human example highlights the continuing entanglement between process and structure in evolutionary studies. Even as structures of significance for human evolution become more and more grounded in the social domain created by complex cognitive and technological capabilities, they remain grounded in processes of descent and modification. Human evolutionary studies, drawing from the diverse historical fields of thought—archaeology, anthropology, anatomy, genetics—each with their own realms of data, are uniquely positioned to interrogate questions at this intersection.

References

Ackermann RR, Cheverud JM (2004) Detecting genetic drift versus selection in human evolution. Proc Natl Acad Sci 101(52):17946–17951

Aiello LC, Wheeler P (1995) The expensive-tissue hypothesis: the brain and the digestive system in human and primate evolution. Curr Anthropol 36(2):199–221

Aiello LC (2007) Notes on the implications of the expensive tissue hypothesis for human biological and social evolution. In: Roebroeks W (ed) Guts and brains. Leiden University Press, Leiden, pp 17–28

Aiello LC, Key C (2002) Energetic consequences of being a Homo erectus female. Am J Hum Biol 14(5):551–565

Aiello LC, Wells JCK (2002) Energetics and the evolution of the genus Homo. Annu Rev Anthropol 31(1):323–338

Akey JM (2009) Constructing genomic maps of positive selection in humans: where do we go from here? Genome Res 19(5):711–722

Alcock J (2017) The behavioral sciences and sociobiology: a Darwinian approach. In: Delisle RG (ed) The Darwinian tradition in context: research programs in evolutionary biology. Springer, Cham, pp 37–60

Antón SC (2003) Natural history of Homo erectus. Am J Phys Anthropol 122(S37):126–170
Antón SC, Potts R, Aiello LC (2014) Evolution of early Homo: an integrated biological perspective. Science 345(6192):1236828
Armelagos GJ, Goodman AH, Jacobs KH (1991) The origins of agriculture: population growth during a period of declining health. Popul Environ 13(1):9–22
Attwell L, Kovarovic K, Kendal JR (2015) Fire in the Plio-Pleistocene: the functions of hominin fire use, and the mechanistic, developmental and evolutionary consequences. J Anthropol Sci 93:1–20
Baab KL (2008) The taxonomic implications of cranial shape variation in *Homo erectus*. J Hum Evol 54(6):827–847
Barkan E (1996) The politics of the science of race: Ashley Montagu and UNESCO's anti-racist declarations. Race and other misadventures: essays in honor of Ashley Montagu in his ninetieth year. General Hall, New York
Berger LR, de Ruiter DJ, Churchill SE, Carlson KJ, Dirks P, Kibii JM (2010) Australopithecus sediba: a new species of Homo-like australopith from South Africa. Science 328(5975):195–204
Berger LR (2013) The mosaic nature of Australopithecus sediba. Science 340(6129):163–165
Berger LR et al (2015) *Homo naledi*, a new species of the genus Homo from the Dinaledi Chamber, South Africa. elife 4:e09560
Blute M (2017) Three modes of evolution by natural selection and drift: a new or an extended evolutionary synthesis? Biol Theor 12(2):67–71
Boivin N et al (2013) Human dispersal across diverse environments of Asia during the Upper Pleistocene. Quat Int 300:32–47
Bocquet-Appel JP (2011) When the world's population took off: the springboard of the Neolithic Demographic Transition. Science 333(6042):560–561
Bowler PJ (2017) Alternatives to Darwinism in the early twentieth century. In: Delisle RG (ed) The Darwinian tradition in context: research programs in evolutionary biology. Springer, Cham, pp 195–218
Bramanti B et al (2009) Genetic discontinuity between local hunter-gatherers and central Europe's first farmers. Science 326(5949):137–140
Braun DR et al (2009) Raw material quality and Oldowan hominin toolstone preferences: evidence from Kanjera South, Kenya. J Archaeol Sci 36(7):1605–1614
Bruner E, Holloway RL (2010) A bivariate approach to the widening of the frontal lobes in the genus *Homo*. J Hum Evol 58(2):138–146
Bruner E (ed) (2014) Human paleoneurology, vol 3. Springer, New York
Bunn HT (2001) Hunting, power scavenging, and butchering by Hadza foragers and by Plio-Pleistocene Homo. Meat-eating and human evolution. Oxford University Press, Oxford
Caspari R, Lee SH (2004) Older age becomes common late in human evolution. Proc Natl Acad Sci USA 101(30):10895–10900
Caspari R (2009) 1918: three perspectives on race and human variation. Am J Phys Anthropol 139 (1):5–15
Cerling TE et al (2011) Diet of Paranthropus boisei in the early Pleistocene of East Africa. Proc Natl Acad Sci 108(23):9337–9341
Cordain L et al (2005) Origins and evolution of the Western diet: health implications for the 21st century. Am J Clin Nutr 81(2):341–354
Darwin C (1859) On the origin of species by means of natural selection, or the preservation of favoured races in the struggle for life. John Murray, London
Darwin C (1871) Sexual selection and the descent of man. John Murray, London
Dean MC, Smith BH (2009) Growth and development of the Nariokotome youth, KNM-WT 15000. In: Grine FE, Fleagle JG, Leakey RE (eds) The first humans – origin and early evolution of the genus Homo. Springer, Dordrecht, pp 101–120
Delisle RG (2017) From Charles Darwin to the evolutionary synthesis: weak and diffused connections only. In: Delisle RG (ed) The Darwinian tradition in context: research programs in evolutionary biology. Springer, Cham, pp 133–168

Dennell R, Roebroeks W (2005) An Asian perspective on early human dispersal from Africa. Nature 438(7071):1099–1104
Demes B, Creel N (1988) Bite force, diet, and cranial morphology of fossil hominids. J Hum Evol 17:657–670
DeSilva JM, Lesnik JJ (2008) Brain size at birth throughout human evolution: a new method for estimating neonatal brain size in hominins. J Hum Evol 55(6):1064–1074
DeSilva JM (2011) A shift toward birthing relatively large infants early in human evolution. Proc Natl Acad Sci 108(3):1022–1027
Dunsworth HM et al (2012) Metabolic hypothesis for human altriciality. Proc Natl Acad Sci 109 (38):15212–15216
Ellison PT (2009) On fertile ground: a natural history of human reproduction. Harvard University Press, Cambridge
Esposito M (2017) The organismal synthesis: holistic science and developmental evolution in the English-speaking world, 1915–1954. In: Delisle RG (ed) The Darwinian tradition in context: research programs in evolutionary biology. Springer, Cham, pp 219–242
Excoffier L, Schneider S (1999) Why hunter-gatherer populations do not show signs of Pleistocene demographic expansions. Proc Natl Acad Sci 96(19):10597–10602
Falk D (1987) Hominid paleoneurology. Annu Rev Anthropol 16(1):13–28
Ferring R et al (2011) Earliest human occupations at Dmanisi (Georgian Caucasus) dated to 1.85–1.78 Ma. Proc Natl Acad Sci 108(26):10432–10436
Finlayson C (2005) Biogeography and evolution of the genus Homo. Trends Ecol Evol 20 (8):457–463
Fuentes A (2016) The extended evolutionary synthesis, ethnography, and the human niche: toward an integrated anthropology. Curr Anthropol 57(S13):S13–S26
Futuyma DJ (2015) Can modern evolutionary theory explain macroevolution? In: Serrelli E, Gontier N (eds) Macroevolution. Springer, Cham, pp 29–85
Gabunia L, Vekua A (1995) A plio-pleistocene hominid from Dmanisi, East Georgia, Caucasus. Nature 373(6514):509
Gabunia L et al (2001) Dmanisi and dispersal. Evol Anthropol 10(5):158–170
Gibbons A (2007) The first human. Anchor, New York
Gilbert SF, Bosch TCG, Ledón-Rettig C (2015) Eco-Evo-Devo: developmental symbiosis and developmental plasticity as evolutionary agents. Nat Rev Genet 16(10):611–622
Glantz M, Van Arsdale AP, Temirbekov S, Beeton T (2016) How to survive the glacial apocalypse: Hominin mobility strategies in late Pleistocene Central Asia. Quat Int. https://doi.org/10.1016/j.quaint.2016.06.037
Gould SJ (2002) The structure of evolutionary theory. Harvard University Press, Cambridge
Grabowski M (2016) From bigger brains to bigger bodies: the correlated evolution of human brain and body size. Curr Anthropol 57(2):174–196
Haak W, Forster P, Bramanti B, Matsumura S, Brandt G, Tänzer M, Villems R, Renfrew C, Gronenborn D, Alt KW, Burger J (2005) Ancient DNA from the first European farmers in 7500-year-old Neolithic sites. Science 310(5750):1016–1018
Hare B, Tomasello M (2005) Human-like social skills in dogs? Trends Cogn Sci 9(9):439–444
Harmand S, Lewis JE, Feibel CS, Lepre CJ, Prat S, Lenoble A, Boës X, Quinn RL, Brenet M, Arroyo A, Taylor N, Clément S, Daver G, Brugal JP, Leakey L, Mortlock RA, Wright JD, Lokorodi S, Kirwa C, Kent DV, Roche H (2015) 3.3-million-year-old stone tools from Lomekwi 3, West Turkana, Kenya. Nature 521(7552):310
Hawks J et al (2007) Recent acceleration of human adaptive evolution. Proc Natl Acad Sci 104 (52):20753–20758
Henry AG, Brooks AS, Piperno DR (2011) Microfossils in calculus demonstrate consumption of plants and cooked foods in Neanderthal diets (Shanidar III, Iraq; Spy I and II, Belgium). Proc Natl Acad Sci 108(2):486–491
Hovers E, Braun DR (eds) (2009) Interdisciplinary approaches to the Oldowan. Springer, Dordrecht

Jablonka E, Lamb MJ (2014) Evolution in four dimensions, revised edition: Genetic, epigenetic, behavioral, and symbolic variation in the history of life. MIT Press, Cambridge
Kaifu Y (2006) Advanced dental reduction in Javanese Homo erectus. Anthropol Sci 114 (1):35–43
Laland K, Wray GA, Hoekstra HE (2014) Does evolutionary theory need a rethink? Nature 514 (7521):161
Laland K, Uller T, Feldman MW, Sterelny K, Müller GB, Moczek A, Jablonka E, Odling-Smee J (2015) The extended evolutionary synthesis: its structure, assumptions and predictions. Proc R Soc B 282(1813):1–14
Laland K (2017) Schism and synthesis at the royal society. Trends Ecol Evol 32(5):316–317
Langley MC, Street M (2013) Long range inland–coastal networks during the Late Magdalenian: evidence for individual acquisition of marine resources at Andernach-Martinsberg, German Central Rhineland. J Hum Evol 64(5):457–465
Lee SH, Wolpoff MH (2003) The pattern of evolution in Pleistocene human brain size. Paleobiology 29(2):186–196
Leigh SR (2006) Brain ontogeny and life history in Homo erectus. J Hum Evol 50(1):104–108
Leonard WR, Robertson ML (1992) Nutritional requirements and human evolution: a bioenergetics model. Am J Hum Biol 4(2):179–195
Levit GS, Hossfeld U (2017) Major research traditions in 20th century evolutionary biology: the relations of Germany's Darwinism with them. In: Delisle RG (ed) The Darwinian tradition in context: research programs in evolutionary biology. Springer, Cham, pp 169–194
Lieberman L (1989) A discipline divided: Acceptance of human sociobiological concepts in anthropology. Curr Anthropol 30(5):676–682
Livingstone FB, Dobzhansky T (1962) On the non-existence of human races. Curr Anthropol 3:279–281
Loison L, Herring E (2017) Lamarckian research programs in French biology (1900–1970). In: Delisle RG (ed) The Darwinian tradition in context: research programs in evolutionary biology. Springer, Cham, pp 243–270
Lordkipanidze D, Vekua A, Ferring R, Rightmire GP, Zollikofer CPE, Ponde de León MS, Agusti J, Kiladze G, Mouskhelishvili A, Nioradze M, Tappen M (2006) A fourth hominin skull from Dmanisi, Georgia. Anat Rec 288(11):1146–1157
Lordkipanidze D, Jashashvili T, Vekua A, Ponce de León MS, Zollikofer CPE, Rightmire GP, Pontzer H, Ferring R, Oms O, Tappen M, Bukhsianidze M, Agusti J, Kahlke R, Kiladze G, Martinez-Navarro B, Mouskhelishvili A, Nioradze M, Rook L (2007) Postcranial evidence from early Homo from Dmanisi, Georgia. Nature 449(7160):305–310
Lordkipanidze D, Ponce de León MS, Margvelashvili A, Rak Y, Rightmire GP, Vekua A, Zollikofer CPE (2013) A complete skull from Dmanisi, Georgia, and the evolutionary biology of early Homo. Science 342(6156):326–331
Lovejoy CO (1981) The origin of man. Science 211(4480):341–350
Lovejoy CO (2009) Reexamining human origins in light of *Ardipithecus ramidus*. Science 326 (5949):74e1–74e8
Lowe WH, Kovach RP, Allendorf FW (2017) Population genetics and demography unite ecology and evolution. Trends Ecol Evol 32(2):141–152
Lynn CD, Glaze AL, Evans WA, Reed LK (eds) (2017) Evolution education in the American South: culture, politics, and resources in and around Alabama. Palgrave Macmillan, New York
MacCord K, Maienschein J (2017) Cells, development, and evolution: teeth studies at the intersection of fields. In: Delisle RG (ed) The Darwinian tradition in context: research programs in evolutionary biology. Springer, Cham, pp 289–308
Marks J (2008) Race across the physical-cultural divide in American anthropology. In: Kublick H (ed) A new history of anthropology. Blackwell, Hoboken, pp 242–258
Matsuzawa T (2008) Primate foundations of human intelligence: a view of tool use in nonhuman primates and fossil hominids. In: Matsuzawa T (ed) Primate origins of human cognition and behavior. Springer, Tokyo, pp 3–25

Mayr E (1950) Taxonomic categories in fossil hominids. Cold Spring Harb Symp Quant Biol 15:109–118

McBrearty S, Brooks AS (2000) The revolution that wasn't: a new interpretation of the origin of modern human behavior. J Hum Evol 39(5):453–563

McHenry HM, Coffing K (2000) Australopithecus to Homo: transformations in body and mind. Annu Rev Anthropol 29(1):125–146

McPherron SP, Alemseged Z, Marean CW, Wynn JG, Reed D, Geraads D, Bobe R, Béarat HA (2010) Evidence for stone-tool-assisted consumption of animal tissues before 3.39 million years ago at Dikika, Ethiopia. Nature 466(7308):857–860

Nielsen R, Akey JM, Jakobsson M, Pritchard JK, Tishkoff S, Willerslev E (2017) Tracing the peopling of the world through genomics. Nature 541(7637):302–310

Nowell A, Davidson I (2010) Stone tools and the evolution of human cognition. University Press of Colorado, Boulder

O'Connell JF, Hawkes K, Blurton Jones NG (1999) Grandmothering and the evolution of *Homo erectus*. J Hum Evol 36(5):461–485

Organ C, Nunn CL, Machanda Z, Wrangham RW (2011) Phylogenetic rate shifts in feeding time during the evolution of *Homo*. Proc Natl Acad Sci 108(35):14555–14559

Pigliucci M (2007) Do we need an extended evolutionary synthesis? Evolution 61(12):2743–2749

Pigliucci M (2008) Is evolvability evolvable? Nat Rev Genet 9(1):75–82

Pigliucci M (2017) Darwinism after the modern synthesis. In: Delisle RG (ed) The Darwinian tradition in context: research programs in evolutionary biology. Springer, Cham, pp 89–104

Plutzer E, Berkman M (2008) Trends evolution, creationism, and the teaching of human origins in schools. Public Opin Q 72(3):540–553

Pobiner B (2016) Accepting, understanding, teaching, and learning (human) evolution: obstacles and opportunities. Am J Phys Anthropol 159(S61):S232–S274

Pontzer H et al (2010) Locomotor anatomy and biomechanics of the Dmanisi hominins. J Hum Evol 58(6):492–504

Pontzer H (2012) Ecological energetics in early Homo. Curr Anthropol 53(S6):S346–S358

Potts R, Teague R (2010) Behavioral and environmental background to 'Out-of-Africa I' and the arrival of *Homo erectus* in East Asia. In: Fleagle JG (ed) Out of Africa I. Springer, Dordrecht, pp 67–85

Racimo F, Sankararaman S, Nielsen R, Huerta-Sánchez E (2015) Evidence for archaic adaptive introgression in humans. Nat Rev Genet 16(6):359–371

Rightmire GP (1981) Patterns in the evolution of Homo erectus. Paleobiology 7(2):241–246

Rightmire GP (2004) Brain size and encephalization in early to Mid-Pleistocene Homo. Am J Phys Anthropol 124(2):109–123

Roebroeks W, Villa P (2011) On the earliest evidence for habitual use of fire in Europe. Proc Natl Acad Sci 108(13):5209–5214

Rose MD (1984) A hominine hip bone, KNM-ER 3228, from East Lake Turkana, Kenya. Am J Phys Anthropol 63(4):371–378

Roseman CC (2016) Random genetic drift, natural selection, and noise in human cranial evolution. Am J Phys Anthropol 160(4):582–592

Ruff CB, Trinkaus E, Holliday TW (1997) Body mass and encephalization in Pleistocene *Homo*. Nature 387(6629):173

Sahlins MD (1976) The use and abuse of biology: an anthropological critique of sociobiology. University of Michigan Press, Ann Arbor

Sandgathe DM, Dibble HL, Goldberg P, McPherron SP, Turq A, Niven L, Hodgkins J (2011) On the role of fire in Neandertal adaptations in Western Europe: evidence from Pech de l'Azé IV and Roc de Marsal, France. PaleoAnthropology 216–242

Sankararaman S, Mallick S, Patterson N, Reich D (2016) The combined landscape of Denisovan and Neanderthal ancestry in present-day humans. Curr Biol 26(9):1241–1247

Schrenk F, Kullmer O, Bromage T (2015) The earliest putative Homo fossils. In: Henke W, Tattersall I (eds) Handbook of paleoanthropology. Springer, Berlin, pp 2145–2165

Sémah F, Saleki H, Falguères C, Féraud G, Djubiantono T (2000) Did early man reach Java during the Late Pliocene? J Archaeol Sci 27(9):763–769

Shipman P, Walker A (1989) The costs of becoming a predator. J Hum Evol 18(4):373–392

Slatkin M, Racimo F (2016) Ancient DNA and human history. Proc Natl Acad Sci 113 (23):6380–6387

Skoglund P, Malmström H, Raghavan M, Stora J, Hall P, Willerslev E, Gilbert TP, Götherström A, Jakobsson M (2012) Origins and genetic legacy of Neolithic farmers and hunter-gatherers in Europe. Science 336(6080):466–469

Skoglund P et al (2014) Genomic diversity and admixture differs for Stone-Age Scandinavian foragers and farmers. Science 344(6185):747–750

Smith BD, Zeder MA (2013) The onset of the Anthropocene. Anthropocene 4:8–13

Smocovitis VB (2012) Humanizing evolution: anthropology, the evolutionary synthesis, and the prehistory of biological anthropology, 1927–1962. Curr Anthropol 53(S5):S108–S125

Snodgrass JJ, Leonard WR, Robertson ML (2009) The energetics of encephalization in early hominids. In: Hublin JJ, Richards M (eds) The evolution of hominin diets. Springer, Dordrecht, pp 15–29

Spoor F, Leakey MG, Gathogo PN, Brown FH, Antón SC, McDougall I, Kiarie C, Manthi FK, Leakey LN (2007) Implications of new early Homo fossils from Ileret, east of Lake Turkana, Kenya. Nature 448(7154):688–691

Stedman HH, Kozyak BW, Nelson A, Thesier DM, Su LT, Low DW, Bridges CR, Shrager JB, Minugh-Purvis N, Mitchell MA (2004) Myosin gene mutation correlates with anatomical changes in the human lineage. Nature 428(6981):415–418

Steudel-Numbers KL (2006) Energetics in Homo erectus and other early hominins: the consequences of increased lower-limb length. J Hum Evol 51(5):445–453

Stiner MC, Munro ND, Surovell TA, Tchernov E, Bar-Yosef O (1999) Paleolithic population growth pulses evidenced by small animal exploitation. Science 283(5399):190–194

Strier KB (2016) Why biological anthropologists should identify as anthropologists: The meaning of membership in the AAA and the AAAS. Am J Phys Anthropol 161(1):3–5

Stringer C (2012) Lone survivors: how we came to be the only humans on earth. Macmillan, New York

Suwa G, Asfaw B, Haile-Selassie Y, White T, Katoh S, Woldegabriel G, Hart WK, Nakaya H, Beyene Y (2007) Early Pleistocene Homo erectus fossils from Konso, southern Ethiopia. Anthropol Sci 115(2):133–151

Swisher CC, Curtis GH, Jacob T, Getty AG, Suprijo A (1994) Age of the earliest known hominids in Java, Indonesia. Science 263(5150):1118–1122

Szathmáry E (2015) Toward major evolutionary transitions theory 2.0. Proc Natl Acad Sci 112 (33):10104–10111

Tappen M (2009) The wisdom of the aged and out of Africa I. In: Shea J, Lieberman D (eds) Transitions in prehistory: essays in honor of Ofer Bar Yosef. Oxbow, Oakville, pp 24–41

Teaford MF, Ungar PS, Grine FE (2002) Paleontological evidence for the diets of African Plio-Pleistocene hominins with special reference to early Homo. Human diet: its origin and evolution. Bergin & Garvey, London, pp 143–166

Tennie C, Call J, Tomasello M (2009) Ratcheting up the ratchet: on the evolution of cumulative culture. Philos Trans R Soc Lond B Biol Sci 364(1528):2405–2415

Tomasello M (1999) The human adaptation for culture. Annu Rev Anthropol 28(1):509–529

Ungar PS, Grine FE, Teaford MF (2006) Diet in early Homo: a review of the evidence and a new model of adaptive versatility. Annu Rev Anthropol 35:209–228

Van Arsdale AP, Lordkipanidze D (2012) A quantitative assessment of mandibular variation in the Dmanisi hominins. PaleoAnthropology 134–144

Van Arsdale AP, Wolpoff MH (2013) A single lineage in early Pleistocene Homo: size variation continuity in early Pleistocene Homo crania from East Africa and Georgia. Evolution 67 (3):841–850

Van Schaik CP, Deaner RO, Merrill MY (1999) The conditions for tool use in primates: implications for the evolution of material culture. J Hum Evol 36(6):719–741
Villmoare B, Kimbel WH, Seyoum C, Campisano CJ, DiMaggio EN, Rowan J, Braun DR, Arrowsmith JR, Reed KE (2015) Early Homo at 2.8 Ma from Ledi-Geraru, Afar, Ethiopia. Science 347(6228):1352–1355
Walker A (1981) Dietary hypotheses and human evolution. Philos Trans R Soc Lond B Biol Sci 292(1057):57–64
Walker A, Leakey RE (1993) The Nariokotome Homo erectus skeleton. Harvard University Press, Cambridge
Wall JD, Brandt DYC (2016) Archaic admixture in human history. Curr Opin Genet Dev 41:93–97
Want SC, Harris PL (2002) How do children ape? Applying concepts from the study of non-human primates to the developmental study of 'imitation in children. Dev Sci 5(1):1–14
Washburn SL (1951) Section of anthropology: the new physical anthropology. Trans NY Acad Sci 13(7), Series II):298–304
Wilkins J, Schoville BJ, Brown KS, Chazan M (2012) Evidence for early hafted hunting technology. Science 338(6109):942–946
Whiten A, Goodall J, McGrew WC, Nishida T, Reynolds V, Sugiyama Y, Tutin CEG, Wrangham RW, Boesch C (1999) Cultures in chimpanzees. Nature 399(6737):682–685
Wilson EO (1975) Sociobiology: the new synthesis. Harvard Belknap, Cambridge
Wolpoff MH (1975) Some aspects of human mandibular evolution. In: McNamara JA (ed) Determinants of mandibular form and growth. Craniofacial growth series monograph. Center for Human Growth and Development, Ann Arbor, vol 4, pp 1–64
Wolpoff MH (1999) Paleoanthropology. McGraw-Hill Humanities, Social Sciences & World Languages, Boston
Wood B (1992) Origin and evolution of the genus Homo. Nature 355(6363):783
Wood B, Collard M (1999) The human genus. Science 284(5411):65–71
Wray GA, Hoekstra HE (2014) Does evolutionary theory need a rethink?-COUNTERPOINT No, all is well. Nature 516:161–164
Zaim Y, Ciochon RL, Polanski JM, Grine FE, Bettis EA, Rizal Y, Franciscus RG, Larick RR, Heizler M, Eaves KL, Marsh HE (2011) New 1.5 million-year-old Homo erectus maxilla from Sangiran (Central Java, Indonesia). J Hum Evol 61(4):363–376
Zeder M (ed) (2006) Documenting domestication: new genetic and archaeological paradigms. University of California Press, Berkeley
Zhu RX, Potts R, Xie F, Hoffman KA, Deng CL, Shi CD, Pan YX, Wang HQ, Shi RP, Wang YC, Shi GH, Wu NQ (2004) New evidence on the earliest human presence at high northern latitudes in northeast Asia. Nature 431(7008):559–562

Part II
Crossing the Boundaries: Between Non-Darwinian and Darwinian

From Charles Darwin to the Evolutionary Synthesis: Weak and Diffused Connections Only

Richard G. Delisle

Abstract The understanding of "Darwinism" is replete with difficulties, keeping in mind that intellectual movements have no essence. Undoubtedly, Charles Darwin offered to evolutionary biology a plethora of notions used in a novel way. Yet, the inconsistent utilization of these notions within numerous complementary and competing research programs in twentieth-century evolutionary biology makes the traceability of Darwinism in time a difficult, if not unrewarding, enterprise. Furthermore, if attempts to define Darwinism have often focused on finding similarities between Charles Darwin and post-1900 developments, the task of searching for conceptual and theoretical dissimilarities between Darwin and presumably well-identified Darwinians has been less frequently addressed. This is precisely what is to be attempted in this chapter, which will analyze the profound differences between Darwin's own views and those of the founding members of the Evolutionary Synthesis, including Julian S. Huxley, Theodosius Dobzhansky, Bernhard Rensch, George Gaylord Simpson, and Ernst Mayr. It will be shown that connections binding Darwin and some of these "synthesists" are rather weak and diffused, especially when their significant use of rhetorical arguments is bypassed, revealing that a "reticulate model" of science is more appropriate than a "multilevel model" when it comes to analyzing the development of evolutionary biology. Since "Darwinism" as a label hides more than it reveals, it is suggested that it might be best to abandon it altogether. The chapter will conclude by revisiting some notions included in the traditional historiography: the "eclipse of Darwinism," the "hardening of the Evolutionary Synthesis," and an "Expanded Evolutionary Synthesis."

Keywords Darwinism • Evolutionary synthesis • Eclipse of Darwinism • Hardening of the evolutionary synthesis • Expanded evolutionary synthesis • Multilevel model of science • Reticulate model of science • Charles Darwin • Theodosius Dobzhansky • Julian S. Huxley • Ernst Mayr • Bernhard Rensch • George Gaylord Simpson

---

R.G. Delisle (✉)
Departments of Liberal Education and Philosophy, University of Lethbridge, Lethbridge, AB, Canada
e-mail: richard.delisle@uleth.ca

R.G. Delisle (ed.), *The Darwinian Tradition in Context*,
https://doi.org/10.1007/978-3-319-69123-7_7

1 Introduction

Why is it that today we speak of "Darwinism," yet not of "Einsteinism"? Is this a pure historical accident of labeling? Or is it because we see Charles Darwin as more immediate to us than other scholars in the past? Michael Ruse (2011: 10) states:

> I expect that in fifty years or a hundred years we will still have the theory of the *Origin* [*of Species*] around… [P]recisely because it does not stand still, but remakes itself and grows and changes by virtue of the fact that it gives such a terrific foundation. Is Darwinism past its sell-by-date? Not by a long chalk yet!

Let us uncritically accept Ruse's assessment for a moment. Which theory, then, is believed to be flexible enough to have persisted through time under the same label, as also elegantly argued in Timothy Shanahan (2017), John Alcock (2017), David Depew (2017), and Massimo Pigliucci (2017)? An important historical fact should be recalled here: the familiarity of Darwin's name among scholars is largely related to the fate of his ideas in twentieth century science. While Darwin's books of the 1860s and 1870s became widely known, especially following the initial publication of the *Origin of Species* in 1859 (Desmond and Moore 1991: 477–481), the central ideas it contained were often deemed insufficient to explain biological evolution at large, hence the notion of an "eclipse of Darwinism" in the late nineteenth and early twentieth centuries (Bowler 1983, 1988). Darwin's name really came of age in twentieth-century evolutionary biology, when a series of empirical and theoretical developments between the 1920s and the early 1960s in areas such as population genetics, zoology, and paleontology seemed to show that evolution could be consistent with key concepts thought to be already contained in Darwin's work (Fisher 1930; Wright 1931; Haldane 1932; Dobzhansky 1937, 1951; Huxley 1942; Mayr 1942, 1963; Simpson 1944, 1953; Rensch 1960a [1947]; Stebbins 1950). These developments eventually came to be known as the Evolutionary Synthesis.

Consequently, Darwin's place in this intellectual story became ever more solidified when the ongoing professionalization of the fields of history and philosophy of biology—which really made itself felt beginning with the 1959 Centennial celebrating the publication of the *Origin of Species* (Appleman 1959; Fleming 1959; Loewenberg 1959, 1965; Smith 1959, 1960; see also Smocovitis 1999)—generated an abundant literature. From there, several avenues were exploited. First, a tradition was launched that reflected upon the historiography of Darwin studies dubbed the "Darwin industry" (Ruse 1974, 1996a; Greene 1975; Churchill 1982; Oldroyd 1984; La Vergata 1985; Lenoir 1987). Second, intellectual bridges were sought to connect Darwin himself with the Evolutionary Synthesis (Provine 1971, 1978, 1985; Mayr 1980, 1982a, 1991; Hull 1985; Greene 1981, 1986; Gayon 1992 [1998]; Depew and Weber 1989, 1995; Ruse 1996b, 2009; Gould 2002). Third, analyses appeared exploring the explanatory structure of the Darwinian theory of evolution (Beckner 1959; Goudge 1961; Ruse 1973, 1979; Hull 1974; Caplan 1978; Tuomi 1981; Rosenberg 1985).

On this view, a mid-nineteenth century scholar like Darwin was pulled onto the side of the twentieth century. To say it differently, there was little continuity between Darwin himself and the Evolutionary Synthesis; a sort of historical break occurred between the two main events of Darwinism, the *Origin of Species* and the Evolutionary Synthesis. Scholars of the post-1930 period had to reach back in time to *re-connect* with Darwin. Are we sure, then, that the Darwin of the twentieth century matches the Darwin of the nineteenth century and that the two key events marking the development of Darwinism are aligned with one another? This is the main question to be investigated in this paper, the answer being more in the negative than in the affirmative. As Jacques Roger (1985: 813) remarked:

> [I]n the many meetings and colloquia that took place in 1959, Darwin was unanimously declared to have been right in his explanation of evolution... Things are more complicated today... It is no longer certain that 'Darwin was right'; or at least we feel that it is more necessary than ever to go back to 'what Darwin really said' and thought, to put the synthetic theory in historical perspective, and to examine the extent to which it gave a distorted idea of Darwin himself.

This investigative line is as relevant today as when Roger proposed it. It must be acknowledged at the outset that an intellectual movement may be interrupted more or less completely by a lack of sufficient adherents, only later to be reactivated by a new generation of scholars. Intellectual continuity does not require generational continuity. Yet, as far as the intellectual content of Darwinism is concerned, the extent to which self-proclaimed darwinians actually walk in the footsteps of Darwin does matter. Whereas it is expected that new generations will continue to develop Darwin's views, it is also expected that enough intellectual continuity should exist across generations if all participating scholars are to be considered part of a single evolving intellectual movement.

The main thesis of this paper can be summarized thus: intellectual connections between Darwin and the Evolutionary Synthesis are weaker and more diffuse than usually assumed in the historiography. It will be argued that this is the case because the actual accomplishments of both Darwin and the proponents of the Evolutionary Synthesis are not what they have claimed to be. This latter group is hereafter referred to as the "synthesists," a group that includes Theodosius Dobzhansky, Julian Sorell Huxley, Ernst Mayr, Bernhard Rensch, and George Gaylord Simpson. Beyond self-proclamations made by Darwin and those of the synthesists, whether taken together and individually, there exists another evolutionary biology not easily reconcilable with the idea of a dominant evolving intellectual thread. To support that thesis, our analytical framework will articulate four issues in evolutionary studies: (1) reductionism versus holism; (2) the tree of life approached either from a horizontal or a vertical perspective; (3) the influence of external versus internal agencies in evolution; and (4) the central versus the peripheral positioning of the mechanisms in evolutionary explanations. Before applying this analytical framework in the second half of this chapter, it will be necessary to first examine in some depth the importance of rhetorical arguments in Darwin.

2 Beyond Rhetorical Arguments in the *Origin of Species*

The following considerations about Darwin would require a book-length exposition (currently in preparation), but the following sketch should be suggestive enough for present purposes. A cursory look at the Table of Contents of the *Origin of Species* (1859) immediately reveals at least two main parts: (1) a group of chapters (Chaps. 1–5) concerned with topics relevant to the rise of variation and the action of selective processes in the context of places in the economy of nature; (2) a group of chapters (Chaps. 9–13) surveying evolution in various fields such as paleontology, biogeography, systematics, morphology, embryology, and comparative anatomy (rudimentary organs). Darwin's earlier *Sketch* (1842) and *Essay* (1844) already largely manifested the same twofold division (Darwin 1909). Some have remarked that by so doing Darwin wanted to follow the best epistemological prescriptions of his time—as held by contemporaries like John Herschel and William Whewell—which consisted in showing, in the first part, the effectiveness and empirical foundation of natural selection as a real cause (vera causa) or mechanism of biological evolution and, in the second part, the unifying and explanatory power of this cause when applied against apparently unrelated disciplines belonging to distinct classes of facts (consilience of inductions), such as paleontology, morphology, embryology (Ruse 1975, 2000; Hodge 1977, 2000). Darwin's explanatory structure, therefore, could be conceived as having at its core natural selection acting on biological variations, in addition to unifying and explaining series of facts descriptively found in various biological disciplines (Ruse 1979: 198, 2009: 18). This explanatory structure positioning the cause in a privileged hierarchical position relative to other components or fields is called the "multilevel model" of science.

Whereas the explanatory structure just outlined constitutes the background of the *Origin of Species*—the argument goes—the foreground of Darwin's theory would articulate a series of specific notions: (1) relative adaptation; (2) variation; (3) demographic pressure and competition; (4) selection; (5) common ancestry and divergence; (6) a self-regulated economy of nature; (7) evolutionary contingency.

Darwin largely managed to summarize his entire theory in the visual form of a diagram in the *Origin*, also known as the Queer Diagram. In it, Darwin depicts the evolution of two main lineages over time (A and I), each being characterized by an ongoing succession of forms—some being short-lived, others surviving longer—genealogically bound in common ancestry and fanning out from each other. A key explanatory feature of the diagram is found in its analogy with what we would call today "fractal geometry": just as patterns are similar at all levels in a fractal object, Darwin's two main lineages conform to the same pattern-process of divergence throughout taxonomic levels. Darwin thus used the exact same diagram to explain evolution at all levels (Darwin 1859: 116–126, 331–332, 412–413, 420–422, 431–432). More than anything else, the Queer Diagram constitutes a powerful propaganda machine. It seems to encapsulate in a remarkably compact fashion the entirety of Darwin's theory: the differential action of natural selection, a segmented economy of nature, the principles of common ancestry and divergence, and the process of adaptation and extinction.

This being said, I argue that everything outlined thus far is more properly classified as part of the *rhetorical* component of the *Origin*, than of its actual accomplishments. It is difficult for twentieth and twenty-first century scholars to read Darwin critically, influenced as they are with modern notions of evolutionary biology. They tend to read more than is really present in Darwin's work, thus buying into his rhetoric while smoothing over tensions and contradictions of his theory. It is not good enough, however, to follow Darwin's self-proclaimed goals and realizations. The whole question can be approached differently by turning to a critical investigation of the *Origin*'s argumentative structure: instead of searching for Darwin's intent, we may want to evaluate his actual achievements. I am not here proposing to apply a normative approach to the *Origin*; rather, I am simply suggesting we confront Darwin's own accomplishments against his own claims by paying attention to the internal consistency of his arguments. The reader willing to go beyond Darwin's rhetoric in the *Origin of Species* encounters a strangely dislocated book. To see this more clearly, let us focus on five different parts of the book.

2.1 Chapters 1–5 and 8

In Chaps. 1–5 and 8 of the *Origin* (1859), evolution is described as a selective process that generates an observable pattern (divergence) seen today at low taxonomic levels, with the assistance of the analogy between artificial selection and natural selection. However, the careful reader of the *Origin* finds in these chapters entities at low taxonomic levels strongly bound in complex reproductive networks, as if life was caught up in these tight networks. Darwin unconvincingly tried to impose his theory upon this factual background, hoping to show that closely related forms bound in common reproductive networks could be segregated from one another in a divergence process. By so doing, *he himself provided* considerable counterfactual evidence:

1. Darwin identified several reasons as to why life forms are prevented from changing, thus from segregating: resistance to change for a number of generations (p. 7); reversion to the parental condition (pp. 14–15, 25–26, 152–154, 159); interbreeding being favored in nature over close inbreeding (pp. 70–71, 96, 104–106, 248–250, 253).
2. Darwin was unable to extract life from the complex web of reproductive networks, networks he explicitly described as being composed of "strains" and "sub-breeds" (pp. 31–32, 96, 267), that is, of non-monolithic entities too often hidden under rigid and abstract taxonomic categories such as "varieties" and "species."

3. Against the widespread opinion of breeders (pp. 28–29), and against the lack of good evidence as explicitly recognized by Darwin himself (pp. 17, 40), it was held that many domesticated forms are not of polyphyletic origin but rather of monophyletic origin (pp. 16–20, 23–28).
4. Darwin established that the family level constituted the absolute sterility level, meaning that forms belonging to the genus, species, and variety levels are fertile to a lesser or greater degree (pp. 22–23, 248, 253, 255–257, 261, 267–272). That lack of clear reproductive separation at low taxonomic levels renders the segregation and the speciation process more difficult to conceive. As if the issue was not already muddled enough, Darwin alluded to the fact that breeders were able to create, through artificial selection, forms separated at the generic level (i.e., pigeons) and still fully fertile with one another (pp. 22–23, 26, 445).
5. Darwin recognized that natural selection could not select for a negative feature such as "sterility," that feature being a mere by-product accompanying other features that are, themselves, positively selected (pp. 245, 260, 269). In addition, continued Darwin, the relationship between the phenotype and fertility is an imperfect one, with quite dissimilar forms being able to reproduce and quite similar ones being unable (p. 269). If this is the case, on what exactly is natural selection working to segregate forms?
6. Darwin wavered between two distinct models of speciation, each presenting a major disadvantage for his theory (pp. 102–108): (1) the sympatric model of forms living in close proximity allows for plenty of competition to instigate evolutionary change but such a high level of contiguity among closely related forms renders the segregation of reproductive networks difficult; (2) the allopatric model of forms offers an easy way to break free from common reproductive networks geographically, but comes with a lack of competition among isolated entities.
7. In his haste to promote an ongoing divergence process all the way up to the highest taxonomic levels, Darwin entirely overlooked the fact that the evolutionary drive fed by the competitive exclusion principle he postulated among very closely related forms (pp. 75–76, 110) could only get ever weaker as forms became more separated from each other, taxonomically speaking. With the exhaustion of such a drive, the pattern-process of evolution should look more like a random walk than divergence.
8. Darwin was so committed to a theory assuming a pan-divergent pattern-process of evolution that he overlooked the obvious distinction between two distinct competitive contexts he himself introduced and which should lead to two distinct evolutionary outcomes: (1) the competitive exclusion principle involving very closely related forms (pp. 75–76) and (2) the entangled bank involving unrelated forms (pp. 71–75). Whereas the former context might conceivably explain divergence at low taxonomic levels (assuming that one can come up with the proper conceptual tools to explain how forms can be segregated from reproductive networks), the latter context is more susceptible to generate a random walk rather than divergence.

Confronted with so many difficulties, contradictions, and counterfactual evidence, the least that can be said is that much imagination would be required of the reader of Chaps. 1–5 and 8 of the *Origin* for him or her to follow Darwin in his theoretical views centered around the principle of divergence. In all objectivity, Darwin managed to establish a fairly convincing case for theories based on hybridism (polyphyletism and reticulate evolution), theories not entirely inconsistent with some amount of divergence, but in opposition to a theory entirely based on the principle of divergence.

Furthermore, as much as Darwin may have wanted his theory to comply with the ideal of a science founded on a vera causa—a non-conjectural cause known to really exist (natural selection) that can produce the postulated effects, that is, an ongoing divergence all the way up the taxonomic scale—the case was hardly convincing. This is true to such an extent that, when Darwin presented the Queer Diagram in the context of low taxonomic entities, knowledge from his own writing would have required him, in a large part at least, to visually represent complex, semi-continuous networks of strains caught up in reproductive nexuses of various sizes and taxonomic levels (varieties, species, genera), or entities under an evolutionary drive to which a random walk is imposed upon them. Instead, Darwin elected to force his way visually, as it were, by presenting neat, clear, and segregated evolutionary lines through irreversible speciations disposed along an ascending taxonomic scale. For the careful reader, however, this perfectly neat and symmetrical pattern-process of evolution remains largely unfounded.

2.2 *Chapters 6 and 7*

With Chaps. 6 and 7 of the *Origin* (1859), the reader is presented with an entirely different view of evolution. The approach shifts from a horizontal view based on entities that are part of reproductive networks at low taxonomic levels, (Chaps. 1–5, 8) to a vertical one. Natural selection is now depicted as a brute force or wild power to which nothing can resist; transitional forms across evolutionary grades and taxonomic levels are imagined. This is a vertical and directional view of evolution. Darwin applies in these two chapters a thick coat of rhetorical paint, resorting to speculative evolutionary scenarios and blurring the fundamental conceptual difference, so crucial for his theory, between "homologies" and "analogies." Tracking evolutionary links and gradations everywhere among currently living forms, Darwin became ever more ambitious, moving from transitional features observed within the same organisms (pp. 179–180, 190, 210), to transitions between closely related forms (pp. 180, 219–235), and eventually to groups separated at high taxonomic levels (pp. 181–183, 216–218). In his desire to continually widen the taxonomic scope of his analysis, Darwin entirely neglected the fact that he went from cases reasonably explained by common ancestry and divergence (homologies), to cases more easily explained by parallel evolution (analogies), and ultimately to cases entirely beyond biology for the latter exclusively appealed to our imagination. Here are three cases at hand found in these chapters:

1. Among cases suggestive of phylogenetic sequences proposed by Darwin involving very closely related forms are those appealing to the squirrels, the petrels, the ants, and the bees (pp. 180, 184–185, 219–235).
2. Among cases presented by Darwin which should have required of him a serious confrontation with the issues of analogies and parallel evolution (although he refrained from doing so) are: the flying abilities among mammals (lemurs, flying lemurs, and bats) and birds (Apteryx, ostriches, ducks, penguins); the parasitic nesting behavior in birds (ostriches, cuckoos); and the complexification of eyes among widely dispersed taxonomic groups (pp. 181–182, 186–188, 216–218).
3. Finally, the case of the complexification of eyes was taken by Darwin into new explanatory territory by adopting a two-part argumentative strategy: a selection of various levels of complexity seen among disperse taxonomic groups of living animals, followed by an entirely hypothetical reconstruction of a directional evolution of the eye's architecture based on an eye-telescope analogy (pp. 186–189).

I would argue that Darwin's theory in Chaps. 6 and 7 constitutes a truly different theory from the one presented in the rest of the *Origin of Species*. The rationale for promoting this radical thesis is based on the aforementioned explanatory shift from an horizontal approach of evolution to a vertical one, a shift striking at the heart of the kind of theory Darwin intends to offer, as seen in the modified relationship between the "Unity of Type" (descent) and the "Conditions of Existence" (selective forces). Indeed, by gradually widening the taxonomic scope of his quest in Chaps. 6 and 7, Darwin moved away from a theory delicately balancing "descent" and "natural selection," as visually represented in the Queer Diagram, to a different theory in which natural selection overruns descent. Under the new theory, the notions of common ancestry and divergence are jeopardized by the ease with which natural selection can forge analogical features and parallel evolution, thus blurring affiliative traces. Furthermore, the neat pattern-process of divergence that Darwin tried so painstakingly to impose in Chaps. 1–5 and 8 over the blurred reality of complex reproductive networks is replaced in Chaps. 6 and 7 by a pattern-process based on linear or vertical evolution: a directional drive freely climbing the levels of complexity.

2.3 Chapters 9 and 10

With Chaps. 9 and 10 of the *Origin* (1859)—the paleontological chapters—the reader is confronted with yet another picture of reality, one giving the impression that evolution is largely at a standstill (pp. 293, 313–316, 323–324, 331–332). One key implicit notion organizes knowledge in these chapters: all past forms are believed to be directly aligned with or falling in between currently living forms (pp. 329–330). Darwin's ultimate conceptual tool in this conquest of the past is convergence backward in time. This idea implies that evolutionary divergence in each class or phylum has reached its maximum today, meaning that it took the entire history of life just to get from the original species of each class (progenitor) to the several orders seen today in that same class. Such a deployment saw the extinction of many lower taxonomic entities (species, genera, families, and orders), by none at higher levels (class/phylum). In the *Origin*, Darwin made clear that he thought entities of all taxonomic levels already existed by the Silurian period (p. 338), although he remained vague as to whether or not all classes had already deployed in full by that early geological time (except when resorting to entirely hypothetical scenarios when discussing the Queer Diagram). Evolution is seen, therefore, as an extremely slow-to-deploy process of distinct and independent classes (Delisle 2014). Also flowing from Darwin's key notion that divergence today is at a maximum (relative to the past) is the assumption that there is no need to leave explanatory room for genuine evolutionary novelties, since variability's past is believed to be either aligned with or falling in-between extant forms. For Darwin, the present always reveals more than the geological past.

This explains why Darwin rushed to assimilate past forms to currently living ones, in an epistemological quest that avoided the recognition of distinct and entirely independent faunas and floras at various geological periods. Against proponents of a series of successive, sudden, and largely independent groups of forms in time, Darwin opposed the view of a single, unified, gradual, and continuous network binding the past to the present (p. 489). As the floor under his feet opened up to reveal the deep past uncovered in paleontological investigations, Darwin rushed to incorporate it into the present (pp. 303–305). Darwin thus assimilated to living counterparts Tertiary monkeys, Secondary mammals and crustaceans, and Pre-Secondary fish. In fact, it is striking in Chaps. 9 and 10 how Darwin mentions only a limited number of precise fossil groups—such as trilobites, *Lingula*, *Nautilus*, ganoid fish, mastodons, *Megatherium*, *Toxodon*—often preferring to refer to general designations such as ancient mammals, birds, reptiles, mollusks, crustaceans, ammonites, shells, and cephalopods, thus blurring to a significant extent the potential difference between past and present forms. Darwin admitted that he was faced with a Silurian temporal barrier: before that barrier we know nothing of life; after that barrier life is largely organized around well-defined groups as we know them today (pp. 302–310). It is not that Darwin entirely overlooked the differences between some fossil groups and their possible extant counterparts (pp. 313, 323, 429). It is simply that he recognized some evolutionary change within each class while simultaneously holding that the variability seen among extant groups is more than enough to organize that of the past. It is difficult to see how real evolutionary novelties can arise from the view presented in Chaps. 9 and 10. For Darwin, the past is entirely under the epistemological control of the present, the former being a pale and incomplete representation of the latter.

2.4 Chapters 11 and 12

Another picture of evolution—a fourth one—is presented to the reader of Chaps. 11 and 12 of the *Origin* (1859). Here, Darwin is busy explaining the geographical distribution of currently living forms by appealing to assumed common ancestors and unique centers of dispersal that are traceable to fairly recent geological periods. In so doing, Darwin was implicitly guided by an *ideal model* that consisted in adjusting or calibrating a series of explanatory variables in such a way as to keep a symmetrical or proportional relation between them: a certain quantity of geological time is proportional to a certain level of taxonomic separation; a certain distance in geographical space is proportional to a certain amount of selective pressure; a certain level of taxonomic separation is proportional to a certain distance in geographical space; a certain amount of selective pressure is proportional to a certain quantity of geological time. Presumably, in these two chapters at least, Darwin thought that the dual core of this theory—composed of the principle of divergence and the principle of gradualism-gradation—would best be supported by maintaining the proportionality of such explanatory variables: the gradual pattern-process of divergence being accompanied by proportional amounts of evolutionary change (as seen in taxonomic levels), geological time, geographical space covered, and level of selective pressure.

Unsurprisingly, few case studies could really comply with this ideal model of proportionality; after all, evolutionary contingency requires much more explanatory flexibility than that. Knowing this, Darwin did introduce some explanatory flexibility when attempting to match the theory and the evidential basis. The ideal model of proportionality was thus used by Darwin as a default position: only when facts did not match the model did Darwin proceed to adjust some of the explanatory variables. At this precise epistemological junction, it becomes clear that Darwin refused to commit himself towards a *strong contingency thesis*, adopting instead a *weak contingency thesis*. Darwin could not embrace evolutionary contingency to its full extent precisely because the core of his theory (divergence + gradation) as presented in these chapters prevented him from questioning its validity.

But why did Darwin assume in the first place that divergence and gradualism are universal? Apparently, he fell victim to his own ontological commitment, using the present time as the only valid time horizon to investigate the evolutionary past. The principles of divergence and gradualism-gradation are, for him, the two conceptual tools used to bind the past to the present: (1) divergence being at its maximum today, all past forms must fall in-between extant forms; (2) the living world being tightly knit through innumerable gradations, the past must be intertwined with the present: *Natura non facit saltum* (pp. 194, 206, 210, 243, 460, 471).

Under such assumptions, Darwin's theory was significantly blind to that past by introducing a partial and distorting view of it. In short, Darwin's theory lacked explanatory flexibility when it came to accommodating the full contingency of evolution, forcing it under an interpretative framework that is too rigid. This explains why Darwin's framework showed signs of explanatory strain in the exposition of some of his case studies, necessitating him to resort to speculative and ad hoc explanations. To save the core of his theory (divergence + gradation) while trying to create a match between the theory and the evidential basis, Darwin displaced the explanatory stress by redirecting it towards more debatable explanations. The various explanatory variables contained in Darwin's interpretative framework in Chaps. 11 and 12 have the following characteristics:

1. An inflexible and irrefutable theoretical core: the principle of divergence (including unique center of dispersal, common ancestry, and monophyletism) and the principle of gradualism-gradation. These are the principles that Darwin tries to save at all costs and without which his strict *actualistic method* would plainly be ineffective, thus jeopardizing his whole theory.
2. Two explanatory variables applied in a preferred state of polarity: short geological time periods (over long periods) and low taxonomic levels (over high levels). It should be noted that Darwin feels more comfortable applying these two variables in the just-described states of polarity throughout the *Origin*. This is unsurprising for someone thinking about the past mainly from the perspective of the present.
3. Two explanatory variables exploited in highly variable ways: geographical distribution (locally confined or widely dispersed) and level of selective pressure (low or high). Darwin is never shy about modulating the state of these two variables in his evolutionary explanations, as if migration even over long distances was never a problem and as if postulated selective pressure could be modified at will to fit the case at hand.
4. Speculative and ad hoc explanatory variables imagined in order to make the predictions of the theory fit the evidential basis: floating icebergs, strong winds, complete absence of selective pressure, sunken lands, authority arguments. If the worse comes to worst, Darwin ultimately turns to this bag of evolutionary tricks to save the explanation.

Equipped with this interpretative framework, Darwin tackled several case studies, such as the alpine plants in the Northern Hemisphere (pp. 365–369), life forms in the Galapagos Islands (pp. 397–403), and a worldwide dispersal from around the North Pole (pp. 369–382). Let us consider this latter case only. The dual core of Darwin's theory was respected under two conditions: while the notion of a unique center of dispersal incarnated de facto the principle of divergence, the principle of gradualism-gradation was seen in a clinal distribution radiating from that center in direction of the south in both the Old and New Worlds: identical forms found around the common Arctic region, different yet closely related varieties found in the Subarctic regions of both continents, different yet closely related species found in the Temperate regions of both continents, and a host of allied forms randomly distributed in the rest of the world, such as in South America, the Cape of Good Hope, Ceylon, Java, and Australia (pp. 369–376). The other explanatory variables were kept proportional to each other in an attempt to comply with Darwin's *ideal model*. For instance, the more forms migrated away from the northern cradle under the impetus (selective pressure) of a shifting climate, the further away related forms living in the Old and New Worlds are from each other, spatially speaking, but also taxonomically speaking (from perfect identify, to distinct varieties, to distinct but related species). Darwin fine-tuned his scenario with some speculations in an attempt to match the ideal model with the evidential basis (lack of space prevents us from going into details here).

More interesting to us is how Darwin was forced to deal with two main categories of facts going beyond the explanatory power of his northern cradle scenario. In the first category, biogeographical facts in the southern hemisphere could not be incorporated in his original scenario, as some extant forms could not be affiliated with northern ones. Darwin had, therefore, to resort to postulating a more modest and more localized migration from the vicinity of the South Pole in direction of the north (pp. 377–382). So doing implied explaining how similar forms could have spread to widely isolated areas such as Antarctica, Australia, South Africa, and South America, thus putting some strain on his theory. Darwin turned to more speculative and bolder explanations in order to try to address that issue. An extraordinary means of dispersal was postulated (drifting icebergs), relying for that matter on an appeal to authority, in the form of Charles Lyell's supportive opinion (p. 381). Furthermore, it was imagined that the migrational process could have benefitted from the exploitation of some Antarctic lands and some unspecified southern lands intermediately positioned to formerly housed forms responsible for colonizing lands as far apart as New Zealand and South America (p. 399).

The second category of facts going beyond the explanatory power of Darwin's northern cradle scenario concerned the presence of a group of quite distinct and scattered species of plants currently restricted to the southern hemisphere (pp. 381–382). Unable to make sense of these plants by the previous two scenarios (northern cradle or southern cradle), Darwin turned to a third, vague, and entirely new scenario that imagined that places like Kerguelen Land, New Zealand, and Fuegia could have been colonized by speculative means, such as the aforementioned drifting icebergs and by frankly ad hoc explanations like unknown sunken islands.

Interestingly, this third scenario reveals Darwin's inability to think about extant forms having no equivalence in the world of today. In all other case studies, Darwin ensured that the horizontality of the present time (recent migration from other regions) would be the organizing principle of the time dimension. This is precisely what went wrong with this unexpected and scattered group of quite distinct species of plants restricted to the southern hemisphere. It is as if they had made a sudden irruption out of an obscure past, weakening the core of Darwin's theory on two counts: (1) quite distinct species of plants not falling under the purview of the principle of divergence (convergence backward in time starting from extant forms); (2) a disruption in the postulated gradation and contiguity of forms (the principle of gradualism-gradation) owing to their combined distinctness and random distribution.

In contradistinction to Darwin's *horizontal approach* in this case study, a *vertical approach* would have been more appropriate to make sense of the evolutionary past of such isolated or unusual species emerging from an obscure evolutionary past. Apparently, Darwin's theory was powerless to confront the full contingency of evolution. His commitment towards a *weak contingency thesis* rather than a *strong contingency thesis* deprived him of the necessary explanatory flexibility that would have allowed him to face the complex reality of evolution. When Darwin explicitly or implicitly claims repeatedly in the *Origin* that "there is no law of development" (pp. 313–315, 318, 331–332, 343, 351, 408–409), meaning that biological evolution can only be understood as complying with evolutionary opportunism—each lineage having its own independent evolutionary fate—it should be remembered that this statement is partly rhetorical, as Darwin's theory does not allow for such a full recognition.

2.5 Chapter 13

Finally, the reader is offered a fifth approach to evolution in Chap. 13 of the *Origin* (1859). In it, Darwin intends to use the field of systematics as the backbone of evolutionary biology, a field assumed to have registered a pan-divergent view of evolution, as seen in clusters of forms regrouped under ever more inclusive clusters along the taxonomic scale. The principle of divergence was projected backward in time (convergence) in an attempt to embrace the entire evolutionary history of the distinct and independent classes. Fields like morphology, embryology, and comparative anatomy (rudimentary organs) were believed to support this vision of evolution, each bringing something specific to a unified and all-encompassing evolutionary field. It might have been assumed by Darwin that the fields of systematics, morphology, embryology, and comparative anatomy were organized in such a way as to comply with the epistemological ideal of a science based on a "consilience of inductions," whereby the confirmation of a theory rests on independent classes of facts extracted from various disciplines. This is not what the careful reader of Chap. 13 actually finds, however; instead, the neat explanatory unity of Darwin's theory is disrupted in two main ways.

First, Darwin had to explicitly and vigorously combat the two extreme faces of the action of natural selection over time, realities jeopardizing his theory: perfect imitation caused by evolutionary analogies (pp. 414–415, 425, 427–428) and the rise of entirely novel or "ancestorless" evolutionary features (p. 421); two realities blurring or erasing the traces of affiliation or divergence. Darwin's self-assurance that his theory could easily trace descent was betrayed when he thought he could distinguish between homologous and analogous species belonging to the phylogenetic groups of the marsupials and of the rodents (pp. 429–430). To impose more order than is in fact present in evolution, Darwin founded his theory on three key assumptions: (1) more often than not the selective process will retain the most divergent forms of all, driving less divergent ones to extinction (p. 412); (2) the taxonomic *arrangement* currently seen among living forms was preserved throughout the entire evolutionary history of the groups (p. 421); and (3) evolution today is at its maximum state of divergence, with all past forms being either aligned with or falling in between currently existing forms (p. 329). Having thus locked up evolution in a rigid interpretative framework, Darwin did not leave enough conceptual room for a subscription to the *strong contingency thesis*, as mentioned above. The contrast between the genuine complexity of evolution—which also includes reticulate evolution, polyphyletism, parallel evolution (analogies), and deleted traces of affiliation—and Darwin's rigid interpretative framework is striking to the careful reader of Chap. 13.

The second way in which the neat explanatory unity of Darwin's theory was dissolved is seen in the nature of the relationship he established between various disciplines. Whereas the section on systematics deployed in full the conceptual panoply accompanying Darwin's theory centered around the principle of divergence (pp. 411–434); the sections devoted to morphology (pp. 434–439), embryology (pp. 439–450), and comparative anatomy (pp. 450–456) merely supported the notion of common ancestry. As we know, "divergence" and "common ancestry" are not necessarily equivalent. This being the case, most of Chap. 13 provided little support to the ideal of a science founded on "consilience of inductions," considering that several key explanatory components of Darwin's theory received no help from them. On the contrary, the sections on morphology, embryology, and comparative anatomy brought tensions, contradictions, or inconsistencies to what we might presume constitutes Darwin's ideal of science. For instance, the field of morphology was exploited by Darwin in a very general way only, not being precise enough to generate phylogenetic connections within members belonging to the same class (p. 333). Furthermore, after having claimed that the field of embryology was more useful for probing the evolutionary past than the field of morphology (p. 449), Darwin overlooked significant counterfactual evidence he himself noted in embryology. Instead, he adopted a theoretical posture that locked evolution within a rigid view by embracing an embryological theory based on two assumptions (pp. 443–450): (1) changes appear at later embryological stages than at earlier ones; (2) changes appear at corresponding embryological ages within lineages. Again, evolution was put in a straitjacket, depriving it of the evolutionary flexibility it would require to accommodate the *strong contingency thesis*. Finally, the field of comparative anatomy (rudimentary organs) encountered difficulties at Darwin's hands. It will suffice to mention two of them: (1) difficulties in distinguishing between truly ancestral features and features modulated under the laws of organogenesis (pp. 453–454); (2) ad hoc explanations introduced to distinguish between rudimentary organs produced under domestication and those in a state of nature, undermining the legitimacy of any such features altogether (p. 454).

The *Origin of Species* truly constitutes a challenging piece of work for a careful reader when taken as a whole. Beyond Darwin's self-proclamations in favor of a theory of evolution believed to be unified and all-encompassing, it is fairly clear that the Queer Diagram serves more to conceal profound tensions and inconsistencies than to reveal its unity and coherence. Jean Gayon (2009: 282) rightly points out that such a Diagram was, for Darwin, "a heuristic device, a plausible bet" about what evolution is all about, a bet that will be challenged from several perspectives during the twentieth century and after. I would like to add here that the first to challenge Darwin was Darwin himself in the *Origin*. The reader willing to go beyond Darwin's rhetoric encounters a book displaying at least five independent sets of issues or pictures. While some are squarely incompatible with one another, others are less than clearly related to each other:

1. Life forms being caught up in complex web of reproductive networks at low taxonomic levels and from which it is difficult to segregate.
2. Natural selection being a wild and brute force overrunning the pattern-process of divergence through a directional ascent of taxonomic levels and evolutionary grades, thus blurring the homology–analogy distinction and overstressing the "Conditions of Existence" (natural selection) over the "Unity of Type" (descent).
3. Evolution being at a standstill and lacking in novelties, as seen in the paleontological annals, with a past that falls entirely under the purview of today's biological variability or disparity.
4. The negation of the strong contingency thesis in an attempt to calibrate explanatory variables along a symmetrical-proportional ideal explanatory model, with some variables being treated as irrefutable (divergence and gradualism-gradation), others being more or less flexible (taxonomic level, geological timeframe, geographical distance, and selective pressure), still others being highly conjectural (deriving icebergs, sunken land, authority arguments).
5. Explanatory tensions, inconsistencies, and contradictions generated by an attempted synthesis between fields like systematics, morphology, embryology, and comparative anatomy. In Darwin's hands, those fields were deprived of the necessary conceptual flexibility to accommodate in full evolutionary contingency.

The extreme pluralism encountered in the *Origin* should be taken in account when Darwin scholars are reflecting upon the changes of the Darwinian movement over time. We need to be careful not to envision Darwin's contribution as being reducible to a sort of neat, compact, and abstract theoretical construct, especially when considering that such an abstraction is largely erected upon Darwin's own rhetoric. To confine Darwin's Darwinism to selective pressures acting on small variations at low taxonomic levels—a process extrapolated to higher taxonomic levels through the divergence pattern—falls short of what is actually contained in the *Origin*.

3 From Darwin to the Evolutionary Synthesis: More Rhetorical Arguments

It is almost irresistible, however, to cling to the narrow view of Darwin's Darwinism, as just described, in light of the definition often provided to characterize the Evolutionary Synthesis (Gould 1980: 119–120; Mayr 1980: 1): (1) that gradual evolution is explained by small genetic changes (variations) oriented by natural selection leading to adaptation; (2) that speciational, evolutionary trends, and macroevolutionary events are consistent with known genetic mechanisms at the microevolutionary level. Not only does the Evolutionary Synthesis seems to be a direct extension of Darwin's theory, but the synthesists themselves give the impression of being Darwin's heirs, as evidenced by titles such as: *Genetics and the Origin of Species* (1937) by Dobzhansky; Mayr's "Darwin's Impact on Modern Thought" (1995); and Simpson's "One Hundred Years Without Darwin Are Enough" (1961; see also Laporte 2000: 78–79), among other suggestive titles.

But just as there was a lot of rhetoric in Darwin's *Origin*, the synthesists also availed themselves of rhetorical declarations when it came to positioning themselves within the Darwinian movement. As argued elsewhere (Delisle 2008, 2009a, b, 2011a, b) and as will be seen further below, behind the superficial unity of the Evolutionary Synthesis lies profound ontological, conceptual, epistemological, and metaphysical oppositions, pointing to identical issues at about the same time in the French tradition (Loison and Herring 2017), the German-speaking tradition (Levit and Hossfeld 2017), and the English-speaking tradition devoted to organicism (Esposito 2017).

Whether or not Darwin (Ruse 1979: 176) and the synthesists (Smocovitis 1996: 171) really intended to emulate Newton in presenting a compact theory centered around a causal core giving meaning to an array of disciplines, the project largely failed. Apparently, the topic of biological evolution is too complex and the amount of relevant information too great to be captured by a simplified view of science. Unsurprisingly, this explanatory complexity was open to significant interpretative issues, well beyond what Orzack (1981) has originally rightly recognized. As far as the structure of the Darwinian theory is concerned, it seems that nothing less than a "reticulate model" of science is capable of accommodating the pluralism observed among the darwinians. As originally suggested by Morton Beckner (1959) and discussed by Ruse (1973: 48–52) and Tuomi (1981), the reticulate model offers a diffused and flexible explanatory structure, with no privileged role given to causal components (i.e., natural selection) over descriptive ones, and no hierarchical organization of disciplines over others (explanatory core vs. explained periphery), in opposition to the "multilevel model" of science. To make this point clearer, let us enter into the specifics of our analysis by applying the analytical framework already presented in the Introductory section of this paper. In order to make the comparison easier and the presentation compact, an abridged style is used in the following sections.

3.1 *Reductionism Versus Holism: The Ultimate Level of Evolution*

It might be assumed that all darwinians placed the ultimate drive of evolution somewhere between the genetic level and the species level. Contrary to received wisdom, however, not only is this issue a contentious one among them, but the pluralism of views far exceeds what is usually presented in the traditional historiography:

Darwin We have seen how Darwin provided fairly strong evidence in favor of life forms at low taxonomic levels being caught up in common reproductive networks. Assuming he could find a way to segregate them from one another, Darwin (1859) was hoping to show that a selection process of variations among individual organisms, subspecies, species, and genera sufficed to explain all of biological evolution when extrapolated in the time dimension and across taxonomic levels.

Mayr Following Darwin's explanation closely, Mayr (1942, 1963) avoided some of Darwin's conceptual pitfalls by relying on notions such as the biological species concept, the polytypic species concept, genetic revolution, the founder effect, and the allopatric/peripatric models of speciation. So doing allowed Mayr to avoid the interpenetration and confusion of low taxonomic levels seen in Darwin when the latter considered variations introduced during sexual reproduction in gametes, in individual organisms, and within species. Mayr thus introduced a clear separation between organismic biology and genetics, with the drive of biological evolution being placed in the lowest taxonomic levels of organismic biology only (selection of genetic features occurring at the individual and species levels).

Rensch Whereas Darwin's and Mayr's views are reductionistic in light of all the existing taxonomic levels (not as reductionistic as population geneticists, though), Rensch's approach is entirely out of scale, as it embraces a "super-reductionism" that places the ultimate drive of evolution in entities housed at the subatomic levels. According to Rensch, all cosmic entities (including complex life forms) are the by-product of what he calls an "epigenetic process" binding them together in a single deterministic causal nexus traceable to fundamental laws (causality, universal constants, principles of symmetry, conservation of energy) and basic cosmic entities (energy/mass, charge, spin, speed, space, time). While Rensch's view incorporates a weak form of emergentism with new and unpredictable features arising through more complicated stages of matter/law integration during the cosmic deployment, he invites us to apply reductionism further still in search of the basic physical level responsible for the biological manifestations. Rensch hints at an ultimate source which fuses matter and laws in a sort of energetic field of protopsychical nature, in line with his panpsychism (Rensch 1960a, 1971, 1974, 1985). For him, cosmic evolution unpacks the inherent potentialities originally contained in its original energetic state and manifested during the ascension of biological complexity. We will expand on Rensch's view as we go along with our analysis (which overlaps significantly with the analysis independently proposed by Levit et al. 2008, 2014; Levit and Hossfeld 2017).[1]

Dobzhansky Just as Rensch is out of scale of the traditional historiography with his super-reductionism, so is Dobzhansky with his "super-holism." Contrary to received wisdom, genetic reductionism for Dobzhansky the geneticist was never the ultimate foundation of his evolutionary thinking. Instead, he envisioned the universe as an ontologically unified and intrinsically dynamic entity, taking in its evolutionary wake its component parts including life, the whole being more than the sum of its parts (Dobzhansky 1955, 1962, 1963, 1965, 1967a, 1968a, 1973). The other main features of Dobzhansky's universe are the following: (1) it is progressive and tends towards improvement; (2) it is irreversible; life cannot go back to its inorganic state nor humankind to its simple organic one; (3) the evolutionary pace is picking up as time passes; (4) freedom is increasing; determinism in inorganic matter is gradually giving way as life forms climb the scale of complexity; (5) cosmic evolution constitutes an open-ended process. We will also expand further on Dobzhansky's view as we go along with our analysis, a view with many similarities to Huxley's (1923, 1953, 1960).

3.2 *Horizontal Evolution Versus Vertical Evolution: Conceiving the Tree of Life*

Darwinism is often thought to be associated with an atomistic ontology, whereby the evolutionary fate of each individual organism or lineage is seen as independent of other such entities, under the selective pressure of separate, unique, unrepeatable, and differential conditions. Again, this view is not shared by a number of darwinians, who have proposed holistic alternatives of the tree of life:

Darwin Other than vague allusions that all life forms are connected through a single or only a few hypothetical common ancestors, Darwin's tree of life is not a unified one. Rather, it is characteristically segmented around isolated classes/phyla, such high taxonomic entities being both closed and permanent in the sense that past forms are entirely interpreted in the light of present ones. As seen above, Darwin reconstructs the past with the help of two principles: (1) all past forms within a class/phylum are believed to be directly aligned with or falling in-between currently living forms; (2) it is assumed that the entire history of life was required for each

[1]It is interesting to note that whereas Rensch's work had received little serious attention from historians and philosophers concerned with the Evolutionary Synthesis that two sets of studies independently produced came to significantly overlapping conclusions (Delisle 2008, 2009a, b, 2011a; Levit et al. 2008, 2014; Levit and Hossfeld 2017). Both sets of studies also independently concluded that the standard view of the Evolutionary Synthesis is jeopardized by the kind of pluralism seen in Rensch's view when also compared to the pluralism of other synthesists, requiring a rethinking of what the Evolutionary Synthesis really was. See also Delisle (2017).

class/phylum to reach current level of diversity. This being the case, little explanatory room is left for recognizing real novelties in the evolutionary past. For Darwin, clearly, to travel in geographical space is to travel in geological time; evolution is captured through the horizontality of extant forms.

Mayr Walking in the footsteps of Darwin, Mayr (1942, 1954, 1959, 1960, 1963, 1974, 1982b, 1992) also subscribed to an evolutionary view largely deprived of historical depth. Focusing on horizontal evolution rather than vertical evolution, Mayr saw the process of cladogenesis (as opposed to anagenesis) as the cornerstone of evolutionary biology, explaining both the diversity of life and the rise of evolutionary novelties. Rigorously applying the principle of actualism, Mayr hoped to show that one could see in the geographical distribution of currently living forms all the gradual steps leading to a full speciation, an explanatory model believed to be complete enough to make sense of the most profound evolutionary changes registered in the history of life. Like Darwin, Mayr assumes that the present reveals all there is to know about the past.

Huxley Whereas Darwin's and Mayr's tree of life lacked unity and/or historical substance, Huxley's view (1923, 1936, 1942, 1947) fully developed both aspects by opposing their horizontality with verticality, that is, a temporal projection forward in time. Huxley's tree of life shares a collective fate by being pulled forward up the scale of complexity during a process that consists in sharing successful novelties across a wide taxonomic range, as especially seen in generalized organisms (and as opposed to specialized ones). Once a particular novelty is widespread—the argument goes—the only way life can go on is through the invention and sharing of more novelties, thus allowing life to climb the evolutionary grades. Huxley organizes the holistic unity of the tree of life around a main emerging trend as seen in six tendencies: (1) an increase in the size of life forms in units of composition or mode of clustering (for instance, unicellular, pluricellular, and communities); (2) an increase in the complexity of life in the division of labor of parts; (3) an increase in the harmonization of parts through a central nervous system; (4) an increase in the autoregulation of life forms, with internal processes reducing the impact of external conditions; (5) an increase in learning from past situations (memories, learning, shared knowledge in social groups); and (6) an increase in mental capacities. Huxley's tree of life finds its ultimate epistemological foundation in the postulation of a progressive universe that, in its wake, brings matter down the inorganic, organic, and mind path. Just as Huxley's holistic stance shares many similarities with Dobzhansky's, so does their view of the tree of life (Dobzhansky 1955, 1956, 1967a, b).

Rensch Even more clearly than Huxley and Dobzhansky, Rensch's (1954, 1959, 1960a, 1967, 1978) tree of life is solidly bound together by a single causal and law-like nexus running through an ascending scale of complexity, which extends from manifestations in microphysics to anthropology. Overlooking the ramification of the tree of life, Rensch focuses instead on the common thread characterizing the ascension of intelligence. Applying its preferred method of investigation—reductionism or ontological stripping—he follows this thread by gradually descending the chain of being through five subhuman levels: (1) the faculties of reasoning and insights in primates and carnivores; (2) the faculty of formulating nonverbal conclusions in birds; (3) the faculty of nonverbal judgements in invertebrates; (4) the faculty of forming mental imagines in lower invertebrates; (5) the faculties of sensation and sensorial reactions in protozoans.

3.3 External Versus Internal Agencies: What Is Driving Evolution?

In the traditional historiography on Darwinism, evolution is believed to be driven by external selective pressures (as opposed to internal agencies), justifying why evolutionary contingency constitutes the right picture for biological evolution. Yet, a number of darwinians will also take important liberties with this conceptual piece:

Darwin Only in the rhetorical Darwin is the evolutionary contingency thesis well-founded. A careful reading of the *Origin of Species* (1859), however, reveals a surprisingly high number of factors constraining its expression: (1) Chapters 1–5 and 8 see life forms inextricably connected in common reproducing networks. (2) Chapters 6 and 7 show natural selection overrunning everywhere the pattern of divergence across evolutionary grades and taxonomic levels in favor of a strongly directional evolution. (3) Chapters 9 and 10 depict evolution to be so slow as to approach evolutionary standstill. (4) Chapters 11 and 12 gave up on the strong contingency thesis in favor of a weaker thesis considering that knowledge is organized around two inflexible and irrefutable principles—divergence and gradualism/gradation—thus preventing Darwin from accommodating a full evolutionary contingency. (5) Finally, Darwin (1859: 118–119, 412) claimed that, in general, the most divergent forms will outcompete all others, thus imposing upon evolution a fairly rigid and preestablished pattern of evolution.

Simpson In a textbook exposition of how evolutionary contingency should express itself, Simpson (1944, 1949) argues that life's opportunism is manifested in the diversity of evolutionary rhythms (fast or slow) and modes (anagenesis and cladogenesis): modalities that vary throughout evolutionary lineages and geological periods. To promote the view that life's history is characterized by a collection of individual and independent histories, Simpson will go so far as to claim that the manifestations of complexity seen in the tree of life are entirely unrelated to each other: there was no cumulative effect of tendencies over geological times; evolutionary branches had to independently rediscover the road to complexity. With this theoretical posture, Simpson (1949) condemned life to poor development only since it is prevented from building upon the spreading of key innovations which, nevertheless, he himself recognizes: the ability to exploit resources more efficiently; the increase in global energy needed to sustain metabolism; the increase in awareness and perception of the environment; the increase in parental care; the control of the environment by humans. Whatever Simpson claimed, he implicitly recognized that portions of the tree of life required sustained accumulations of evolutionary novelties.

Dobzhansky Not entirely denying the ramification of life, Dobzhansky's (1966, 1967b, 1968b, 1974) view is nonetheless more oriented and directional than usually assumed, certainly too much for a genuine evolutionary contingency driven by external agencies. In his ontologically unified, dynamic, and progressive cosmos, the biological portion of this universal evolution found a way to generate a new emergent level of complexity (or transcendence), as seen in humankind. He refers to the biological process employed to get there "groping," that is, the rise of a plethora of forms of which only the more generalized ones are selected (as opposed to the more specialized ones) in order to carry on with the next step of progressive evolution. Apparently, there is something more acting on Dobzhansky's life forms other than mere external opportunistic processes to get them to the next progressive level; this other thing being a directional and progressive cosmos—holistically conceived—acting as a sort of internal drive to biological evolution.

Rensch In a view that stresses evolutionary directionality as much as Dobzhansky's, although showing a more deterministic bent, Rensch holds that the biological portion of cosmic evolution constitutes a deployment gradually unpacking the internal potentialities already largely contained in the original ontological matter (laws/entities). In the biological phase seeing the rise of greater mental powers, which it should be reminded preexisted under a protopsychical aspect during a prebiological phase, Rensch (1960b) refers to the organizing principles as "rules" rather than "laws" in order to insist on a slightly looser determinism. The fact remains, however, that there is little explanatory room for external opportunistic processes in Rensch's understanding of evolution.

3.4 The Mechanisms of Evolution: Central or Peripheral Explanations?

We have intentionally kept discussion of the issue of evolutionary mechanisms for the end. In the traditional historiography, the history of evolutionary biology is sometimes equated with the development of evolutionary theories/mechanisms (including Lamarckism, Darwinism, Orthogenesis, Mutation Theories, Saltationism, and Neo-Darwinism). On such an assumption, the mechanisms are believed to be central to evolutionary biology, as presented in the multilevel model of science. If the reticulate model of science is adopted, however, no privileged place is recognized to them relative to other explanatory components. It is argued here that the views of Darwin and some of the synthesists are much better adapted to the reticulate model than to the multilevel one. For such scholars, the mechanisms of evolution have no meaning in themselves, unless integrated and harmonized in complex explanatory structures taking into account, for instance, the level at which the evolutionary drive is positioned, the interpretation given to the tree of life, and whether or not evolution is confined to the biological realm or expanded to also include the entire universe. To assume that the concept/theory of natural selection comes with an evolutionary view centered around a haphazard, contingent, and nondirectional evolution is to distort what many darwinians believe. For them, natural selection is conceptually neutral and its evolutionary role or function is to be determined by other explanatory components:

Darwin As much as natural selection seems to be central to the *Origin of Species*, the fact remains that Darwin continually modified its application against different factual backgrounds directly related to his various competing interpretations of the tree of life: (1) natural selection being so weak as to have difficulty segregating reproductive networks (Chaps. 1–5 and 8); (2) natural selection being so powerful as to produce strong directional evolution overruning evolutionary grades and taxonomic levels (Chaps. 6–7); (3) natural selection being sufficiently weak so as to require the entire history of life to merely manage to deploy the currently living classes/phyla (Chaps. 9–10 and 13). Apparently, natural selection is not as central to Darwin as it constitutes only one explanatory component in need of harmonization with others. For instance, it could be argued that in light of Chaps. 11 and 12 of the *Origin*, the real explanatory core of Darwin's theory is composed of the principles of divergence and gradation, since both are treated as inflexible, with other variables (such as selective pressure) being used in a very loose way only, in an attempt to make the theory fit the factual reality. To borrow Imre Lakatos's (1978: 48) terminology, I would hold that the dual principle of divergence + gradation represents the irrefutable hard core of Darwin's theory in Chaps. 11 and 12, by methodological decision, with natural selection merely constituting an auxiliary hypothesis, a part of the protective belt whose function is to protect the hard core through a series of adjustments. These adjustments range from a complete lack of selective pressure to a very strong selective pressure. Darwin's competing interpretations of the tree of life is so central to his understanding of evolution that he had to continually adjust the mechanisms accordingly.

It should be noted that the epistemological attitude giving precedence to the tree of life is common in the history of evolutionary biology. A few examples will make that point clearer. Henry Fairfield Osborn (1917, 1932, 1934) subscribed to what he eventually called "aristogenesis" to designate what he thought were linear goal-directed manifestations in the annals of life. Taking this for granted, he kept looking for causal explanations to make sense of them, with limited success (Bowler 1983: 160–161, 174–176). Similarly, Otto Schindewolf's (1993 [1950]) "typostrophism" presented the tree of life as characterized by sudden appearance of biological types (saltationism), followed by a diversification along parallel and directional lines (orthogenesis), often accompanied by overspecialization and extinction, a pattern repeated at several taxonomic levels. Schindewolf searched in macromutations and ontogeny the cause or mechanism for what he considered was an overall pattern in life's history (Reif 1983, 1993). Pattern and process in Schindewolf's view are separate enough from one another that Stephen Jay Gould could write: "I don't think that the typostrophic theory offers much to our search for mechanisms (though it describes a common pattern that we will have to render in other causal ways)" (Gould 1993: xiv). Finally, most instructive are the phases through which the theory of punctuated equilibria went through in the recent decades (Ruse 1988: 35–50, 1992; Turner 2017). At first, the moments of rapid evolutionary change followed by long periods of stasis in the annals of life were presented by its founders as being largely consistent with Darwinism (Eldredge and Gould 1972). Later, the theory of punctuated equilibria was presented as at least requiring a fairly important extension of Darwinism's theoretical scope (Gould 1980, 1982; Eldredge 1985a; Gould and Eldredge 1986, 1993). This shift in theoretical interpretation shows that the phenomenal reality of punctuated equilibrium was never questioned by its founders, preferring instead to modulate their understanding of the causal issues. The tree of life and the evolutionary mechanisms are, therefore, largely uncoupled in this case study, although punctuated equilibrium was originally presented as emerging from a reflection upon the implications accompanying the mechanism of allopatric speciation.[2] Yet, once the causal theory had open the eyes of the founders, so to speak, they were apparently no longer able to close them to the "fact" of punctuated equilibrium, as clearly shown in Gould's quote about Schindewolf's view of evolution.

Simpson Most instructive are the consequences that sometimes accompany the attempt to fit the entire complexity of evolutionary biology into a narrow view exclusively centered on evolutionary mechanisms. As a paleontologist, Simpson (1944: xvii) wanted to forcefully show that past evolutionary patterns were entirely explained by processes known in population genetics, which he took to mean that evolution was not under the action of a single and regular agent pushing life forward but rather driven by as many distinct evolutionary changes as there are opportunistic lineages (evolutionary contingency). This zealous commitment toward the Evolutionary Synthesis, however, forced Simpson into a theoretical trap by depriving him of the tools for explaining how many life forms came to share so many evolutionary novelties and progressive features. This struggle between Simpson's use of the evolutionary mechanisms and his interpretation of the tree of life is revealed as the reader progresses through the chapters of *The Meaning of Evolution* (1949).

Huxley Other scholars had fewer scruples than Simpson when dealing with evolutionary mechanisms. Regardless of the fact that Huxley's book, *Evolution: The Modern Synthesis* (1942), gave the Evolutionary Synthesis its name, his general

[2]Whether or not this line of argument was merely rhetorical to make punctuated equilibrium more palatable to darwinians remains an open question. As much as Eldredge and Gould wanted to avoid the "inductivistic sin," it seems to me that the 1972 paper exposes enough paleontological "facts" taken seriously by them as to at least raise the possibility that they were already convinced of the inductivistic value of the fossil record. Taking seriously the long history of paleontological interpretations helps being more open about inductivism.

conception of evolution which promoted a universal progressionism never fundamentally changed from the 1920s to the 1960s, irrespective of the fact that he gradually modified his commitment toward the evolutionary mechanisms. At first not entirely averse to recognizing the role of orthogenetic factors, macromutations, and the inheritance of acquired characteristics, Huxley (cf. 1923, 1963) eventually adopted a more panselectionist stance. Yet, this shift had no impact on his general view of evolution, as he believed that cosmic evolution and the tree of life had explanatory precedence over the evolutionary mechanisms.

Dobzhansky Similarly, Dobzhansky gave epistemological precedence to the tree of life and cosmic evolution over the mechanisms of evolution in his synthetic effort to elaborate a general view of evolution. This is explicitly expressed in the following words: "The ascertainment of the fact of directedness of the general evolution is not tantamount to its explanation. The fact of directedness had been discovered, it would seem, prematurely, before the causes that bring evolution about even begun to be deciphered" (Dobzhansky 1974: 312). Evolutionary mechanisms can only serve, then, the purpose of justifying a biological fact recognized long ago by paleontology: progress. It is to this end that Dobzhansky conceptualized natural selection as an agent that generates a plethora of forms, of which only the most generalized (progressive) ones are selected to take evolution on to the next level.

Rensch The concept of natural selection was believed to be so neutral, explanatory speaking, that Rensch had no difficulty integrating it in his single deterministic nexus, tightly binding together entities from the subatomic level to the level of higher organisms. In this nomological universe, the "law" of natural selection (as he calls it) is merely a part of a larger selection process active in the deployment of ever more complex and intelligent entities. Actually, Rensch (1971: 282–283) even explicitly speaks of a "cosmic law of selection."

Mayr Of all the darwinians reviewed in this paper, only Mayr is fully committed to an etiological or causal view of biological evolution. His view is entirely about deploying a conceptual apparatus (e.g., the polytypic species, the biological species concepts, etc.) explaining how small genetic variations can ultimately lead to the multiplication of lineages through a selective process. There is no epistemological quest, in Mayr, for a harmonization between the mechanisms of evolution and the tree of life, for instance. His understanding of the tree of life is merely a projection in the past of processes currently observed. Mayr's etiology includes all the explanatory components believed to be relevant for biological evolution, the evolutionary past being dissolved in such an etiology. In that sense, Mayr is much more Darwinian than Darwin himself, for we have seen that in building his theory Darwin took into account other explanatory components, such as his conflicting interpretations of the tree of life.

4 What Is Darwinism? Revisiting the Historiography

Our analysis of the intellectual connections binding Charles Darwin and the Evolutionary Synthesis throws more light on Darwinism. With the mechanisms of evolution being variably exploited by evolutionists, it seems that the multilevel model of science founded on a special explanatory status for a causal core cannot universally account for the multiplicity of approaches adopted. While Mayr, and to a certain extent Simpson, give this causal core predominance, this is not the case for Darwin, Huxley, Dobzhansky, and Rensch. For the last four scholars, the mechanisms are integrated in multidimensional explanations based on a harmonization with other components like the tree of life, the chain of being, the nature of the entities studied (inert matter, life, mind), and methodological choices. In such synthetic explanations, the evolutionary mechanisms are often peripheral: it is the other components that give conceptual meaning to the mechanisms, and not the other way around. In these cases, the reticulate model of science with no preestablished hierarchy with regard to the importance of explanatory components or disciplines is more appropriate when it comes to describing how such scholars proceed with organizing the information. Each scholar establishes his own hierarchical order, an order that cannot be taken for granted by historians and philosophers, since it must be reconstituted.

Confronted with such diffused and diversified intellectual strands among darwinians, the question of "what Darwinism is" becomes ever more challenging. The issue is not only to make sense of what has changed within Darwinism over the years, a vexing question if there is one. Even more problematic is the issue of what unites darwinians both across time and *in contemporaneity* (see also Hull 1985). As proposed elsewhere (Delisle 2009a, b: 377–393), the pluralism among darwinians is so significant that it may be best to classify them as working in distinct research programs and traditions. To put this succinctly, views that hold that evolution is a directional and open-ended cosmic process (Huxley, Dobzhansky), or a directional cosmic process deploying its inherent potentialities until its eventual exhaustion and relapse (Rensch), or reconstructed from the assumption that all there is to know about the evolutionary past and future is entirely revealed by the present (Darwin, Mayr), cannot all be part of the same research entity. Building on a previous proposal and now going beyond it (Delisle 2011a, b), it seems to me that "Darwinism" as a label has outlived its utility, inasmuch as it acts like a blanket, cloaking fundamental issues instead of revealing them.[3]

Evolutionism is perhaps best seen as a broad research area within which scholars work in distinct research entities either running in parallel or consecutively, being

[3]I am conscious of the implications accompanying the solution of getting rid of the label "Darwinism." The label has often served the purpose of being equated with "evolutionary studies" as often seen in titles of books, thus making them more appealing on the book market. As such, the label has a strong sociological utility. But because sociological issues should not have priority over conceptual ones, setting aside this label could help clarify the latter issues.

complementary or contradictory. In that research area, natural selection, for instance, is reduced to a mere "theory" or "concept" that can be incorporated and interpreted in the light of distinct research agendas. Our proposal has the merit of getting rid of the historian's and philosopher's task of weighting "good" and "bad" research programs with a scale evaluating Darwinian purity. It also eliminates the witch hunt for epithets like "ultra-darwinians," "near-darwinians," "pseudo-darwinians," or "anti-darwinians". "Darwinism" cannot simply be equated with natural selection. Since there is much more to the *Origin of Species* than the theory of natural selection (see also Mayr 1985), why should Darwin's name be exclusively associated with natural selection? Similarly, scholars like Huxley, Dobzhansky, and Rensch only paid lip service to the mechanisms of evolution when elaborating their synthetic views of evolution. Unless one wants to exclude them from Darwinism altogether while keeping others—an attempt that would lead to an awkward situation—it seems that the only alternative is to rethink the label itself. By discarding "Darwinism," natural selection can be set apart from other relevant explanatory components in evolutionary biology, allowing evolutionists to freely relate to it or contribute to it, including Darwin himself.

In contradistinction to what is being proposed by Pigliucci (2017) whose analysis struggles to make sense of what he calls "bizarre traditions" in French-, German-, Russian-, and even English-speaking countries, we propose to fully draw the lessons of pluralistic, complementary, and alternative views. Pigliucci's normative approach forecloses the reflection about what evolutionary biology is about, including in its future and in its past (see also Depew 2017, on the foreclosing issue). Knowledge about an allegedly founding figure like Darwin cannot be reduced to a figment of presentism, being confined to a sort of abstract construct that comes in handy for contemporary scholars. The flesh-and-blood Darwin, so to speak, must be allowed to exist, as should all the other evolutionists. The discarding of the label "Darwinism" contributes to breaking or redefining the frontiers between research entities, allowing for a reassessment of the nature of their complex relations. Possible connections in need of further investigation include:

1. French neo-lamarckians (Loison 2010; see also Delisle 2012) and a neo-Darwinian like Rensch share a serious interest in both a physicalistic reductionism and a nomological science, not unlike Jacques Loeb in the early twentieth century (Esposito 2017).
2. Self-organizing principles (Brooks and Wiley 1988; Kauffman 1995) and thermodynamics (Wicken 1987; Depew and Weber 1995) are not incompatible with what Huxley, Dobzhansky, and Rensch had to say in their synthetic views (see especially Huxley 1923: 72–74, 250–261, 1953: 5).
3. Rensch's suggestion to put natural selection in the context of a cosmic law of evolution has been recently investigated in cosmology (Davies 2000; Impey 2007; Smolin 1997).
4. The developmental model of evolution originally founded on the fields of embryology and morphology in the nineteenth century (Coleman 1971; Maienschein 1991; Nyhart 1995) often appealed to a fairly strong directionality

in the tree of life, not unlike the one defended by Huxley, Dobzhansky, and Rensch.
5. To this, it should be added that Rensch himself was no stranger to the morphological research tradition (Levit et al. 2008, 2014; Levit and Hossfeld 2017).
6. Mayr's organicism resisting genetic reductionism (Delisle 2009b: 303–376) is certainly no stranger to the robust organicist tradition described by Maurizio Esposito (2017).
7. The idea which stipulates that specialized living forms are evolutionarily doomed in the long term while generalized ones keep their options open is widespread as seen in Huxley and Dobzhansky among darwinians and Vandel among neo-lamarckians (see Loison and Herring 2017).
8. The un-Darwinian notion of tree of life *à la* Bergson was shared by neo-darwinians and lamarckians alike (Gayon 2008; Herring 2016).
9. The idea of a hierarchy of emergent novelties/properties was promoted by neo-darwinians like Huxley and Dobzhansky, organismic biologists like Woodger, Lillie, and Ritter (Esposito 2017), and holistic thinkers like Jan Smuts, Teilhard de Chardin, and Lloyd Morgan.

The picture that emerges from these considerations is a complex network of research programs or ideas variably connected through commitments towards methodologies, concepts, theories, ontologies, metaphysics, and factual evidence [see also my Preface to this volume and Delisle (2017)]. As made clear in many contributions to this volume, such research programs have no essences. Apparently, evolutionary studies have from the outset been intertwined in such a way as to encompass many topics and disciplines. This picture is incompatible with the impoverished historiography depicting "Darwinism" as a linear development centering on evolutionary mechanisms and going from Darwin to the Evolutionary Synthesis and beyond. Evolutionism as a doctrine, it seems, never was organized around a single dominant intellectual thread.

Let us conclude this paper by briefly revisiting some landmarks often accompanying the historiography:

4.1 *The Eclipse of Darwinism*

Peter Bowler (1983, 1988, 2017) has argued that evolutionism in the decades around the year 1900 often developed out of research programs that owed little to Charles Darwin's selection theory (natural selection), a phase he calls the "eclipse of Darwinism." I would like to expand upon the notion that natural selection was believed to be neither central to evolutionary biology, nor efficient enough as an evolutionary force to explain the phenomenal reality. It should first be mentioned that very few evolutionists entirely discarded the idea that life forms must be sufficiently adapted to their milieu to survive and reproduce. That is why the debate around the value of selection theories is usually couched along an explanatory spectrum ranging from a "strong" role for a selection entirely creating and orienting variations to a "weak" role for a selection merely eliminating the variations or forms, leaving orientation to additional factors (for more subtleties, see Bowler 2017). In this story, presumably, it is believed that the "darwinians" are those which incline more towards natural selection playing a stronger role than a weaker one. Is this really the case? For instance, no matter how much space Huxley (1942) and Dobzhansky (1937; see also Lewontin et al. 1981) devoted to considering selective processes, the fact remains that such mechanisms were intrumentalized in their synthetic views. Are we to believe, then, that Huxley and Dobzhansky are proponents of a strong selection theory? At one level of analysis, the answer is negative, insofar as the evolutionary mechanisms are, for them, conceptually neutral for not being inherently associated with a view based on a haphazard, undirectional, and opportunistic process—preferring instead to give epistemological precedence to the tree of life and cosmic evolution, both believed to be strongly directional. At another level, Huxley and Dobzhansky did contribute to the development of a selection theory, irrespective of whether or not we are content with the way they have instrumentalized it. Assuming we want to retain the historiographical notion of an "eclipse" of strong selection theories, it will be necessary to expand it considerably to include at least a part of what Huxley, Dobzhansky, Rensch, and, to a lesser extent, Simpson, did. So doing, however, will muddle the clear dichotomy between proponents of strong and weak selection theories, perhaps to the point of requiring the abandonment of the notion of an eclipse of Darwinism (see Largent 2009). We may have to live with that solution, given that Darwin himself kept *adjusting*—in Imre Lakatos' sense (1978: 48)—the notion of natural selection to his several competing views of the tree of life in the *Origin of Species*.

4.2 *The Hardening of the Evolutionary Synthesis*

It has been argued by Stephen Jay Gould (1983, 2002) that a number of contributors to the Evolutionary Synthesis between the 1930s and the 1960s went through a constriction phase, a conceptual notion describing a loss of explanatory pluralism ranging from evolutionary mechanisms originally appealing to selective as well as to stochastic processes (among other components) to a panselectionist view that relies almost exclusively on the role of selection as the cause of change. Among other evolutionists, Dobzhansky's work is seen as being at the forefront of this explanatory shift. In light of what has already been said, our comment here will look somewhat anticlimactic. The historiographical notion of a hardened synthesis is entirely a by-product of an etiological view of science in which causes or mechanisms are believed to be central, a view perhaps best referred to as a "mechanism-centered" evolutionary biology. We have seen, on the contrary, that evolutionism comprised many other explanatory components besides the mechanisms of evolution themselves. We have also seen that each evolutionist establishes his or her own hierarchy of components in a synthetic proposal, to the point of sometimes either paying lip service to the evolutionary mechanisms or conceptualizing them in such a way as to comply with preestablished notions. Not only is the notion of a hardening of the Evolutionary Synthesis not of universal application among so-called neo-darwinians, but I would suggest that its historiographical value is entirely dependent upon the recognition that complex evolutionary syntheses are multidimensional research entities requiring on the part of the historian or philosopher the specification at which level the analysis is conducted.

4.3 An Extended Evolutionary Synthesis

The post-1960 period in evolutionary biology is understandably accompanied with frequent pleas for a greater pluralism as far as the understanding of the evolutionary patterns and processes are concerned (Depew and Weber 2013; Eldredge 1985a, b; Gould 1982; Laland et al. 2014; Müller 2007; Pigliucci 2007; Wicken 1987). After all, explanatory room must be allowed for new factual discoveries and conceptual breakthroughs. Yet, just as there was no hardening of the Evolutionary Synthesis in the usual sense of the expression because pluralism was maintained at other explanatory levels, so the plea for an expanded synthesis must also be confronted to this historical complexity. New generations of evolutionists must avoid erecting a straw man called the "Evolutionary Synthesis" in order to better promote their own so-called "novel" ideas. There is still much pluralism in the works Darwin, Huxley, Dobzhansky, and Rensch that has not yet been fully integrated in the historiography (not to mention the names of other evolutionists).

5 Conclusion

The *Origin of Species* is a fascinating book full of intriguing and original suggestions. Its overall explanatory structure, however, is contradictory and dislocated. Darwin does not provide for the evolutionists of the future a unified and coherent view of evolution, but instead offers a whole range of tools and concepts from which one can individually pick, in a piecemeal fashion. And this is exactly what the founders of the "Evolutionary Synthesis" did: they each selected whatever suited them in the elaboration of their separate views, often combining "Darwinian" and "non-Darwinian" notions. That is why the Evolutionary Synthesis is perhaps best conceived as a sociological event rather than a conceptual one (Levit et al. 2008; Cain 2009; Delisle 2011a). There was no unity in the *Origin of Species*, just as there was no real unity in the so-called Evolutionary Synthesis. The intellectual connections binding all the scholars presented in this paper are rather weak and diffused. Apparently, evolutionism offers a kaleidoscopic complexity of research entities and ideas running in parallel and overlapping, thought systems containing several explanatory levels. Only a very inclusive historiography can fully capture the inherent complexity of this research area. Looked at retrospectively, it seems to me that the analyses of John C. Greene (1981, 1999) may well represent the single most important source of inspiration when it comes to adopting the necessary breadth for capturing all that complexity.

Acknowledgments I thank James Tierney (Yale University) for assistance in improving the English version of this paper.

References

Alcock J (2017) The behavioral sciences and sociobiology: a Darwinian approach. In: Delisle RG (ed) The Darwinian tradition in context: research programs in evolutionary biology. Springer, Cham, pp 37–60

Appleman P (1959) The logic of evolution: some reconsoderations. Vic Stud 3:115–125

Beckner M (1959) The biological way of thought. Columbia University Press, New York

Bowler PJ (1983) The eclipse of Darwinism. Johns Hopkins University Press, Baltimore

Bowler PJ (1988) The non-Darwinian revolution. Johns Hopkins University Press, Baltimore

Bowler PJ (2017) Alternatives to Darwinism in the early twentieth century. In: Delisle RG (ed) The Darwinian tradition in context: research programs in evolutionary biology. Springer, Cham, pp 195–218

Brooks D, Wiley E (1988) Evolution as entropy: toward a unified theory of biology, 2nd edn. University of Chicago Press, Chicago

Cain J (2009) Rethinking the synthesis period in evolutionary studies. J Hist Biol 42:621–648

Caplan A (1978) Testability, disreputability, and the structure of the modern synthetic theory of evolution. Erkenntnis 13:261–278

Churchill F (1982) Darwin and the historian. Biol J Linn Soc 17:45–68

Coleman W (1971) Biology in the nineteenth century: problems of form, function, and transformation

Davies P (2000) The 5th miracle: the search for the origin and meaning of life. Touchtone, New York

Darwin C (1859) On the origin of species. John Murray, London

Darwin F (1909) The foundations of the origin of species: two essays written in 1842 and 1844 by Charles Darwin. Cambridge University Press, Cambridge

Delisle RG (2008) Expanding the framework of the holism/reductionism debate in neo-Darwinism: the case of Theodosius Dobzkansky and Bernhard Rensch. Hist Philos Life Sci 30:207–226

Delisle RG (2009a) The uncertain foundation of neo-Darwinism: metaphysical and epistemological pluralism in the evolutionary synthesis. Stud Hist Philos Biol Biomed Sci 40:119–132

Delisle RG (2009b) Les philosophies du néo-darwinisme. Presses Universitaires de France, Paris

Delisle RG (2011a) What was really synthesized during the evolutionary synthesis? A historiographic proposal. Stud Hist Philos Biol Biomed Sci 42:50–59

Delisle RG (2011b) Disputatio [book symposium]: Les philosophies du néo-darwinisme (2009). By: R. Delisle, D. Becquemont, B. Continenza, and M. Morange. Philosophiques 38 (1):263–303

Delisle RG (2012) Review of Loison's Qu'est-ce que le néolamarckisme? (2010). Hist Philos Life Sci 34:504–506

Delisle RG (2014) Evolution in a fully constituted world: Charles Darwin's debts towards a static world in the *Origin of Species* (1859). Endeavour 38:197–210

Delisle RG (2017) Introduction: Darwinism or a kaleidoscope of research programs and ideas? In: Delisle RG (ed) The Darwinian tradition in context: research programs in evolutionary biology. Springer, Cham, pp 1–10

Depew DJ (2017) Darwinism in the 20th century: productive encounters with saltation, acquired characteristics, and development. In: Delisle RG (ed) The Darwinian tradition in context: research programs in evolutionary biology. Springer, Cham, pp 61–88

Depew D, Weber B (1989) The evolution of the darwinian research tradition. Syst Res 6 (3):255–263

Depew D, Weber B (1995) Darwinism evolving : systems dynamics and the genealogy of natural selelction. MIT Press, Cambridge

Depew D, Weber B (2013) Challenging Darwinism: expanding, extending, replacing. In: Ruse M (ed) The Cambridge encyclopedia of Darwin and evolutionary thought. Cambridge University Press, Cambridge, pp 405–411

Desmond A, Moore J (1991) Darwin : the life of a tormented evolutionist. W.W. Norton, London

Dobzhansky T (1937) Genetics and the origin of species. Columbia University Press, New York

Dobzhansky T (1951) Genetics and the origin of species, 3rd edn. Columbia University Press, New York

Dobzhansky T (1955) Evolution, genetics, and man. Wiley, New York

Dobzhansky T (1956) The biological basis of human freedom. Columbia Univesity Press, New York

Dobzhansky T (1962) Mankind evolving. Yale University Press, New Haven

Dobzhansky T (1963) Evolution – organic and superorganic. Rockefeller Inst Rev 1:1–9

Dobzhansky T (1965) Evolution and transcendence. Main Curr Mod Thought 22:2–9

Dobzhansky T (1966) An essay on religion, death, and evolutionary adaptation. Zygon 1:317–331

Dobzhansky T (1967a) The biology of ultimate concern. New American Library, New York

Dobzhansky T (1967b) Creative evolution. Diogenes (58):62–74

Dobzhansky T (1968a) Darwin versus Copernicus. In: Rothblatt B (ed) Changing perspectives on man. University of Chicago Press, Chicago, pp 173–190

Dobzhansky T (1968b) Teilhard de Chardin and the orientation of evolution. Zygon 3:242–258

Dobzhansky T (1973) Ethics and values in biological and cultural evolution. Zygon 8:261–281

Dobzhansky T (1974) Chance and creativity in evolution. In: Ayala F, Dobzhansky T (eds) Studies in the philosophy of biology. University of California Press, Berkeley, pp 307–338

Eldredge N (1985a) Time frames: the rethinking of darwinian evoluiton and the theory od punctuated equilibria. Touchstone Book, New York

Eldredge N (1985b) Unfinished synthesis: biolological hierarchies and modern evolutionary thought. Oxford University Press, Oxford

Eldredge N, Gould SJ (1972) Punctuated equilibria: an alternative to phyletic gradualism. In: Schopf TJM (ed) Models in paleobiology. Freeman, Cooper, San Francisco, pp 82–115

Esposito M (2017) The organismal synthesis: holistic science and developmental evolution in the English-speaking world, 1915–1954. In: Delisle RG (ed) The Darwinian tradition in context: research programs in evolutionary biology. Springer, Cham, pp 219–242

Fisher R (1930) The genetical theory of natural selection. Clarendon Press, Oxford

Fleming D (1959) The centenary of the origin of species. J Hist Ideas 20:437–446

Gayon J (1992 [1998]) Darwinism's struggle for survival: heredity and the hypothesis of natural selection. Cambridge University Press, Cambridge

Gayon J (2008) L'Evolution créatrice lue par les fondateurs de la théorie synthétique de l'évolution. In: Fagot-Largeot A, Worms F (eds) Annales bergsoniennes IV L'Evolution créatrice 1907–2007: épistémologie et métaphysique. Presses universitaires de France, Paris, pp 59–84

Gayon J (2009) From Darwin to today in evolutionary biology. In: Hodge J, Radick G (eds) The Cambridge companion to Darwin, 2nd edn. Cambridge University Press, Cambridge, pp 277–301

Goudge T (1961) The ascent of life. University of Toronto Press, Toronto

Gould SJ (1980) Is a new and general theory of evolution emerging? Paleobiology 6:119–130
Gould SJ (1982) Darwinism and the expansion of evolutionary theory. Science 216:380–387
Gould SJ (1983) The hardening of the modern synthesis. In: Grene M (ed) Dimensions of Darwinism. Cambridge University Press, Cambridge, pp 71–93
Gould SJ (1993) Foreword of O Schindewolf, Basic questions in paleontology. University of Chicago Press, Chicago, pp ix–xiv
Gould SJ (2002) The structure of evolutionary theory. Belknap Press, Cambridge
Gould SJ, Eldredge N (1986) Punctuated equilibrium at the third stage. Syst Zool 35:143–148
Gould SJ, Eldredge N (1993) Punctuated equilibrium comes of age. Nature 366:223–227
Greene JC (1975) Reflections on the progress of Darwin studies. J Hist Biol 8:243–273
Greene JC (1981) Science, ideology, and world view. University of California Press, Berkeley
Greene JC (1986) The history of ideas revisted. Rev Synth (3):201–227
Greene JC (1999) Debating Darwin. Regina Books, Claremont
Haldane JBS (1932) The causes of evolution. Longmans, London
Herring E (2016) Des évolutionnismes sans mécanisme: les néolamarckismes métaphysiques d'Albert Vandel (1894–1980) et Pierre-Paul Grassé (1895–1985). Rev Hist Sci 69(2):369–398
Hodge MJS (1977) The structure and strategy of Darwin's 'long argument'. Br J Hist Sci 10:237–246
Hodge MJS (2000) Knowing about Darwin: Darwin and his theory of natural selection. In: Creath R, Maienschein J (eds) Biology and epistemology. Cambridge University Press, Cambridge, pp 27–47
Hull DL (1974) Philosophy of biological science. Prentice Hall, Englewood Cliffs
Hull DL (1985) Darwinism as a historical entity: a historiographic proposal. In: Kohn D (ed) The darwinian heritage. Princeton University Press, Princeton, pp 773–812
Huxley JS (1923) Essays of a biologist. Chatto & Windus, London
Huxley JS (1936) Natural selection and evolutionary progress. Report of the British Association for the Advancement of Science, pp 81–100
Huxley JS (1942) Evolution: the modern synthesis. Allen & Unwin, London
Huxley JS (1947) A re-definition of progress. Pilot Pap 2:8–25
Huxley JS (1953) Evolution in action. Harper, New York
Huxley JS (1960) Man's new vision of himself. University of Natal, National Conference on Education
Huxley JS (1963) Evolution: the modern synthesis, 2nd edn. Allen & Unwin, London
Impey C (2007) The living cosmos. Random House, New York
Kauffman S (1995) At home in the universe: the search for the laws of self-organization and complexity. Oxford University Press, Oxford
Lakatos I (1978) The methodology of scientific research programmes. Philosophical papers, vol 1. Cambridge University Press, Cambridge
Laland K, Uller T, Feldman M, Sterelny K, Müller G, Moczek A, Jablonka E, Odling-Smee J (2014) Does evolutionary theory need a rethink? yes, urgently. Nature 514:161–164
Laporte L (2000) George Gaylord Simpson: paleontologist and evolutionist. Columbia University Press, New York
Largent M (2009) The so-called eclipse of Darwinism. Trans Am Philos Soc (N Ser) 99(1):3–21
La Vergata A (1985) Images of Darwin: a historiographic overview. In: Kohn D (ed) The darwinian heritage. Princeton University Press, Princeton, pp 901–972
Lenoir T (1987) Essay review: the Darwin Industry. J Hist Biol 20:115–130
Levit GS, Hossfeld U (2017) Major research traditions in 20th century evolutionary biology: the relations of Germany's Darwinism with them. In: Delisle RG (ed) The Darwinian tradition in context: research programs in evolutionary biology. Springer, Cham, pp 169–194
Levit GS, Simunek M, Hossfeld U (2008) Psychoontogeny and psychophylogeny: Bernhard Rensch's (1900–1990) selectionist turn through the prism of panpsychistic identism. Theor Biosci 127:297–322

Levit GS, Hossfeld U, Olsson L (2014) The Darwinian revolution in Germany: from evolutionary morphology to the modern synthesis. Endeavour 38:268–279

Lewontin R, Moore JA, Provine WB, Wallace B (1981) Dobzhansky's genetics of natural populations. Columbia University Press, New York

Loewenberg B (1959) Darwin scholarship of the Darwin year. Am Q 11:526–533

Loewenberg B (1965) Darwin and Darwin studies, 1959–63. Hist Sci 4:15–54

Loison L (2010) Qu'est-ce que le néolamarckisme? Les biologistes français et la question de l'évolution des espèces 1870–1940. Vuibert, Paris

Loison L, Herring E (2017) Lamarckian research programs in French biology (1900–1970). In: Delisle RG (ed) The Darwinian tradition in context: research programs in evolutionary biology. Springer, Cham, pp 243–270

Maienschein J (1991) Transforming traditions in American biology, 1880–1915. Johns Hopkins University Press, Baltimore

Mayr E (1942) Systematics and the origin of species. Columbia University Press, New York

Mayr E (1954) Change of genetic environment and evolution. In: Huxley JS, Hardy AC, Ford EB (eds) Evolution as a process. Allen & Unwin, London, pp 157–180

Mayr E (1959) Where as we? Cold Spring Harb Symp Quant Biol 24:1–14

Mayr E (1960) The emergence of evolutionary novelties. In: Tax S (ed) Evolution after Darwin, vol 1. University of Chicago Press, Chicago, pp 349–380

Mayr E (1963) Animal species and evolution. Belknap Press, Cambridge

Mayr E (1974) The challenge of diversity. Taxon 23:3–9

Mayr E (1980) Prologue: some thoughts on the history of the evolutionary synthesis. In: Mayr E, Provine WB (eds) The evolutionary synthesis: perspectives on the unification of biology. Harvard University Press, Cambridge, pp 1–48

Mayr E (1982a) The growth of biological thought. Belknap Press, Cambridge

Mayr E (1982b) Speciation and macroevolution. Evolution 36:1119–1132

Mayr E (1985) Darwin's five theories of evolution. In: Kohn D (ed) The darwinian heritage. Princeton University Press, Princeton, pp 755–772

Mayr E (1991) One long argument: Charles Darwin and the genesis of modern evolutionary thought. Harvard University Press, Cambridge

Mayr E (1992) Controversies in retrospect. In: Futuyama D, Antonovics J (eds) Oxford surveys in evolutionary biology, vol 8. Oxford University Press, Oxford, pp 1–34

Mayr E (1995) Darwin's impact on modern thought. Proc Am Philos Soc 139:317–325

Müller GB (2007) Evo-devo: extending the evolutionary synthesis. Nat Rev Genet 8:943–949

Nyhart LK (1995) Biology takes form: animal morphology and the German universities, 1800–1900. University of Chicago Press, Chicago

Oldroyd D (1984) How did Darwin arrive at his theory? The secondary literature to 1982. Hist Sci 22:325–374

Orzack S (1981) The modern synthesis is partly wright. Paleobiology 7(1):128–131

Osborn HF (1917) The origin and evolution of life on the theory of action, reaction and interaction of energy. Scribners, New York

Osborn HF (1932) The nine principles of evolution revealed by paleontology. Am Nat 66:52–60

Osborn HF (1934) Aristogenesis: the creative principle in the origin of species. Am Nat 68:193–235

Pigliucci M (2007) Do we need an extended evolutionary synthesis? Evolution 61:2743–2749

Pigliucci M (2017) Darwinism after the modern synthesis. In: Delisle RG (ed) The Darwinian tradition in context: research programs in evolutionary biology. Springer, Cham, pp 89–104

Provine WB (1971) The origins of theoretical population genetics. University of Chicago Press, Chicago

Provine WB (1978) The role of mathematical population geneticists in the evolutionary synthesis of the 1930s and 1940s. Stud Hist Biol 2:167–192

Provine WB (1985) Adaptation and mechanisms of evolution after Darwin: a study in persistent controversies. In: Kohn D (ed) The darwinian heritage. Princeton University Press, Princeton, pp 825–866
Reif W-E (1983) Evolutionary theory in German paleontology. In: Grene M (ed) Dimensions of Darwinism. Cambridge University Press, Cambridge, pp 173–203
Reif W-E (1993) Afterword of O Schindewolf, basic questions in paleontology. University of Chicago Press, Chicago, pp 435–453
Rensch B (1954) The relation between the evolution of central nervous functions and the body size of animals. In: Huxley JS, Hardy AC, Ford EB (eds) Evolution as a process. Allen & Unwin, London, pp 181–200
Rensch B (1959) Trends towards progress of brains and sense organs. Cold Spring Harb Symp Quant Biol 24:291–303
Rensch B (1960a [1947]) Evolution above the species level. Columbia University Press, New York
Rensch B (1960b) The laws of evolution. In: Tax S (ed) Evolution after Darwin, vol 1. University of Chicago Press, Chicago, pp 95–116
Rensch B (1967) The evolution of brain achievements. Evol Biol 1:26–68
Rensch B (1971) Biophilosophy. Columbia University Press, New York
Rensch B (1974) Polynomistic determinationof biological processes. In: Ayala F, Dobzhansky T (eds) Studies in the philosophy of biology. University of California Press, Berkeley, pp 241–258
Rensch B (1978) Psychogenesis from lowest organisms to man. In: Loflin M, Silverberg J (eds) Discourse and inference in cognitive anthropology. Mouton, The Hague, pp 259–280
Rensch B (1985) Biophilosophical implications of inorganic and organismic evolution. Die Blaue Eule, Essen
Roger J (1985) Darwinism today (commentary). In: Kohn D (ed) The darwinian heritage. Princeton University Press, Princeton, pp 813–823
Rosenberg A (1985) Structure of biological science. Cambridge University Press, Cambridge
Ruse M (1973) The philosophy of biology. Hutchinson, London
Ruse M (1974) The Darwin Industry – a critical evaluation. Hist Sci 12:43–58
Ruse M (1975) Darwin's debt to philosophy: an examination of the influence of the philosophical ideas of John F.W. Herschel and William Whewell on the development of Charles Darwin's theory of evolution. Stud Hist Philos Sci 6:159–181
Ruse M (1979) The Darwinian revolution. University of Chicago Press, Chicago
Ruse M (1988) Philosophy of biology today. State University of New York Press, Albany
Ruse M (1992) Is the theory of punctuated equilibria a new paradigm? In: Somit A, Peterson S (eds) The dynamics of evolution: the punctuated equilibrium debate in the natural and social sciences. Cormell University Press, Ithaca, pp 137–167
Ruse M (1996a) The Darwin industry: a guide. Vic Stud 39(2):217–235
Ruse M (1996b) Monad to man: the concept of progress in evolutionary biology. Harvard University Press, Cambridge
Ruse M (2000) Darwin and the philosophers: epistemological factors in the development and reception of the theory of the *Origin of Species*. In: Creath R, Maienschein J (eds) Biology and epistemology. Cambridge University Press, Cambridge, pp 3–26
Ruse M (2009) The history of evolutionary thought. In: Ruse M, Travis J (eds) Evolution: the first four billion years. Belknap Press, Cambridge, pp 1–48
Ruse M (2011) Is Darwinism past its 'sell-by' date? The *Origin of Species* at 150. Stud Hist Philos Biol Biomed Sci 42:5–11
Schindewolf OH (1993) Basic questions in paleontology. University of Chicago Press, Chicago [originally published in 1950 as Grundfragen der Paläontologie]
Shanahan T (2017) Selfish genes and lucky breaks: Richard Dawkins' and Stephen Jay Gould's: divergent Darwinian agendas. In: Delisle RG (ed) The Darwinian tradition in context: research programs in evolutionary biology. Springer, Cham, pp 11–36

Simpson GG (1944) Tempo and mode in evolution. Columbia University Press, New York
Simpson GG (1949) The meaning of evolution. Columbia University Press, New York
Simpson GG (1953) The major features of evolution. Columbia University Press, New York
Simpson GG (1961) One hundred years without Darwin are enough. Teach Coll Rec 60:617–626
Smith S (1959) Evolution: two books and some Darwin marginalia. Vic Stud 3:109–114
Smith S (1960) The origin of 'The Origin'. Adv Sci 16(64):391–401
Smocovitis VB (1996) Unifying biology: the evolutionary synthesis and evolutionary biology. Princeton University Press, Princeton
Smocovitis VB (1999) The 1959 Darwin centennial celebration in America. Osiris 14:274–323
Smolin L (1997) The life of the cosmos. Oxford University Press, Oxford
Stebbins GL (1950) Variation and evolution in plants. Columbia University Press, New York
Tuomi J (1981) Structure and dynamics of Darwinian evolutionary theory. Syst Zool 30(1):22–31
Turner DD (2017) Paleobiology's uneasy relationship with the Darwinian tradition: stasis as data. In: Delisle RG (ed) The Darwinian tradition in context: research programs in evolutionary biology. Springer, Cham, pp 333–352
Wicken JS (1987) Evolution, information and thermodynamics: extending the darwinian program. Oxford University Press, Oxford
Wright S (1931) Evolution in Mendelian populations. Genetics 16:97–159

Major Research Traditions in Twentieth-Century Evolutionary Biology: The Relations of Germany's Darwinism with Them

Georgy S. Levit and Uwe Hoßfeld

Abstract Evolutionary theory has been likened to a "universal acid" (Daniel Dennett) that erodes its way into more and more areas of science. Yet, every single branch of biology has developed this context with its own specific characteristics, which, either through hindering or promoting, has affected the national scientific developments in evolutionary biology. We will argue that the Darwinian theories interacted with national research traditions such that the resulting conceptual body represented an amalgamation of a metatheoretical framework with the "purely empirical" theoretical beliefs such as the theory of natural selection. We will demonstrate this using the example of the German research tradition in evolutionary biology. We will analyse this German tradition comparing it to other major traditions in evolutionary biology such as the English- and Russian-speaking evolutionism. The problem of specific influences constituting the German, English-language (Great Britain and the USA), and Russian-language context of the first and the second Darwinian revolutions will be addressed. In addition, we will introduce a concept of "metaparadigm" reflecting the specificity of German evolutionary theory at the time of the first and the second Darwinian Revolutions.

Keywords National research traditions in evolutionary biology • German evolutionary biology • Metaparadigms • Johann Wolfgang von Goethe • Ernst Haeckel • Adolf Naef • Hans Böker • Bernhard Rensch

G.S. Levit (✉)
Department of Humanities and Social Sciences, ITMO University, St. Petersburg, Russia

Biology Education Research Group, Jena University, Jena, Germany
e-mail: gslevit@corp.ifmo.ru; georgelevit@gmx.net; georg.levit@uni-jena.de

U. Hoßfeld
Biology Education Research Group, Jena University, Jena, Germany
e-mail: uwe.hossfeld@uni-jena.de

R.G. Delisle (ed.), *The Darwinian Tradition in Context*,
https://doi.org/10.1007/978-3-319-69123-7_8

1 Introduction

Evolutionary biology is a scientific discipline of a special kind (Junker and Hoßfeld 2009; Dennett 1995). Its appearance and development was and is closely related to literary insights, religious world views, philosophies, policies and national psychological peculiarities. All sciences are like that, but evolutionary biology is intertwined especially strongly with its social-cultural surroundings dependent on national intellectual traditions. To grasp such national peculiarities, we have introduced the notion of "metaparadigms" (Levit et al. 2014). Metaparadigms relate to a mental tradition detectable by language use rather than by politically determined national borders. For example, we speak about German-language or English-language evolutionary biology. Metaparadigms are not alike famous scientific "paradigms" of Thomas Kuhn (1922–1996) with their seclusion and revolutionary shifts from one steady state to another (Kuhn 2012). Metaparadigms are rather like "strange attractors" in the chaos theory: "A chaotic system has a strange attractor, around which the system oscillates forever without ever repeating itself or settling into a steady state of behaviour" (Boeing 2016). There are certainly some reservations about this comparison and the most important one being the duration of a system, which in the above definition "oscillates forever". While in the case of social-cultural system, an "attractor" appears at a certain time, where it flourishes and then gradually disappears. There are affinities between "metaparadigms" and the concept of "cultural attractor" as well (Buskell 2017), since a metaparadigm is a transformation pattern effecting scientific evolution. A metaparadigm crystallises due to internal and external influences. With regard to internal influences, we understand research traditions (broadly construed) are brought about by whatever epistemological beliefs that are essential for empirical sciences. External influences, however, may include social-political circumstances and national cultural peculiarities. Here, we concentrate on the "internal factors".

Simplified examples of metaparadigms are the Russian bias towards inclusive interpretations of living systems (organism plus its environment) (Levit 2007). Such a bias tended in Russia towards a cooperation model in the biosciences (Todes 1987), by contrast to an emphasis on competition model in the English-speaking world. In German lands, such a metaparadigm has crystallised around the monistic principle, which appeared in biology long before Darwin and persisted through the first and the second Darwinian revolutions. Monism was tightly coupled with typology, whereas the latter was the logical consequence of the first (in this very case). The objective of this chapter is to reconstruct this "strange attractor" of German evolutionary biology and to contrast it with English-speaking and Russian-speaking national traditions in evolutionary theory.

2 Goethe's Dangerous Idea

Both major figures of the first and the second Darwinian revolutions in Germany, Ernst Haeckel and Bernhard Rensch (1900–1990), regarded Johann Wolfgang von Goethe (1749–1832) as one of their most essential predecessors in biology. Citations from Goethe (epigraphs) introduce Haeckel's seminal Darwinian book *Generelle Morphologie* (General Morphology) (Haeckel 1866) as well as *Natürliche Schöpfungsgeschichte* (Natural History of Creation) (Haeckel 1868), a book written for the general reader. In this, Goethe is placed alongside Darwin and Lamarck as a forerunner of the theory of descent (Hoßfeld and Olsson 2003a). Yet, Goethe was not an evolutionist in a modern sense. What Goethe offered to the subsequent generations of bioscientists was a certain world view and methodology of biological—first, biomorphological—research. In the first half of the twentieth century, both Darwinian and non-Darwinian camps in evolutionary morphology and comparative anatomy appealed to Goethe as an authority and as a source of methodological inspiration. In their turn, debates between both camps shaped the agenda of the second Darwinian revolution in the German lands (Junker 2004; Levit and Meister 2006a, b) (Fig. 1).

In Germany, the very beginning of scientific morphology was closely connected to the name of Goethe (Kanaev 1970). This was one reason amongst others for the University of Jena to award Goethe an honorary MD in 1825. As a "low Church",

Foto (im Besitz des Verfassers)

Fig. 1 A drawing with Goethe and Friedrich Schiller, Ex Libris Ernst Haeckel (private collection)

his morphology and comparative anatomy were comparative studies of various organic structures. These studies had also an applied aspect. For example, Goethe championed the establishment of a veterinary school in Jena and frequented, on a daily basis, this school from 1816 until his death (Levit et al. 2016). Remarkably, Goethe tried to involve the veterinary school in fundamental anatomical studies as well, because comparative anatomy was always a priority for him as a science constitutive of his world view. The origin of comparative plant biology can be traced to Goethe as well: Goethe's plant archetype "provided the conceptual basis for much that followed historically" (Niklas and Kutschera 2016) (Fig. 2).

In fact, Goethe looked for differences and similarities between organisms and their parts to arrive at a general doctrine of form, in order to grasp the *idea* of a certain structure by means of empirical studies and scholarly intuition. His methodology was opposed to the mainstream Newtonianism, and arose from the holism of Kant and Spinoza, as well as from Spinoza's hylozoism, especially considering that Spinoza "had dared to fuse hylozoism [...] with monist metaphysics" (Bulman and Ingram 2016: 260). In Ernst Mayr's words, "it was a rebellion against the reductionism and mechanization of Newtonianism" (Mayr 1982: 387). The influence of Kant and his critical stance towards Newtonianism can be traced in

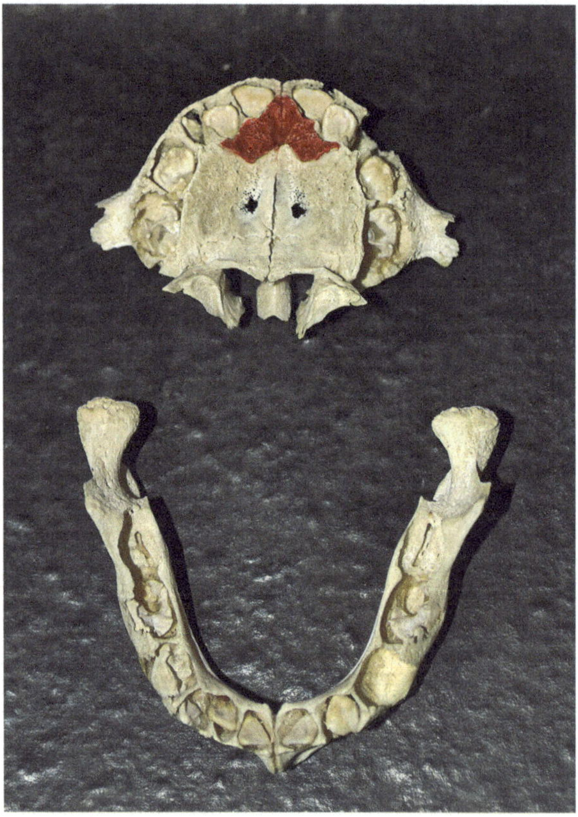

Fig. 2 Goethe's discovery: *Intermaxillary Bone*

Goethe's attitude towards teleology. His objections to teleology are summarised in "*Versuch einer allgemeinen Vergleichungslehre*" (an attempt of a general comparative doctrine 1794) (Goethe 2006: 50–55): "In language similar to his later criticisms of a Newtonian doctrine he finds fault with teleology as a traditional concept of the world"[...] (Tantillo 2002: 96–97). Goethe's "living being" did not come about in accord with external purposefulness and its *Gestalt* is not determined by a "conscious primary force" (*eine absichtliche Urkraft*) (Goethe 2006: 51).

Goethe's universe is an interdependent whole, where organic and inorganic are tightly interlocked. This concept is less metaphysical than a contemporary biologist could expect. Goethe's fish exist in the water and this means the water gets its form and functions within a certain environment. The idea that humans are nature's final purpose is alien to Goethe as well (Tantillo 2002: 99), since the picture he offered was a dynamic one. For example, in describing plants Goethe attempted to reconstruct the ultimate conformation (*Gestalt*) of nature hidden behind the observable. This was the ultimate goal of his structuralist method (Richards 2002: 440, 490). His morphology as a "high Church" had as its subject a moving, emergent and disappearing Gestalt: "The doctrine of Metamorphosis is the clue to all signs of Nature [Zeichen der Natur]" (cited and translated from Jahn 1998: 279). For Goethe, morphology was a fundamental enquiry into the most essential features of life and ultimately of the universe. The "high Church" methodological principles guided empirical research and principles of the "low Church".

Goethe's structuralism followed his concept of the morphological archetype (Williams and Ebach 2008: 29–30), which connected both "high Church" and "low Church" principles. The archetype is, for him, an ideal structure (*Bauplan*) of an organism partly expressed in the basic elements of real organismic organisation: "Thence appears a proposition about an anatomical type, a general entity, which covers (as far as possible) the structures (*Gestalten*) of all animals and allows to specify each animal in a certain system (*Ordnung*)" (Goethe 1795). The search for a vertebrate archetype resulted, for example, in the discovery of the *intermaxillary bone* in man (Goethe [1786] 1988: 111–116). Goethe's intention was to compare various vertebrate "osteological" structures to search for the general vertebrate archetype: "Goethe tried to reach a clear idea of the vertebrate archetype not only from wide induction but also from a study of function. A bone which is not only present in most vertebrates but also obviously serves a very important function is likely—for both these reasons—to belong to the archetype" (Wells 1967). The "archetype" (*Der Typus*) was, for Goethe, a "main thread" running through the labyrinth of Gestalts, a general scheme to be found because of empirical generalisations. In the works of the 1790s, devoted to the structure of animals, Goethe put forward the idea of the archetype as a pattern to be used in comparative morphology, but most importantly he saw the archetype as "a dynamic force actually resident in nature" (Richards 2002: 440), as a potentiality: "... an anatomical archetype will be suggested here, a general picture containing the forms of all animals as potential, one of which will guide us to an orderly description of each animal. [...] The mere idea of an archetype in general implies that no particular animal can be used as our point of comparison; the particular can never serve as a measure for the whole" (Goethe [1794–1795] 1988: 118).

As an epistemological aspect of his morphology, Goethe advocated a cognitive method, which he described in a short essay, *Judgment through intuitive perception* (Goethe 1988: 31–32). There, Goethe undertook a short critical analysis of Kant's thesis that human cognition is restrained by the "discursive judgement" (logical, analytical thought) as opposed to intuitive "viewing of a whole to the parts". In other words, Goethe claimed that "intuitive perception of a whole is a valid scientific method" (Heitler 1998: 65). In *A Study based on Spinoza*, Goethe abandoned empiricism and emphasised that a living thing must be measured by its own gauge. Its essence is spiritual and cannot be found through the senses (Goethe 1988). Goethe's morphological studies certainly contained more than the aspects described above, but they are crucial for our purposes. All these parts are interconnected. Thus, the search for an archetype presupposes the use of intuitive perception, and the presence of the divine in nature was an immanent part of experience. At the same time, Goethe's comparative morphology wasn't only a "Faustian" project aimed at the cognition of the final causes of the Universe, but also an applied science. Goethe himself saw this site as a place for conducting research of both practical and theoretical relevance (Levit et al. 2014).

To summarise, Goethe proposed three methodological principles, which became pivotal for the subsequent German-language evolutionary biology:

1. The first principle was a typological thinking, which Ernst Mayr unfairly equated with essentialism (Winsor 2006; Levit and Meister 2006b). The cornerstone of the typological method was the concept of type as an abstract pattern representing a certain class of phenomena and embodying the norm of this class. The primary objective of pure typology was to create classification systems for living organisms based on structurally explicable (morphological) characters without references to phylogenetic history or causal explanations. At the same time, the typological method, as it is, was quite compatible with the idea of evolution. In other words, a conflict between typological thinking and evolutionary idea was neither a direct nor an inevitable logical consequence of typology.
2. The second major methodological belief Goethe coined was the monistic principle. Goethe never used the term, and he was not a monist in a Haeckelian sense, but within German science history ultimately, it was not Spinoza but Goethe who established a unified view of nature, claiming the unity of nature and God (Kleeberg 2005: 246). The hylozoistic bias of German monism can be traced back to Goethe as well.
3. The third principle championed by Goethe, and highlighted here, was holism, which appeared in both poetic (Goethe 1950: 157) and scientific writings (Meyer-Abich 1949). Goethe's holism was especially inspirational for non-Darwinian biological thought between the second half of the nineteenth and the first half of the twentieth century. As Bernhard Rensch noted: "At the time this idea of wholeness and the allied conception of 'Gestalt' have misled

scholars into inadequate and occasionally almost mystical utterance. Many have sought support from Goethe's words: [...] *No time and no power dismembers Moulded Form which Life develops*" (Rensch 1972: 42).

These principles determined the paths of German evolutionary biology by offering a logically coherent "metaparadigmatic" construction. Yet, the way these principles were applied within Darwinian and non-Darwinian traditions were different.

3 Riot in Goethe's Garden: First Darwinian and Non-Darwinian Revolutions in Germany

It is not our intention here to fully reconstruct the history of Darwinian and non-Darwinian revolutions in Germany and elsewhere (compare: Bowler 1988, 2017). We are going to mention a few crucial figures of the late nineteenth and early twentieth centuries with the objective of demonstrating the crucial influence of Goethean principles on both Darwinian and non-Darwinian camps (which were, at that time, difficult to distinguish).

Haeckel was certainly the most prominent champion of early Darwinism in Germany (Hoßfeld 2016a, pp. 146–180). Quite soon, after the publication of Darwin's seminal *On the Origin of Species* (Darwin 1859), Haeckel began with the serious research along these Darwinian lines. It is now 150 years ago that Haeckel published his first major scientific work, *Generelle Morphologie der Organismen*, in 1866 (Haeckel 1866; Hoßfeld 2016b). Here, he, for the first time, started to formulate his Biogenetic law, which he later developed further in a monograph on calcareous sponges (Haeckel 1872). Neither *General Morphology* nor the calcareous sponges monograph were ever translated into other languages and reached a limited audience even in the German-speaking lands (Olsson et al. 2017). The popularisation of Haeckel's ideas followed in 1868 when a collection of lectures that he had held at Jena University (where he was the first professor of zoology) were published as *Natural History of Creation* (Haeckel 1868). This popular science book became a bestseller and was also translated into many different languages (Hoßfeld 2010, 2016a) (Fig. 3).

The *Generelle Morphologie der Organismen* consists of a first volume called "The General Anatomy of Organisms" and a second volume called "General Developmental History". The subtitle is *General principles of the organic form-science, founded mechanically through the theory of descent as reformed by Charles Darwin*. The first volume was dedicated to Haeckel's teacher, the anatomist Carl Gegenbaur (1826–1903), and the second volume to the "founders of the theory of descent", Darwin, Goethe and Lamarck. This book is the key to Haeckel's later work, its goal being to apply Darwin's theory to biology in general but especially to morphology. Haeckel presents here his first ideas on the relationship between ontogeny and phylogeny and introduces a system of the existing groups of

Fig. 3 Ernst Haeckel sitting on the beach of Rapallo and painting, winter 1903/1904 (Archive of the Ernst Haeckel House, Jena)

organisms based on genealogy rather than the old typological or idealistic concepts (Hoßfeld et al. 2016).

Another important aspect of the book is Haeckel's attempt to establish a promorphology—a general theory of basic forms—in the first volume. The second volume can be viewed as the first attempt to establish evolutionary morphology and evolutionary embryology as new fields of research. In the seventh "book" (one chapter within the second volume), Haeckel also formulates his ideas for a biological anthropology based on Darwin's theory of evolution. In Haeckel's view, evolution is a universal phenomenon affecting everything from inorganic matter to man. He believed in the unity of body and soul and the unity of spirit and matter. This monism guided Haeckel's work from the *Generelle Morphologie* to his last book on *Crystal souls* (Haeckel 1917).

Monism and evolutionary theory were for Haeckel parts of the same research program labelled the "monistic doctrine of evolution" (*monistischen Entwicklungslehre*). At the core of the monistic world view was the idea that "all sciences exploring humans and their soul activities [and especially so-called humanities] ... as well as special fields of zoology can be interpreted as natural sciences" (our translation from Haeckel 1904: 48). The strong connection between the concepts of evolution and monism can be seen on an example of Haeckel's work *Monism and the Link between Religion and Science. The Creed of a Natural Scientist* (Haeckel 1892). In this printed lecture, Haeckel confessed that "our monistic idea of God" is compatible with natural sciences and recognises "the spirit of God on all things"; God cannot be seen as a "personalised being" anymore, i.e. "an individual with a constrained spatial and temporal extension" (Haeckel 1892). Furthermore, Haeckel claimed that "*the Truth, the Good, and the Beautiful* are the three noble divinities before which we kneel" (our translation, original italics, Haeckel 1892: 35–36). There will be new altars built in the twentieth century, Haeckel argued, to celebrate the "trinity of monism" (Nöthlich et al. 2006; Weber and Hoßfeld 2006) (Fig. 4).

The affinity of Haeckel's monistic world view to Goethe's conceptual heritage is easy to see: "Firstly, it was Goethe whose *Naturphilosophie* served as an interpretive pattern within which Haeckel moves; it is not an accident that his *General Morphology* and each of its chapters are introduced with quotes from Goethe" (our translation from Kleeberg 2005: 114). There were typological elements in Haeckel's doctrine as well. In the second half of the nineteenth century, the

Fig. 4 Entry of the Phyletic Museum in Jena (1910) with a handwritten note by Haeckel. One can see a Goethe citation on a ceiling ledger (1908): "Wer Wissenschaft und Kunst besitzt, hat auch Religion, wer jene beiden nicht besitzt, der habe Religion" (Who science has and art he has religion too, who neither of them owns religion is his due) [Archiv des Ernst-Haeckel-Hauses, Jena]

theoretical landscape of morphology and evolutionary theory was dominated by the Jena school, i.e. by Haeckel and his senior friend Carl Gegenbaur (1826–1903). They succeeded in moving the centre of gravity in morphological research to comparative phylogenetic studies (Hoßfeld and Olsson 2003b). At the same time, their concepts appear contradictory from the modern viewpoint. Gegenbaur failed in making the methodology of evolutionary morphology either consistently evolutionary (historical) or consistently Darwinian. Although the results of his research were presented in phylogenetic terminology, the way he posed the problems was significantly typological (Starck 1965; Coleman 1976). Haeckel's Darwinism was accompanied by a strong typological bias as well (Levit and Meister 2006b; Hoßfeld 2010). Thus, Haeckel as well as Gegenbaur along with their direct successors failed in creating a consistent evolutionary morphology. "Typological thinking" survived in their concepts. As Di Gregorio suggested for Haeckel: "The old wolf had survived in sheep's clothing" (Di Gregorio 1995). Haeckel's famous embryos can serve as an example. As Hopwood explained: "Haeckel's synthesis recalls Goethe's much more ambitious intuition of the 'original plant' from accumulated observations, and Haeckel, who with a bit more talent might have become an artist, was as strongly committed to aesthetic judgment in science" (Hopwood 2006).

Haeckel's "oecologie" can be interpreted as a by-product of the revolution in biology he began in 1866 with his *General Morphology* (Kutschera 2016). "Oecologie" was for him a branch of physiology replacing the tasks and subject matter of a discipline formerly known as "economy of nature". Insofar, Haeckel successfully re-introduced the research programme of *Naturgeschichte* into his post-Darwinian monist project. The holistic attitude of Alexander von Humboldt (1769–1859) certainly influenced Haeckel: "Haeckel similarly emphasized a unique form of holism, describing the unity of nature in his philosophy of monism" (Grim and Tucker 2014: 65). To this end, Haeckel created his pro-Darwinian theoretical system along the lines of Goethe's methodological principles.

Yet, German non-Darwinians saw themselves as Goethe's disciples as well. Sometimes they were even more explicit in advertising their affinity with his ideas. At the core of non-Darwinian currents in Germany was an alternative vision of morphology. An outstanding German morphologist, Dietrich Starck (1908–2001), argued that after the death of Gegenbaur, in 1903, comparative anatomy in Germany began to fade (Starck 1977; Hoßfeld and Junker 1998, 1999). In addition, Ghiselin pointed out that morphology took no real part in the Evolutionary Synthesis and existed in "another world" in relation to the rest of biology (Ghiselin 1982: 181). Yet, the "failure" of morphology was a "failure" only from the retrospective standpoint of the Evolutionary Synthesis, which asseverated the triumph of Darwinian (historical) method (Ghiselin 1969). In contrast, from the structuralist viewpoint it was a time of rebirth and the (re)emergence of a "true" idealistic methodology, which at that time seemed to represent an effective alternative to the self-contradictory evolutionary morphologies (Rieppel 2011, 2012, 2016; Rieppel et al. 2013). In the German-speaking world, idealistic morphology had a great influence during the whole first half of the twentieth century and to a certain extent even after the Second World War. This differed substantially from the situation in Anglo-American morphology (Reif 1983). All idealistic morphologists proceeded from the same initial idea that the organism is a structural phenomenon and the purpose of comparative morphological studies must be an exact mental reconstruction of the fundamentals, the typical elements of this structure.

The first comparative-morphological theories were clearly based on the typological-teleological methodology and this made possible not only morphological but also physiological studies as well. In the early twentieth century, the theoretical landscape experienced such a serious influence of typologists, especially in morphology and paleontological studies, that one can talk about the Renaissance of the idealistic morphology in the German biological sciences (Levit and Meister 2006b). Almost simultaneously several biologists declared themselves to be adherents of typology. However, as distinct from the early typology, this new movement, which became known as "idealistic morphology", consciously opposed typological method to the method of evolutionary morphology. Correspondingly, they stood in opposition to Darwinism. This movement was represented by Edgar Dacqué (1878–1945), Wilhelm Troll (1897–1978), Wilhelm Lubosch (1875–1938), Adolf Naef (1883–1949), Otto H. Schindewolf (1896–1971), Adolf Remane (1898–1976) and many others (Rieppel 2011, 2012, 2016; Rieppel et al. 2013; Williams and Ebach 2008). At the same time, idealistic morphology hardly represented a kind of methodological monolith opposed to the Darwinian evolutionary morphology (Starck 1980). All of them followed the basic principles of typology but interpreted the results of typological classification differently (Levit et al. 2008a; Levit and Meister 2006b). There are also examples of non-Darwinian, but not primarily typological theories, explicitly referring to Goethe as their inspiration. One such theory was a neo-Lamarckian holism of the German anatomist and zoologist Hans Böker (1886–1939) (Fig. 5).

In 1924, Böker wrote a paper entitled *Begründung einer biologischen Morphologie* (Justification of Biological Anatomy), where he declared his Lamarckian research programme by stating that species "vary before our very eyes by means of inheritance of acquired features" (Böker 1924: 20). Böker, like many biologists of his time, believed he could create a new "evolutionary synthesis". He was aware of the mutationist and selectionist research programmes but maintained that they were unable to deliver the whole truth (Böker 1937: iv; Hoßfeld 2002) (Fig. 6).

Böker was opposed to the search for "separate features" and proposed the holistic research programme combining idealistic morphology, genetics, evolutionary morphology (Lamarckian version), functional explanations, ecology and even ethology (Hoßfeld 2002). He called his doctrine "comparative biological anatomy" and proceeded from the assumption that the organism is a kind of "construction" that consists of parts while being confronted with its environment. In our context, it is important to note that Böker saw himself as part of a tradition greatly influenced by Goethe (Böker 1932). Furthermore, later representatives of German holism such as Adolf Meyer-Abich (1893–1971) fully realised the role of Goethe for the growth of holistic methodology (Meyer-Abich 1949, 1970). In summary, all idealistic

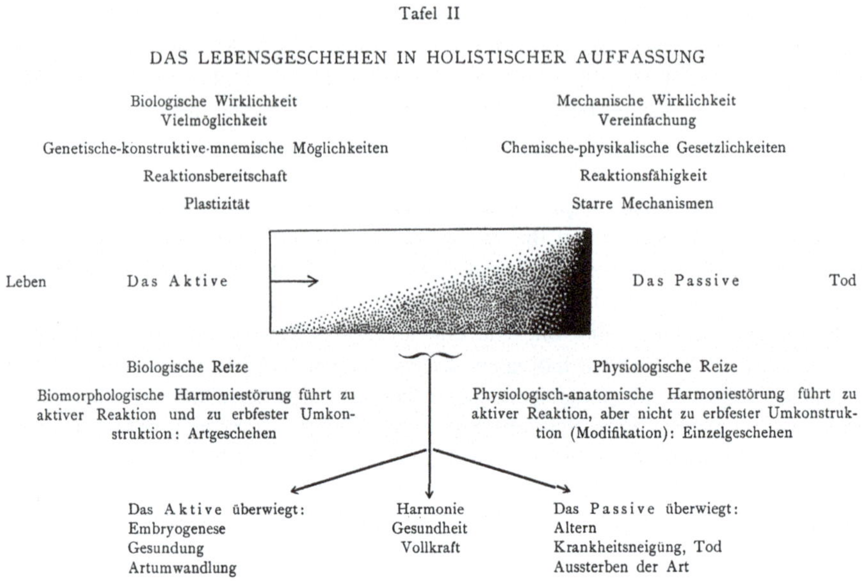

Fig. 5 The scheme by Hans Böker representing the life processes from the holistic viewpoint (from: Böker, H.: Form und Funktion im Lichte der vergleichenden biologischen Anatomie. Folia Biotheoretica 1, Ser. B: 27–41, hier S. 38, Tafel II). According to Böker, a new scientific discipline "comparative biological anatomy" would breathe a new life into the entire anatomy and, ultimately, into biology by bringing into account not only phylogenetic research (genesis) but life appearances along with environmental conditions

Fig. 6 Portrait of Bernhard Rensch (1954) given as a gift to Gerhard Heberer in 1960 (private archive of U.H.)

morphologists explicitly referred to Goethe's principles as the source of their inspiration. Holists, like Böker, saw themselves as Goethe's disciples as well. Considering the influence of Haeckel's version of Darwinism in German lands, one can say that both Darwinian and non-Darwinian pre-Synthetic evolutionary currents were under strong influence of Goethean ideas.

4 Second Darwinian Revolution in German Lands and Its Specificity

The growth of the Evolutionary Synthesis in Germany is a complex topic. Thomas Junker in his seminal tome *The Second Darwinian Revolution* (2004) names thirty biologists who directly influenced the growth of the Modern Synthesis in Germany. Here, we pick up on two names characteristic for both the "evolution" of evolutionary theory and our specific objectives. Ludwig Plate (1852–1937) was one of the most influential pre-Synthetic (but post-Haeckelian) Darwinians (Levit and Hoßfeld 2006), whereas Bernhard Rensch was the major "architect" of the Modern Synthesis in German lands (Reif et al. 2000; Junker and Hoßfeld 2009). The choice of Plate is not an accident. Plate was not only Haeckel's immediate successor in Jena but also the greatest influence on the early Bernhard Rensch (1900–1990). For example, in the first edition of Rensch's most crucial "synthetic" book, *Neuere Probleme der Abstammungslehre* (Evolution above the Species Level) of 1947, Rensch cited his fellow "synthetic Darwinian" Ernst Mayr only five times in contrast to 23 references to Plate. Even Darwin, with 19 references, appears behind Plate in this central "synthetic" book by Rensch. In the late third edition, Rensch (1972) increases Mayr's citations to 14, but Plate is still in the lead with 18 references. Plate's example shows as well how difficult it was at that time to distinguish "proper Darwinians" from "non-Darwinians" (Delisle 2017).

Both Plate's empirical and theoretical works had an enormous impact during his lifetime and are still cited in the morphological literature (e.g. Reynolds 2002). He was translated into Russian early on (Plate 1928). Plate campaigned for a revival of the "original Darwinism" (so-called old-Darwinism) combining selectionism with neo-Lamarckian ideas and was seen by many contemporaries worldwide as a proper advocate of Darwinism (Levit and Hoßfeld 2006; Hoßfeld and Levit 2011). An American evolutionary biologist like Vernon Kellogg (1867–1937) claimed in *The American Naturalist* that Plate "takes the real standpoint of Darwin" (Kellogg 1909). A prominent Russian biologist, geographer and anti-Darwinist Leo S. Berg (1876–1950) saw Plate as his main scientific opponent (Berg 1926). The American palaeontologist Henry F. Osborn (1857–1935), who sought a compromise between selectionist and neo-Lamarckian methodologies, likewise honoured Plate with the title "prominent selectionist" (Osborn 1926). Another of Plate's contemporaries, the Swedish anti-Darwinian historian of science Erik Nordenskiöld (1872–1933) claimed that Plate's *Selektionsprinzip* (1913) contains "all that can be adduced in modern times in defence of the old Darwinism. And as its champion Plate has done a great service, thanks to his wealth of knowledge, his strong convictions, and his honesty" (Nordenskiöld 1928: 572).

Even from our contemporary view, "old-Darwinism" in its fully established and explicit form cannot be reduced to any other theoretical school. The specificity of this theory lay in combining the "standard" Darwinian factors of evolution (mutation, recombination, geographic isolation, natural selection) with the neo-Lamarckian and orthogenetic mechanisms in order to define the exact role of all these mechanisms in evolutionary process proceeding from the whole complex of biosciences including genetics. Old-Darwinians legitimately insisted that they follow the initial ideas of Darwin, who assumed some roles for Lamarckian mechanisms as well as for the auxiliary hypothesis of constraints. The very idea of combining various evolutionary mechanisms was widespread at that time within various cultural contexts (see Loison and Herring 2017).

In addition to Darwin, Haeckel and himself, Plate counted Richard Semon (1859–1919), Wilhelm Roux (1850–1924), Richard von Hertwig (1850–1937), Fritz v. Wettstein (1895–1945), Berthold Hatschek (1854–1941), Jan Paulus Lotsy (1867–1941), Franz Weidenreich (1873–1948) and the future "co-architect" of the Evolutionary Synthesis, Bernhard Rensch, amongst the old-Darwinians. According to Plate, old-Darwinism exactly followed the initial ideas of Darwin while at the same time adapting and processing all healthy and empirically verifiable scientific achievements. Plate aimed to combine all fruitful theoretical approaches (Lamarckism, selectionism, orthogenesis) with the most innovative field of experimental biology. The core of Plate's evolutionary theory can be grasped into two theses:

1. Darwinism is a "stochastic theory" taking into account variations occurring by chance in the individuals of a certain species (Plate 1913: 222).
2. However, the harmonic modification of various features is more easily conceivable from the Lamarckian standpoint (Plate 1913: 224).

In Plate's later works (Plate 1932–1938), we find all the basic factors of evolution later adapted by the Evolutionary Synthesis. Thus, Plate claimed that random mutations and recombination deliver the bulk of raw material for evolution. Natural selection and geographical isolation perform a major role in evolution (Plate 1933: 1045). Also, what is now known as "population thinking" is of great importance for Plate as he analyses the "laws of populations" with some mathematics (Plate 1933: 1047–1052). Yet, Plate also admitted other evolutionary mechanisms going beyond the basic tenets of the Synthetic Theory of Evolution. Plate accepted both macro- and directed mutations, orthogenesis and the inheritance of acquired characters. As to Plate's general "philosophical" standpoint, he "distanced himself from what he saw as the atheism and antireligious politics of monism, but not necessarily from the scientific agenda", Gliboff argues. Plate "continued to consider himself a monist, but emphasized a unity of nature that could include aspects of the divine and need not entirely exclude his Christian and Germanic identity" (Gliboff 2012).

The same strategy of combining up-to-date evolutionary biology with the monist methodology was applied by Bernhard Rensch. Towards the end of the 1930s, Bernhard Rensch (1900–1990) turned from Lamarckism and orthogenesis to selectionism and became one of the key figures in the making of the Synthetic Theory of Evolution (STE) (Levit et al. 2008b; Delisle 2008). He contributed to the "Darwinisation" of biological systematics through the criticism of various anti-Darwinian movements in the German lands, but more importantly he promoted a Darwinian macro-evolution in accord with the principles of gradualism (Reif et al. 2000; Junker and Hoßfeld 2009). In the course of time, Rensch's version of the STE developed into an all-embracing metaphysical conception based on a kind of Spinozism situated within the same tradition as Goethe's hylozoism and Haeckel' monism.

Most astonishing in Rensch's case is the continuity in topics, methodology and empirical generalisations, despite the shift in his views from the "old-Darwinian Synthesis" to the "Modern Synthesis" (Delisle 2008). This continuity in Rensch's theoretical system can be explained, in part at least, by the guiding role of general methodological principles, which underlay his entire system, explicitly or implicitly. For example, Rensch's "philosophy" became an asylum for the concept of orthogenesis which Rensch (as well as other champions of the STE) banned from his evolutionary theory. Unable to explain the directionality of evolution in terms of empirically based science, he "pre-programmed" the occurrence of human-level intelligence by a sophisticated monist philosophy combined with a supposedly naturalistic evolutionary biology. At the core of Rensch's "philosophy" is the idea that the only indisputable objects for a researcher are his or her own psychic phenomena resulting from the immediate experiences: perceptions, imaginations, feelings and thoughts. Only an analysis of these experiences makes it possible to develop concepts of extra-mental reality, which appears to be visible and testable phenomena [Gestalt]. Thanks to physics, Rensch continues, we know that matter consists of atoms, elementary particles, and waves. Finally, matter appears to be "the ultimate something", which will perhaps in the future be described only in terms of interactions of various forces, causal chains and fundamental constants. Rensch appeals to the *reductive realism* of the German philosopher and psychologist Theodor Ziehen (1862–1950), one of the most cited authors in Rensch's works, who posed the question of the suitability of "matter" as a scientific term. Ziehen is known as an author of psychophysiological epistemology (e.g. Ziehen 1898), a philosophical current akin to Haeckel's monism. Rensch did not just share the philosophy of Ziehen; he showed it was absolutely crucial for his whole theoretical system. Rensch labelled Ziehen's epistemology a "monistic principle" (Rensch 1971: 29). As any kind of philosophical monism, the "monistic principle" constitutes an ultimate, ontologically definable reality, which cannot be multiplied or decomposed into further elements, thus representing the very foundation of the Universe and providing it with the elements of an individualised whole. In other words, monism implies elements of holism. Rensch was looking for this type of universal principle (Levit et al. 2008b).

Along these lines, Rensch created his concept of *psychophysical identism* (Rensch 1988: 36). Already in the first edition of his major "synthetic" publication, the *Abstammungslehre* (Rensch 1947), written during the Second World War, Rensch "presented this world view for the first time", although at that time Rensch employed the awkward term *hylopsychism*. Again, psychological identity was not for Rensch a "philosophy" supplementing his biology; rather, it is the core of his world view and his scientific methodology. Rensch championed an all-embracing evolutionism and selectionism. Natural selection was for him the major source of lawfulness in evolution and it was "possible to characterize evolutionary regularities [*Regelhaftigkeiten*] as laws [*Gesetzlichkeiten*]" (Rensch 1991: 107). Considering that natural selection continuously selects the better variants, evolution appears to be channelled into tight pathways, i.e. inevitably proceeds in certain directions. Here, we can observe a long echo of Ludwig Plate's concept of "orthoselection". Although elementary evolutionary events appear to be random, evolution towards human intelligence and evolution of intelligence itself takes place along invisible rails.

Rensch's view on the inevitability of evolution towards human-level intelligence is in sharp contrast to most other leading Synthetic and "post-Synthetic" evolutionists (T. Dobzhansky, G. Simpson, F. Ayala, E. Mayr, J. Monod and many others), which tend to claim that "there is no indication in the geological record that the evolution of intelligence is at all inevitable" (Barrow and Tipler 1986: 133). Rensch, on the contrary, insisted that the origin of humans from ape-like ancestors "was presumably a lawfully determined [*gesetzmäßig bedingter*] process" (Rensch 1991: 225). He did not reduce his concept of "lawfulness" to vulgar determinism and instead coined the notion of "polynomic determination". Polynomic determination implies that the whole range of biological, physical, chemical, social and other natural laws control the entire process of evolution. These overlap and interact, bringing about seemingly stochastic events, which, in fact, can be explicated in terms of lawful processes. In other words, for him randomness does not play the central role in organic and cultural evolution, such that life would certainly occur on other planets with comparable chemical-physical conditions and evolve in a comparable way to evolution here on Earth (Rensch 1991: 108).

The combination of anthropocentric progressionism and pantheism championed by Rensch along with selectionism was certainly in sharp contrast to what other influential selectionists thought. Rensch's major trick was to make consciousness into the imaginary object of pre-organic determined evolution and natural selection. However, the postulation of the pre-phenomenal nature of matter had, as a corollary, that every particle of perceivable reality became supplied with a tiny particle of intelligence. As intelligence is now an essential attribute of the Universe, the evolution of the Universe implies a "pre-programmed" movement in the direction of human-like intelligence. To make his concept compatible with the natural-scientific world view, Rensch concealed this obviously teleological concept into the concept of universal selectionism because selectionism was already widely accepted to be a respectable "teleology-free" concept. In other words, Rensch's anthropocentric determinism is dressed up as universal selectionism. Yet, what he did was to fragment teleology and place it into the interior of things, thus making it subject to natural selection. It is sophisticated and camouflaged but still a kind of typology very close to Mayr's notion of selectionism: there is an "interior of things" (to use the expression of Teilhard de Chardin), the hidden side of evolution getting explicit on its last evolutionary stages. Finally, as a proponent of a "cosmic view of evolution" (Delisle 2009), Rensch championed a kind of holism as well. Overall, Rensch developed his theoretical principles along the methodological principles developed by Goethe and Haeckel, although on a qualitatively different theoretical level.

5 Comparative Remarks and Conclusions

Leading German evolutionists were persistent in applying the three basic principles introduced by Goethe—Monism, Typology and Holism—to their theoretical systems. These principles were not unknown in Russia and English-speaking countries (see Esposito 2017) but did not play such a prominent role as they played in the German lands. It was a set of interconnected methodological principles that survived over centuries despite of political revolutions and paradigmatic shifts in science. German-, Russian- and English-speaking traditions were literally "infected" by holistic thinking, but the roots, theoretical context and, correspondingly, methodological consequences of holism were different in different language realms. For example, Russian holism was initially tightly coupled with both the Darwinian and non-Darwinian traditions but ultimately allied with the mainstream Darwinism. The roots of Russian holism are in the environmental thinking, in approaching organisms as parts of ecosystems. One can detect this bias beginning with early Darwinians and ending with mature representatives of the Modern Synthesis such as Schmalhausen and Timofeeff-Ressovsky (Levit et al. 2006; Levit and Hoßfeld 2009). Russian holism is an outcome of empirical studies, which demonstrated close connection between organisms and their environments. An "environmentalist holism" was characteristic for both Darwinian and non-Darwinian doctrines. Leo Berg's "Nomogenesis" accompanied by the theory of landscape zones may serve as an example. In its extreme manifestation, Russian holistic tradition brought about the biosphere theory. Note that there is only one biosphere on Earth and, maybe, in the whole universe and so in describing the biosphere and its evolution, one is not in the search for a certain "type" (Levit and Hoßfeld 2005). German evolutionary holism (Haeckel, Meyer-Abich, Böker, etc.) takes its roots in typological thinking developed by Goethe. The objective of the German-language tradition was to describe essential morphological features of a certain class of phenomena, i.e. a type (archetype). This explains why German holism in German lands allied with non-Darwinian theories and ultimately developed into a self-sustaining theoretical current in biology.

The second German bias in evolutionary biology, monism, was well known in both Russia and English-speaking countries but found a few followers.

Characterising Haeckel's "Monistic Creed", his contemporary, a zoologist of Stanford University, David Starr Jordan (1851–1931) wrote in *Science*:

> I have myself not the slightest objection to 'Monism' as philosophy. As a dogma it is certainly more attractive than many others which have been brought like lightning from the clouds, as a stimulus to creeping humanity. My objection lies against the use of the divining rod in connection with the microscope. These instruments do not yield homologous results (Jordan 1895).

Jordan's publication was a response to the book review published in *Science* by William Keith Brooks (1848–1908) of Johns Hopkins University under the title *The Tyranny of the Monistic Creed*. The latter claimed that "the monistic confession of faith has led to the discounting of the possibilities of future discovery, and that it has thus obstructed progress" (Brooks 1895). Jordan and Brooks expressed the critical position of many American natural scientists towards monism. Monism had even less currency in Britain:

> Here the term was used by philosophers and psychologists in the sense of mind and matter being two sides of the same coin. But this usage does not seem to have been taken up outside certain academic circles, and many associated debates over religious issues took place without the actual term 'monism' being used (Bowler 2012).

Certainly, monism had some influence within US American and British philosophy. A German–American champion of panpsychism, Paul Carus (1852–1919), the founding editor (from 1888 till 1919) of the journal *The Monist*, generated discussions around monism in the English-speaking world. Yet, Carus cannot serve as an example of a direct influence of monism of the Haeckelian kind on mainstream biological doctrines. As Peter Bowler summarised it:

> Haeckel's naturalistic monism was certainly discussed in Britain and America, especially following the translation of his *Riddle of the Universe* in 1900. But one gets the distinct impression that even some of those who discussed it sympathetically wanted to divorce their version of monism from Haeckel's strident naturalism (Bowler 2012).

The concept of monism, Bowler concludes, is not a very useful category for analysing British intellectual life around 1900 (Bowler 2012). Monist currents in Russian evolutionary biology were arguably even less visible. Igor Polyanski (Polyanski 2012) maintains that monism was central for politically relevant philosophical debates in Russia and that Vladimir I. Lenin (1870–1924) devoted his famous *Materialism and Empiriocriticism* (Lenin 1909) predominantly to the criticism of the Russian *empiriomonism* of Alexander Bogdanov (1873–1928). Bogdanov, in turn, was presumably quite familiar with Haeckel's major works; there is no direct evidence that he read Haeckel, but the indirect evidence is compelling (Adams 1989). Yet, the influence of empiriomonism is overestimated. In his *Materialism and Empiriocriticism*, Lenin names Bogdanov explicitly: "I personally know so far of only one empirio-monist in literature—a certain A. Bogdanov".[1] Lenin's criticism of Bogdanov was a reflection on an internal

[1] https://www.marxists.org/archive/lenin/works/1908/mec/four5.htm

discussion of a quite marginal political current (the RSDRP). On further inspection, in the recently published comparative history of evolutionary theory by Edouard I Kolchinsky, there are no references to Bogdanov at all (Kolchinsky 2014).

Despite terminological influences, Bogdanov's empiriomonism is crucially different from Haeckelian monism, since it represented an amalgam of theoretical elements borrowed from Spencer, Haeckel, Ostwald and especially Marx (Krementsov 2011). At the core of Boganov's epistemology is Marxist materialism: "To Bogdanov, the school of Empiriocriticism was not a rejection of materialism but rather its revival. Materialism, he maintained, had lost its original meaning, the concept of matter having shed its sensuous nature and having turned into a vague abstraction" (Bolls 1981).

Including some Haeckel into a theoretical framework was not something extraordinary. After all, Haeckel's major works were translated into Russian and Haeckelian monism was well known and relatively widespread: "By the 1890s, many Russian naturalists had adopted Haeckelian monism as a basic philosophy of their investigations" (Krementsov 2011: 35). The problem, however, is that a large number of these "many naturalists" were marginalised for evolutionary biology or had a status of exotic "visionaries". The panpsychistic monism of the rocket scientist Konstantin Tsiolkovski (1857–1935) can serve as an example. Being a school teacher in a provincial town, he had a few possibilities to influence the Russian intellectual landscape. Besides, a closer look at Tsiolkovski's theoretical heritage shows that his monism, although certainly influenced by Haeckel, differs from it in crucial points. For example, an antireligious motivation of Haeckelian naturalism is plainly incompatible with Tsiolkovsky's deism (Lytkin 2012).

Leading evolutionists such as Alexander Kowalevsky (1840–1901), Elie Metschnikoff (1845–1916) or Kliment Timiryazev (1843–1920) were critical not only of Haeckel's monism but also of Haeckel's speculative theories in general. Haeckel, by contrast, appreciated Kowalevsky's work very much. In his *Anthropogenie*, Haeckel wrote: "The most significant germ histories in the recent time were those of Kowalevsky" (Haeckel 1874: 49). It is astonishing in this respect that both Kowalevsky and Metschnikoff were either indifferent or even hostile to Haeckel and his theories. There was no letter exchange between Kowalevsky and Haeckel as there was between Darwin and Haeckel. The Archive of the Ernst Haeckel House in Jena holds not a single letter to Haeckel, neither from Kowalevsky nor from Metschnikoff, although there are more than 100 letters from other Russian correspondents in the Archive (Hoßfeld and Breidbach 2005). This is even more curious considering that Kowalevsky's younger brother, Vladimir, undertook his doctoral work under Haeckel's supervision (Uschmann 1956) and that Haeckel's Gastraea theory was to a significant extent based on Kowalevsky's data. In the 185 letters from Kowalevsky to Metchnikov, we find only seven short mentions of Haeckel (Gaisinovich 1974).

To summarise, Haeckel's monism, as well as other versions of German monism (such as Ostwald's), was well known in the pre-revolutionary Russia, but its relatively marginal place in Russian evolutionary biology cannot be compared to its centrality in Germany. As to the third Goethean principle (typology), it is well

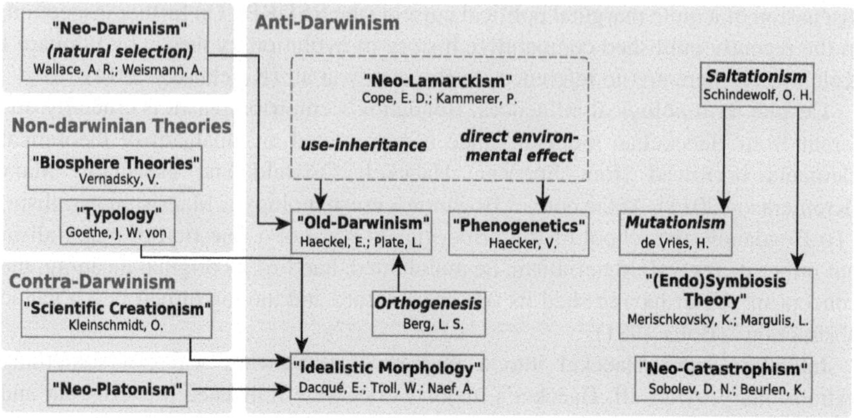

Fig. 7 Scheme: The first decades of the twentieth century became the heyday of alternative evolutionary theories (non- and anti-Darwinian), as they flourished simultaneously in various countries and achieved clarity and conceptual maturity

known that typology under different names (e.g. idealistic morphology) was widespread in German lands before the First World War and even after the Second World War. While English-speaking countries experienced the rapid expansion of the Evolutionary Synthesis, the growth of evolutionary theory in Soviet Union and, partly, in East Germany was distorted by the political repressions associated with the infamous name of Trofim D. Lysenko (1898–1976) and his partisans (Roll-Hansen 2005; Hoßfeld and Olsson 2002) (Fig. 7).

Yet, there was a strong scientific opposition to the Evolutionary Synthesis also in West Germany, which enjoyed all the democratic liberties of the post-war period. Basic principles of the Modern Synthesis became well known in Germany, simultaneously with Great Britain and the USA (Reif et al. 2000). Nevertheless, when Ernst Mayr reported on the "Phylogenetic Symposium" in Hamburg (1956), where he presented the basic principles of the Evolutionary Synthesis, he noted that "all those attending (with the exception of the geneticist de Lattin) argued against the Synthesis" (Mayr 1999). Answering the question "Why then was there so much opposition in Germany?", Mayr gave several reasons. The first reason was the typological or idealistic-morphological tradition, which was in Germany much stronger than in English- or Russian-speaking scientific traditions. The second reason was the "preoccupation of German zoology with phylogeny", which was again connected with the fact that "the students of phylogeny almost without exception adhered to the idealistic morphological philosophy". The third reason was, according to Mayr, the general ignorance of modern genetics by the German biologists, which was again related to the adherence of German morphologists and palaeontologists to the "typological saltationism" (Mayr 1999). In other words, all reasons for the anti-Synthetic resistance in Germany listed by Mayr are ultimately rooted in one and the same theoretical movement: idealistic morphology, which he equates with typology and essentialism.

Alone Mayr's astonishment at the situation in Germany shows that German influence of typological methods in German evolutionary biology was unusually strong. Indeed, already in Darwin's time, typology was not a burning point in British biology: "Because ideal morphology had made so little headway in Great Britain in Darwin's day, Darwin was able to ignore this alternative" (Hull 2010: 135). Idealistic morphology remained barely influential in Great Britain and the USA during both Darwinian revolutions, while it was strikingly prominent in Germany:

> The evolutionary discussions between the German speaking paleontology and biology were interrupted after 1943 [...]. Typological thinking persisted in applied paleontology as well as in systematics and stratigraphy. Therefore there was no reason to study the issue of population genetics. [...] Even after 1945 the German speaking paleontologists continued to exist within a paradigm interpreting evolution as a self-legislating [*eigengesetzlicher*], holistic process. They were unable to change the paradigm (Reif 1999, our translation).

It is not an accident that a prominent historian of German biosciences, Wolf-Ernst Reif (1945–2009), mentioned "typology" and "holism" in one breath. The rigidity of the "paradigm" Reif described was due to the interrelated set of methodological principles highlighted by Goethe, which included typology, holism and monism. In order to emphasise its temporal durability and theoretical complexity, we label this phenomenon a "metaparadigm" (Levit and Hoßfeld 2013).

Acknowledgements This chapter has benefited from the editorial corrections and comments made by Dr. Ian Stewart, University of King's College, Halifax, Canada.

References

Adams MB (1989) "Red Star" another look at Aleksandr Bogdanov. Slavic Rev 48(1):1–15
Barrow J, Tipler F (1986) The anthropic cosmological principle. Clarendon Press, Oxford
Berg LS (1926) Nomogenesis or evolution determined by law. Constable, London
Boeing G (2016) Visual analysis of nonlinear dynamical systems: chaos, fractals, self-similarity and the limits of prediction. Systems 4:37. https://doi.org/10.3390/systems4040037
Böker H (1924) Begründung einer biologischen Morphologie. Z Morphol Anthropol 24:1–22
Böker H (1932) Goethe und die Anatomie. Munch Med Wochenschr 79:457–461
Böker H (1935/1937) Einführung in die vergleichende biologische Anatomie der Wirbeltiere. Gustav Fischer Verlag, Jena, vol I, II
Boll MM (1981) From empiriocriticism to empiriomonism: the Marxist phenomenology of Aleksandr Bogdanov. Slavon East Eur Rev 59(1):41–58
Bowler PJ (1988) The non-Darwinian revolution. Reinterpreting a Historical Myth. Johns Hopkins University Press, Baltimore
Bowler PJ (2012) Monism in Britain: biologists and the Rationalist Press Association. In: Weir TH (ed) Monism: science, philosophy, religion, and the history of a worldview. New York, Palgrave-Macmillan, pp 179–196
Bowler PJ (2017) Alternatives to Darwinism in the early twentieth century. In: Delisle RG (ed) The Darwinian tradition in context: research programs in evolutionary biology. Springer, Cham, pp 195–218
Brooks WK (1895) The Tyranny of the monistic creed, a review. Science 1(14):382–384
Bulman WJ, Ingram RG (2016) God in the enlightenment. Oxford University Press, Oxford
Buskell A (2017) What are cultural attractors? Biol Philos 32(3):377–394
Coleman W (1976) Morphology between type concept and descent theory. J Hist Med 31:149–175

Darwin C (1859) On the origin of species by means of natural selection. John Murray, London
Delisle R (2008) Expanding the framework of the holism/reductionism debate in neo-Darwinism: the case of Theodosius Dobzhansky and Bernhard Rensch. Hist Philos Life Sci 30(2):207–226
Delisle R (2009) The uncertain foundation of neo-Darwinism: metaphysical and epistemological pluralism in the evolutionary synthesis. Stud Hist Philos Biol Biomed Sci 40:119–132
Delisle RG (2017) From Charles Darwin to the evolutionary synthesis: weak and diffused connections only. In: Delisle RG (ed) The Darwinian tradition in context: research programs in evolutionary biology. Springer, Cham, pp 133–168
Dennett D (1995) Darwin's dangerous idea. Penguin Group, London
Di Gregorio MA (1995) A wolf in sheep's clothing: Carl Gegenbaur, Ernst Haeckel, the vertebral theory of the skull, and the survival of Richard Owen. J Hist Biol 28:247–280
Esposito M (2017) The organismal synthesis: holistic science and developmental evolution in the English-speaking world, 1915–1954. In: Delisle RG (ed) The Darwinian tradition in context: research programs in evolutionary biology. Springer, Cham, pp 219–242
Gaisinovich AE (ed) (1974) I.I. Metchnikoff: Pisma (1863–1916). Nauka, Moscow
Ghiselin MT (1969) The triumph of the Darwinian method. University of Chicago Press, Chicago
Ghiselin MT (1982) The failure of morphology to assimilate Darwinism. In: Mayr E, Provine W (eds) The evolutionary synthesis. Belknap Press, Cambridge
Gliboff S (2012) Monism and morphology at the turn of the 20th century. In: Weir TH (ed) Monism: science, philosophy, religion, and the history of a worldview. New York, Palgrave-Macmillan, pp 135–158
Goethe JW (1795) Erster Entwurf einer allgemeinen Einleitung in die vergleichende Anatomie. Sämtliche Werke 12:120–155
Goethe JW (1950) Sämtliche Werke. Die Faustdichtungen. Artemis, Zürich, Bd 5
Goethe JW (1988) The metamorphosis of plants. In: Miller D (ed) Scientific studies. Suhrkamp, New York
Goethe JW (2006) Philosophische und Naturwissenschaftliche Schriften. Elibron Classics, Chestnut Hill
Grim J, Tucker ME (2014) Ecology and religion. Island Press, Washington
Haeckel E (1866) Generelle Morphologie der Organismen. 2 Bde. I. Allgemeine Anatomie der Organismen. II: Allgemeine Entwickelungsgeschichte der Organismen. Verlag von Georg Reimer, Berlin
Haeckel E (1868) Natürliche Schöpfungsgeschichte. Verlag von Georg Reimer, Berlin
Haeckel E (1872) Die Kalkschwämme. Eine Monographie, 3 Bde. Georg Reimer, Berlin
Haeckel E (1874) Anthropogenie oder Entwickelungsgeschichte des Menschen. Wilhelm Engelmann, Leipzig
Haeckel E (1892) Der Monismus als Band zwischen Religion und Wissenschaft. Glaubensbekenntnis eines Naturforschers. Strauss, Bonn
Haeckel E (1904) Die Lebenswunder. Gemeinverständliche Studien über Biologische Philosophie. Ergänzungsband zu dem Buche über die Welträthsel. Alfred Kröner Verlag, Stuttgart
Haeckel E (1917) Kristallseelen: Studien über das anorganische Leben. Kröner Verlag, Leipzig
Heitler W (1998) Goethean science. In: Seamon D, Zajonc A (eds) Goethe's Way of Science. State University of New York Press, New York
Hopwood N (2006) Pictures of evolution and charges of fraud Ernst Haeckel's embryological illustrations. Isis 97:260–301
Hoßfeld U (2002) Konstruktion durch Umkonstruktion: Hans Bökers vergleichende biologische Anatomie der Wirbeltiere. Verh Gesch Theor Biol 9:149–169
Hoßfeld U (2010) Ernst Haeckel. Biographienreihe absolute. Orange Press, Freiburg i. Br
Hoßfeld U (2016a) Geschichte der biologischen Anthropologie in Deutschland. 2. Aufl. Franz Steiner Verlag, Stuttgart
Hoßfeld U (2016b) 150 Jahre Haeckel'sche Biologie. Blätter zur Landeskunde Thüringens, Landeszentrale für politische Bildung Erfurt, Heft 114
Hoßfeld U, Breidbach O (2005) Haeckel-Korrespondenz. VWB-Verlag, Berlin

Hoßfeld U, Junker T (1998) Dietrich Starck zum 90. Geburtstag. Int Z Gesch Ethik Naturwiss Tech Med 6(3):129–147

Hoßfeld U, Junker T (1999) Morphologie und Synthetische Theorie. In: Junker T, Engels E-M (eds) Die Entstehung der Synthetischen Theorie: Beiträge zur Geschichte der Evolutionsbiologie in Deutschland 1930–1950. zgl. Verhandlungen zur Geschichte und Theorie der Biologie 2, VWB-Verlag, Berlin, S 227–240

Hoßfeld U, Levit GS (2011) Ludwig Plate: Haeckel Nachfolger mit vielen Facetten. Bio unserer Zeit 41(6):412–413

Hoßfeld U, Olsson L (2002) From the modern synthesis to Lysenkoism, and back? Science 297 (5578):55–56

Hoßfeld U, Olsson L (2003a) The road from Haeckel. The Jena tradition in evolutionary morphology and the origin of "Evo-Devo". Biol Philos 18(2):285–307

Hoßfeld U, Olsson L (2003b) The history of comparative anatomy in Jena – an overview. Theor Biosci 122(2/3):109–126

Hoßfeld U, Levit GS, Olsson L (2016) Haeckel reloaded: 150 Jahre "Biogenetisches Grundgesetz". Bio unserer Zeit 46(3):190–195

Hull DL (2010) Science as a process: an evolutionary account of the social and conceptual development of science. The University of Chicago Press, Chicago

Jahn I (ed) (1998) Geschichte der Biologie. Gustav Fischer Verlag, Jena

Jordan DS (1895) Haeckel's Monism. Science 1(22):608–610

Junker T (2004) Die zweite Darwinsche Revolution. Geschichte des synthetischen Darwinismus in Deutschland 1924 bis 1950. Basilisken Presse, Marburg

Junker T, Hoßfeld U (2009) Die Entdeckung der Evolution: Eine revolutionäre Theorie und ihre Geschichte. 2. Aufl. Wissenschaftliche Buchgesellschaft, Darmstadt

Kanaev II (1970) Geothe as a naturalist. Nauka, Leningrad. (in Russian)

Kellogg V (1909) Poulton and plate on evolution. Am Nat 43(509):317–320

Kleeberg B (2005) Theophysis: Ernst Haeckels Philosophie des Naturganzen. Böhlau Verlag, Köln (u.a.)

Kolchinsky E (2014) Evolutionary theory unity in the 20th century divided world. Nestor-Historija, St. Petersburg

Krementsov N (2011) A Martian stranded on earth: Alexander Bogdanov, blood transfusions, and proletarian science. Chicago University Press, Chicago

Kuhn T (2012) The structure of scientific revolutions, 50th anniversary edition. University of Chicago Press, Chicago

Kutschera U (2016) Haeckel's 1866 tree of life and the origin of eukaryotes. Nat Microbiol 1:16114

Lenin VI (1909) Materialism and empirio-criticism. Critical comments on a reactionary philosophy. Zveno, Moscow. (in Russian)

Levit GS (2007) The roots of Evo-Devo in Russia: is there a characteristic "Russian tradition"? Theor Biosci 4:131–148

Levit G, Hoßfeld U (2005) Die Nomogenese: Evolutionstheorie jenseits von Darwinismus und Lamarckismus. Verh Gesch Theor Biol 11:367–388

Levit GS, Hoßfeld U (2006) The forgotten "Old Darwinian" synthesis: the evolutionary theory of Ludwig H. Plate (1862–1937). NTM 14:9–25

Levit GS, Hoßfeld U (2009) From molecules to the biosphere: Nikolai V. Timoféeff-Ressovsky's (1900–1981) research program within the Totalitarian landscapes. Theor Biosci 128 (4):237–248

Levit GS, Hoßfeld U (2013) A bridge-builder: Wolf-Ernst Reif and the Darwinisation of German palaeontology. Hist Biol 25(2):297–306

Levit GS, Meister K (2006a) "Methodological ideologies" in the German-language morphology. Jahrb Eur Wiss 2:35–62

Levit GS, Meister K (2006b) The history of essentialism vs. Ernst Mayr's "essentialism story": a case study of German idealistic morphology. Theor Biosci 124(3–4):281–307

Levit GS, Hoßfeld U, Olsson L (2006) From the "modern synthesis" to cybernetics: Ivan Ivanovich Schmalhausen (1884–1963) and his research program for a synthesis of evolutionary and developmental biology. J Exp Zool B Mol Dev Evol 306B(2):89–106

Levit GS, Meister K, Hoßfeld U (2008a) Alternative evolutionary theories: a historical survey. J Bioecon 10(1):71–96

Levit GS, Simunek M, Hoßfeld U (2008b) Psychoontogeny and psychophylogeny: the selectionist turn of Bernhard Rensch (1900–1990) through the prism of panpsychistic identism. Theor Biosci 127:297–322

Levit GS, Hoßfeld U, Olsson L (2014) The Darwinian revolution in Germany: from evolutionary morphology to the modern synthesis. Endeavour 38(3–4):268–278

Levit GS, Hoßfeld U, Reinhold P (2016) Meilensteine aus 200 Jahren *Thierarzneykunst* in Jena (1816–2016). DVG Verlag, Gießen

Loison L, Herring E (2017) Lamarckian research programs in French biology (1900–1970). In: Delisle RG (ed) The Darwinian tradition in context: research programs in evolutionary biology. Springer, Cham, pp 243–270

Lytkin VV (2012) Natural-scientific basis of forming "the cosmic philosophy" by K. E. Tsiolkovsky. Mir mauki, kultury, obraazovanija 2(33):452–454 (in Russian with Engl. abstr.)

Mayr E (1982) The growth of biological thought. The Belknap Press of Harvard University Press, Cambridge

Mayr E (1999) Thoughts on the evolutionary synthesis in Germany. In: Junker T, Engels E-M (eds) Die Entstehung der Synthetischen Theorie: Beiträge zur Geschichte der Evolutionsbiologie in Deutschland. VWB-Verlag, Berlin, pp 19–30

Meyer-Abich A (1949) Biologie der Goethe-Zeit. Hippokrates-Verlag, Stuttgart

Meyer-Abich A (1970) Die Vollendung der Morphologie Goethes durch Alexander von Humboldt. Ein Beitrag zur Naturwissenschaft der Goethezeit. Vandenhoeck u. Ruprecht, Göttingen

Niklas K, Kutschera U (2016) From Goethe's plant archetype via Haeckel's biogenetic law to plant Evo-Devo. Theor Biosci 136(1-2):49–57

Nordenskiöld E (1928) The history of biology. Tudor, New York

Nöthlich R, Weber H, Hoßfeld U, Breidbach O, Krauße E (2006) "Substanzmonismus" und/oder "Energetik": Der Briefwechsel von Ernst Haeckel und Wilhelm Ostwald (1910–1918). Zum 100. Jahrestag der Gründung des Deutschen Monistenbundes. VWB, Berlin

Olsson L, Levit GS, Hoßfeld U (2017) The "biogenetic law" in zoology: from Ernst Haeckel's formulation to current approaches. Theor Biosci 136(1–2):19–29

Osborn HF (1926) Modern aspects of evolution. Nature 118:617–618

Plate L (1913) Selektionsprinzip und Probleme der Artbildung. Ein Handbuch des Darwinismus, 4. Auflage. Verlag von Wilhelm Engelmann, Leipzig

Plate L (1928) Evoliutzionnaja teorija (Kompanejz: Perevod NS). Gosudarstvennoje Izdatelstvo, Moskva-Leningrad

Plate L (1932) Vererbungslehre. Mit besonderer Berücksichtigung der Abstammungslehre und des Menschen. Mendelismus. Gustav Fischer Verlag, Jena, Bd I

Plate L (1933) Vererbungslehre. Mit besonderer Berücksichtigung der Abstammungslehre und des Menschen. Sexualität und Allgemeine Probleme. Gustav Fischer Verlag, Jena, Bd II

Plate L (1938) Vererbungslehre. Mit besonderer Berücksichtigung der Abstammungslehre und des Menschen. Spezielle Genetik einiger Nager. Gustav Fischer Verlag, Jena, Bd III

Polyanski I (2012) Between Hegel and Haeckel. Monistic worldview, Marxist philosophy and biomedicine in Russia and East Germany? In: Weir TH (ed) Monism: science, philosophy, religion, and the history of a worldview. Palgrave-Macmillan, New York pp, pp 197–222

Reif W-E (1983) Evolutionary theory in German paleontology. In: Green M (ed) Dimensions of Darwinism. Cambridge University Press, Cambridge

Reif W-E (1999) Deutschsprachige Paläontologie im Spannungsfeld zwischen Makroevolution und Neodarwinismus (1920–1950). Verh Theor Biol 2:151–188

Reif W-E, Junker T, Hoßfeld U (2000) The synthetic theory of evolution: general problems and the German contribution to the synthesis. Theor Biosci. 119:41–91

Rensch B (1947) Neuere Probleme der Abstammungslehre. Die Transspezifische Evolution. Ferdinand Enke Verlag, Stuttgart

Rensch B (1971) Biophilosophy. Columbia University Press, New York
Rensch B (1972) Neuere Probleme der Abstammungslehre. Die Transspezifische Evolution, 3. Aufl. Ferdinand Enke Verlag, Stuttgart
Rensch B (1988) Probleme genereller Determiniertheit allen Geschehens. Parey, Berlin
Rensch B (1991) Das universale Weltbild: Evolution und Naturphilosophie, 2. Auflage. WBG, Darmstadt
Reynolds PD (2002) The scaphopoda. Adv Mar Biol 42:137–236
Richards RJ (2002) The romantic conception of life: science and philosophy in the age of Goethe. The University of Chicago Press, Chicago
Rieppel O (2011) Wilhelm Troll (1897–1978): idealistic morphology, physics, and phylogenetics. Hist Philos Life Sci 33(2011):321–342
Rieppel O (2012) Adolf Naef (1883–1949), systematic morphology and phylogenetics. J Zool Syst Evol Res 50:2–13
Rieppel O (2016) Phylogenetic systematics. Haeckel to Hennig. CRC Press, Boca Raton
Rieppel O, Williams DM, Ebach MC (2013) Adolf Naef (1883–1949): on foundational concepts and principles of systematic morphology. J Hist Biol 46(3):445–510
Roll-Hansen N (2005) The Lysenko effect: the politics of science. Humanity Books, New York
Starck D (1965) Vergleichende Anatomie der Wirbeltiere von Gegenbaur bis heute. Verh Deutsch Zool Ges Jena 1965:51–67
Starck D (1977) Tendenzen und Strömungen in der vergeleichenden Anatomie der Wirbeltiere im 19. Und 20. Jahrhundert. Nat Mus 197:93–120
Starck D (1980) Die idealistische Morphologie und ihre Nachwirkungen. Med J 15:44–56
Tantillo AO (2002) The will to create: Goethe's philosophy of nature. University of Pittsburgh Press, Pittsburgh
Todes DP (1987) Darwin's Malthusian metaphor and Russian evolutionary thought, 1859–1917. Isis 78(4):537–551
Uschmann G. (1956) Die promotion von W.O. Kowalevsky in Jena, ein Beitrag zur Geschichte der evolutionistischen Paläontologie. EHH, G. Uschmanns Nachlass B 11:15–17 (Hefter: Vorträge III)
Weber H, Hoßfeld U (2006) Stichwort "Monismus". Nat Rundsch 59(9):521–522
Wells GA (1967) Goethe and the intermaxillary bone. Br J Hist Sci 3(4):348–361
Williams DM, Ebach MC (2008) Foundations of systematics and biogeography. Springer, New York
Winsor M (2006) The creation of the essentialism story: an exercise in metahistory. Hist Philos Life Sci 28:149–174
Ziehen T (1898) Psychophysiologische Erkenntnistheorie. Gustav Fischer, Jena

Alternatives to Darwinism in the Early Twentieth Century

Peter J. Bowler

Abstract Julian Huxley claimed that the period around 1900 experienced an 'eclipse of Darwinism' when natural selection was rejected in favour of alternative mechanisms of evolution. These included the Lamarckian theory of the inheritance of acquired characteristics and the belief that variation was directed by forces internal to the organism. Mendelism undermined the credibility of these earlier ideas, but they survived in some areas of biology well into the twentieth century. Mendelism itself derived in part from the theory of evolution by sudden, discrete jumps or saltations.

This chapter describes these non-Darwinian theories and notes the relationships between them. It also identifies the motivations that encouraged biologists to prefer them and describes the evidence they presented. The role of the debate over 'form' and 'function' is stressed, along with the suggestion that much of the debate was driven by disputes over the nature of variation and its role in evolution. The bulk of the chapter consists of a detailed outline of the ways in which the non-Darwinian theories survived into the early twentieth century.

Keywords Acquired characters • Lamarckism • Mutation theory • Neo-Lamarckism • Non-adaptive evolution • Orthogenesis • Saltationism

1 Introduction

The evolutionary debates of the late nineteenth century had been conducted using evidence primarily from areas such as morphology, palaeontology and field studies. Major disagreements had arisen over how the process of development had occurred. Darwin's own thinking contained components that could be developed in different ways (see Delisle 2017), but the theory of natural selection came under increasing pressure from a number of alternatives, initiating what Julian Huxley (1942: 22–26) called an 'eclipse of Darwinism' (Bowler 1983). Around 1900, the increasing level

---

P.J. Bowler (✉)
School of History and Anthropology, Philosophy and Politics, Queen's University, Belfast, Northern Ireland, UK
e-mail: p.bowler@qub.ac.uk

R.G. Delisle (ed.), *The Darwinian Tradition in Context*,
https://doi.org/10.1007/978-3-319-69123-7_9

of experimental work in the life sciences, sometimes referred to as the 'revolt against morphology', intensified the crisis (Allen 1975). The new science of genetics initially added to the problems facing all of the alternatives, Darwinism included.

With hindsight we know that the dispute between the early geneticists and the Darwinians would ultimately be resolved, leading to what Huxley (1942) called the 'modern synthesis' (Mayr and Provine 1980). The difficulties facing alternatives such as the Lamarckian theory of the inheritance of acquired characteristics gradually intensified. But in the first decade of the twentieth century, no one could have predicted the outcome of the debate, and authoritative surveys such as that of Kellogg (1907) still argued that the Darwinian theory faced serious problems. Non-Darwinian ideas continued to play a significant role into the 1930s, especially in those areas of the life sciences least influenced by genetics. By the 1950s, the critics of Darwinism had been largely marginalized, although outside the scientific community there was increased opposition on both religious and ideological grounds. Depew (2017) shows how non-Darwinian ideas influenced the development of the modern synthesis, while this chapter focuses on those naturalists who still saw non-Darwinian mechanisms as the primary cause of evolution.

When the author of this chapter first began to study the 'eclipse of Darwinism' in the 1980s, the triumph of the modern Darwinian theory made it easy to dismiss the alternatives as blind alleys into which scientists had been led temporarily. It was necessary to argue that, right or wrong, those theories had played so significant a role that historians who ignored them would not produce a balanced view of how evolutionism actually developed. Hindsight was not a valid reason for dismissing non-Darwinian theories as a trivial side issue.

In recent decades, our interpretation of this episode has been transformed by the emergence of evolutionary developmental biology. This has reopened issues once marginalized by genetics and the modern Darwinian synthesis. Some enthusiasts see 'evo-devo' as reintroducing a role for non-selectionist factors such as Lamarckism, while even those sceptical of this view acknowledge that the older theories were not as wide of the mark as was once claimed. Esposito (2017) traces some of the developments that have prefigured the rise of evo-devo (see also Gissis and Jablonka 2011; Laublichler and Maienschein 2007).

2 Conceptual Issues

There were three major non-Darwinian positions: the Lamarckian theory of the inheritance of acquired characteristics, orthogenetic theories based on the notion of directed variation, and saltationist theories which assumed that new species appeared suddenly through discontinuous 'leaps'. But this simple division conceals a multitude of complexities and it will be useful to identify the key conceptual issues over which the protagonists of the theories disagreed, both with the Darwinians and among themselves.

Two crucial areas can be identified. The first centres on whether evolution is directed by external factors such as adaptation to the environment, or by internal forces directing individual variation in a manner independent of external constraints. This issue interacts with a second disagreement over the nature of the variations seen as the raw material of evolution. Are those variations directed or undirected, and if they are directed is that direction controlled by factors internal or external to the organism? The debates were often perceived as a dispute over the role of form and function in the shaping of the organism's morphology (Russell 1916). If internally controlled variation directs evolution, it will determine form irrespective of the demands of functional adaptation. If adaptation is crucial, form must follow the demands of function. These disagreements were perceived differently in rival research traditions and the various national intellectual and scientific contexts (as shown for instance in Levit and Hossfeld 2017; Loison 2017).

Darwin assumed that variation is undirected in the sense that a range of slight modifications are available within the population. Something causes individuals to differ among themselves, but he did not believe that the cause predetermined the appearance of only one (or a small number) of new characters. The sheer width of variation available in a population meant that this factor could not in itself direct evolution—evolution is more or less open-ended. The only way that a direction can be imposed is by selection. Some external factor—the environment or the human breeder—allows only certain variants to breed and suppresses all the rest. Natural evolution is adaptive because only variants fitted to the environment will pass their characters on to future generations.

Lamarckism and orthogenesis both imply that variation is directed along determined channels but disagree on whether the direction is imposed by factors external or internal to the organism. Lamarckism sees new characters acquired by the organism in the course of its life as evolutionarily significant variation—significant precisely because they can be passed on to the next generation. A Darwinian or a geneticist could in principle accept that individuals can acquire new characters but would dismiss them as irrelevant because they cannot be inherited. Lamarckians took it for granted that the new characters they focused on were developed by the organism as it accommodated itself to the demands of the environment, as in the popular if misleading image of the giraffe stretching its neck. Evolution was necessarily adaptive, so Lamarckism and Darwinism supplied alternative explanations of the same phenomenon, although each had particular types of adaptation it found easier to accommodate.

This superficial agreement should not blind us to the deep conceptual gulf between the two positions. For Lamarckians, variation was conceived as an addition to the development of the organism, a new stage added in the adult phase of life. To be inherited the acquired character had to be pushed back into the process of ontogeny, so that ontogeny was, in effect, the summation of all the characters acquired by previous generations. Lamarckism was associated far more closely than Darwinism with the belief that ontogeny recapitulates phylogeny (Gould 1977). For Darwinians, the open-endedness of variation made more sense if the new characters were seen as distortions of the original ontogeny rather than additions to it, making it less likely that ancestral adult characters were recapitulated in the embryo.

More seriously for the wider debate, the Lamarckians portrayed natural selection as a purely negative process—it simply eliminated the vast majority of variations produced within the population. The survival of the organism depended on luck—if it was born with a maladaptive character, then nothing it did would prevent it being killed off in the struggle for existence. In the Lamarckian theory, organisms were active agents able to respond positively to environmental challenge, acquiring new characters that gave them a better chance in life but also shaped the future of their species.

This claim was crucial for those concerned about evolutionism's religious, moral and social implications. Some Lamarckians became vitalists, seeing the organism as an agent imbued with a power of choice that lifted it above the status of a material system. This made the theory easier to reconcile with the hope that evolution expressed a divine purpose. Others focused on the moral and social implications. Exponents of free-enterprise individualism and their opponents who favoured state-controlled education both seized on the idea that the individual can be shaped by its environment to argue that Lamarckism offered a way of improving the human race. This diversity of applications has resulted in conflicting interpretations of the theory's influence among historians.

For any theory of adaptive evolution, the concept of specialization offers a way of imagining a form of pre-direction in the results. Once a tendency to specialize for a particular way of life has been established, it will be beneficial for future generations to continue the trend as long as the environment remains suitable. Darwinists accepted this point, but some Lamarckians took it to heart and argued that their theory offered a better way of explaining the apparently directed trends seen in the fossil record. Orthogenesis too sought to explain the pre-directed nature of evolutionary trends but did so by rejecting the role of adaptation altogether. If variation was directed along predetermined channels, then a species would continue to evolve along the path marked out for it whether the results were adaptive or not. Evolution was independent of adaptive constraints and might even produce maladaptive features. The direction of variation was assumed to be controlled by the process of ontogeny. The developmental forces that produced the adult organism could somehow push further along the same path. For this reason, orthogenesis, like Lamarckism, could easily be linked to the belief in recapitulation.

For those Lamarckians who imagined that specialization would impose a trend on the acquisition of new characters, it was possible to see a link with orthogenesis. An adaptive trend leading to increased specialization might gain a kind of 'momentum' that would carry on beyond the point of maximum fit with the environment, producing overdeveloped characters that were eventually maladaptive.

By denying or limiting the role of adaptation, this approach also tended to subvert another key component of Darwinism: the image of the 'tree of life'. Because he did not believe that variation constrained the process of natural selection, Darwin could see how a species could divide when exposed to different environmental conditions. This is why biogeography played such a role in his thinking—migration explained how and why the divergences took place. Visualizing the overall pattern of evolution as an ever-branching tree was an obvious extension of this approach. As a consequence, Darwinians assumed that when a group of species shared a character, it must have been inherited from a recent common ancestor. By contrast, the orthogenetic approach encouraged naturalists to see evolution as a process in which parallel lines could advance in the same direction because they were driven by the same variation trend. They would thus independently develop the same characters, so the possession of a common character was not evidence of divergence from a common ancestor. Evolution exhibited a tree-like structure overall, but parallelism implied that each major branch consisted of a series of independent lineages developing through the same predetermined pattern of development.

Saltationism also minimized the role of adaptation by seeing forces internal to the organism as the primary agent producing new species. But instead of treating ontogeny as a positive directing agent, the saltationists focused instead on the discontinuity of variation. Saltations or sudden leaps would sometimes produce a range of forms with entirely new characters which were the true source of new species. Since the new characters were not formed by the gradual addition of small individual differences, there was no opportunity for the environment to impose any constraints on their production. Saltationism thus challenged both the Darwinians and the Lamarckians by denying two key principles, those of continuity and utility.

In principle, the supporters of orthogenesis could have accepted that their variation trends proceeded by a series of discrete steps. But most of the saltationists who became active around the turn of the century favoured a model in which a whole range of characters could suddenly appear, immediately fragmenting the species into several new subspecies. While denying the role of adaptation in the production of the new forms, this approach did at least retain the model of evolution as an ever-branching tree. Hugo De Vries even tried to work out an accommodation with Darwinism by arguing that natural selection would eventually eliminate most of the new forms.

3 Lamarckism

The term 'Lamarckism' was used in the late nineteenth century to denote what had been just a single component of the evolutionary theory advanced by J. B. Lamarck. This was the inheritance of acquired characteristics or use inheritance—the idea that if an animal modified its bodily structure by adopting new habits, the modifications would be passed on to future generations and could thus accumulate to allow the species to adapt to a changed environment. Lamarck's writings had embedded this into a theory that in many other respects was unacceptable in the post-Darwinian world (Hodge 1971). But many later naturalists adopted the idea without reading Lamarck's own writings. Alpheus Packard (1901) provided the first detailed account of Lamarck's life in English. Darwin himself had allowed a limited role for the inheritance of acquired characteristics, but by the end of the century there were many who saw this as the more significant mechanism of adaptive evolution. The term 'neo-Lamarckism' was coined by Packard in 1889 to denote this position and also came into wide use (Bowler 1983, chaps. 4 and 6).

Lamarckism was never a unified movement because its basic concept (the inheritance of acquired characteristics) could be supported by different arguments and given different implications. This creates problems for historians who tend to focus on a particular interpretation and are then reluctant to accept that others were really seen as 'Lamarckian' at the time. Some think Lamarckism is defined by commitment to vitalism and teleology. This was indeed an important Lamarckian movement, but there were also materialists and naturalistic thinkers who endorsed the inheritance of acquired characteristics.

Other historians have assumed that Lamarckism was an ideologically progressive movement linked to a rejection of 'social Darwinism' and race theory. Arthur Koestler (1971) praised the Lamarckian experiments of Paul Kammerer (discussed below), not realizing that some of his supporters held racist views. Lamarckism was linked to progressive political views by Kammerer but also by the Soviets during the repression of genetics under T. D. Lysenko. The latter episode reminds us that the theory has a darker side all too often ignored by historians. We need to recognize the breadth of support for the basic Lamarckian mechanism in wider culture and accept that many positions helped to keep interest in this non-Darwinian idea alive in science.

3.1 Lamarckism and Vitalism

In the late nineteenth century, the Lamarckian position was taken up by scientists and other thinkers who distrusted the materialistic implications of Darwinism. The author Samuel Butler campaigned against the selection theory and presented Lamarckism as a morally preferable view of evolution because it allowed animals to play an active role in shaping the future of their species. This position was defended in the twentieth century by the playwright George Bernard Shaw, who linked it to Henri Bergson's philosophy of 'creative evolution' (Shaw 1921: preface; Bergson 1911). Bergson's claim that living animals are driven by a non-material 'élan vital' or life force was an influential contribution to a revival of vitalist thinking that influenced a number of biologists and psychologists, many of whom were also tempted by the Lamarckian view of evolution. Loison and Herring (2017) note how it shaped the thinking of the later French Lamarckians. The position had a strong attraction for religious thinkers who could present it as being compatible with the belief that evolution is the unfolding of a purposeful divine plan. The Anglican churchman Charles Raven, for instance, promoted this view and was later a supporter of the teleological evolutionism of Pierre Teilhard de Chardin (Raven 1927, 1962; see Bowler 2001: 137–146, 277–286).

There was also a brief resurgence of vitalist thinking within the life sciences at the turn of the century. A number of biologists and psychologist gave credence to this approach, although as the century progressed they became increasingly out of touch with the latest developments. Nevertheless, several senior figures were able to keep up the impression that opposition to materialism was still active. The embryologist E. W. MacBride (1914: chap. 18) linked his support for Lamarckism to the recapitulation theory. Although originally opposed to the vitalism, he later wrote popular works linking Lamarckism to a rejection of materialism and the belief that evolution exhibited divine purpose (MacBride 1924, 1927; Bowler 2001: 144–145). The anatomist Frederic Wood Jones—known for his theory that humans and apes had evolved in parallel from a more primitive Primate ancestor—came out in open support of Lamarckism later in his career (Jones 1942). The psychologist William McDougal was one of the most prominent opponents of materialism and published experimental evidence for a Lamarckian effect (McDougall 1927).

There were other scientists of the older generation who opposed materialism but were more cautious over the link with Lamarckism. J. Arthur Thomson's survey of theories of heredity (1907) recognized that the evidence for the inheritance of acquired characters was suspect, but in his later career he wrote many popular works supporting an organicist (if not openly vitalist) approach and insisting that animal choice must play a role in directing evolution. He occasionally hinted that the Lamarckian effect could not be ruled out altogether (e.g. Thomson 1934, II: 993, 1010). The psychologist Conwy Lloyd Morgan's vision of 'emergent evolution' saw animals as having active mental powers. Along with James Mark Baldwin and Henry Fairfield Osborn, he had earlier proposed the mechanism of 'organic selection' (also known as the Baldwin effect) in an effort to reconcile the Darwinian and Lamarckian positions. Baldwin argued that characters acquired in response to a newly adopted habit were not necessarily inherited, but they gave the species a chance to adapt to new conditions and then directed the course taken by natural selection (Richards 1987: chaps. 8 and 10; Weber and Depew 2003).

3.2 *Lamarckism and Progress*

An image of Lamarckism equally popular among those who disliked the moral implications of Darwinism linked it to the ideology of social progress. Reacting against the laissez-faire policies of 'social Darwinism', many reformers sought to use state-controlled education to modify people's habits and encourage cooperation for the common good. If acquired characters can be inherited, the new habits would eventually become instincts biologically implanted in an improved human race. This vision of social progress emerged in the post-Darwinian period and continued to be popular in the new century. It was endorsed, for instance, by Paul Kammerer (1924), whose defence of Lamarckism was later praised by Arthur Koestler (1971). In Koestler's version of history, the Lamarckian project was eliminated from orthodox science because it threatened the materialistic worldview of genetics and Darwinism (for a more nuanced view, see Gliboff 2006, 2011).

Kammerer committed suicide in 1926 when his experimental support for Lamarckism was discredited. At that time he was about to move to Soviet Russia, a move that would have highlighted the link between Lamarckism and the Marxist version of the progressive ideology. The Soviets were attracted to the possibility that their social programme might have a permanently beneficial effect on the human race but were also looking for anything that might improve their wheat supply. In the 1930s, the agronomist T. D. Lysenko gained Stalin's support for a breeding programme based on Lamarckian principles. Eventually genetics and Darwinism were dismissed as manifestations of capitalist ideology and many geneticists were purged. The episode has often been seen as an illustration of what goes wrong when politicians interfere with science (e.g. Joravsky 1970). More recent studies take a less critical approach, pointing out that Lysenko's work was not completely out of touch with contemporary plant breeders' thinking and even suggesting that it anticipated modern evolutionary developmental biology (Roll-Hansen 1985, 2011; Graham 2016).

The publicity centred on Lamarckism's links with reformist ideologies has obscured the theory's wide appeal to harsher social programmes including support for unrestrained capitalism and racial inequality. In the late nineteenth century, Herbert Spencer's political philosophy argued that free enterprise, not government sponsored education, was the best way of encouraging people to acquire new characters that would allow the race to progress. His followers' willingness to emphasize the role of competition has allowed them to be described as 'social Darwinists' (Hofstadter 1959) despite the fact that Spencer invoked the Lamarckian effect to explain how individual responses to the challenge of competition were passed on to future generations. Coupled with the popular view that Lamarckism is primarily a theory favoured by idealists, this has led many historians of the social sciences to deny that Spencer can have been a Lamarckian. Nevertheless, he was seen as one of the most influential voices supporting the inheritance of acquired characteristics (Bowler 2015). Although Spencer's influence had waned in Britain by the turn of the century, in America he continued to inspire many life scientists well into the new century (Ruse 1996).

The assumed link between Lamarckism and reformist ideology has also deflected attention from the theory's use by advocates of race science and eugenics. Nineteenth-century Lamarckians such as the palaeontologist E. D. Cope argued that some races were less 'mature' than others. E. W. MacBride called for restrictions on the breeding of the Irish on the grounds that the Lamarckian effect which had adapted them to an inferior environment worked too slowly for there to be any hope of reversing the process in the modern world (Bowler 1984). Even Bernard Shaw called for a eugenics programme that would prevent those incapable of acquiring new characters from reproducing (Hale 2006).

3.3 The Experimental Defence of Lamarckism

Whatever the moral and social concerns of Lamarckism, there were still a number of biologists who sought hard evidence. As the life sciences became more dependent on laboratory work, the need to provide actual demonstrations of the inheritance of acquired characteristics became acute. In the late nineteenth century, there had been numerous efforts, but all were dogged by controversy over their adequacy. In the early decades of the new century, there was still no shortage of efforts being made, but the critics' suspicions could not be allayed. The highly respected surveys of theories of heredity by J. Arthur Thomson (1907) and of evolution theories by Vernon Kellogg (Kellogg 1907) already expressed considerable scepticism.

French biologists had never been enthusiastic about Darwinism, and here evolutionism emerged to a large extent as a by-product of the physiological tradition established by Claude Bernard. The interaction between the individual organism and its environment was seen as the source of any new characters entering the population (Loison 2010, 2011; Loison and Herring 2017). Experimental evidence for the inheritance of acquired characters was provided by C. E. Brown-Séquard and others and this work continued in the new century. Yet, French biologists became increasingly frustrated by the difficulty of rendering the evidence unambiguous. Their problems were as much conceptual as experimental. As many critics pointed out, the Lamarckian effect implied that the long-established character of the species had no power to restrict the development of the individual while demanding that any new features acquired would immediately be incorporated into the species' future inheritance. Demonstrating the acquisition of new characters was easy, but showing that they were genuinely transmitted to future generations by heredity proved impossible. As scepticism mounted, biologists such as Felix Le Dantec and Maurice Caullery began to suggest that the Lamarckian effect operated only in lower organisms and had largely disappeared by the later stages in the ascent of life.

Elsewhere there were also increasingly desperate efforts to provide experimental proof of the Lamarckian effect (Blacher 1982; Bowler 1983: 99–103; Burkhardt 1980 and for a contemporary survey Robson and Richards 1936: 30–42). Considerable excitement was aroused by the experiments on amphibians by the Austrian biologist Paul Kammerer (Gliboff 2006). In what Arthur Koestler (1971) later dubbed 'the case of the midwife toad', Kammerer's evidence was discredited in a sustained critique led by the geneticist William Bateson. Whatever Koestler's protestations of Kammerer's innocence, he was something of an outsider to the scientific community, and there were genuine concerns that his work would not stand up to scrutiny.

Kammerer's death came shortly before a planned move to the Soviet Union, where there were sustained efforts to defend Lamarckism, culminating with the work of T. D. Lysenko (Joravsky 1970; Roll-Hansen 1985, 2011; Graham 2016). Lysenko's studies of the 'vernalization' of wheat (freezing the seeds to advance the period of germination) seemed to vindicate the Lamarckian effect. His work gained the support of Stalin, with consequences noted above, but was dismissed by Western geneticists. The rush to discredit Lysenko during the Cold War did, however, conceal the fact that his approach was in line with established traditions in agronomy, and his work on graft hybrids was taken more seriously even by those who stood aside from the rest of his career (e.g. Blacher 1982).

A variety of other experimental proofs were offered. The psychologist William McDougall (1927) claimed to have shown that rats trained to run a maze could pass the knowledge on to their offspring as an inherited instinct. In America, the palaeontologist Henry Fairfield Osborn helped to set up an experimental programme designed to provide evidence for Lamarckism (Cook 1999). Charles R. Stockard claimed to find inherited defects in rats induced by the effects of alcohol. These were reported in a symposium on the inheritance of acquired characteristics held by the American Philosophical Society (Stockard 1923). Here, the emphasis seems to have switched to finding evidence of damage to the germ plasm or genes—hardly the kind of effect that had encouraged the earlier generation of Lamarckians. There had always been strong support for the inheritance of acquired characteristics among microbiologists and pathologists and many of the twentieth-century experiments focused on lower organisms.

By the 1930s, even surveys unsympathetic to Darwinism admitted that the evidence was increasingly dubious (e.g. Robson and Richards 1936). There were, perhaps, short-term effects such as the 'Dauemodifikations' of V. Jollos or the effects of 'damaged' genes noted above. But as genetics expanded its influence, the Lamarckians found it increasingly difficult to suggest plausible ways in which the effects they claimed to demonstrate could actually operate. One idea was that hormones could somehow influence the activity of the genes. More plausibly, there were efforts to suggest that work on cytoplasmic inheritance would challenge the dogmatism of chromosome-centred genetics (Sapp 1987). But as the synthesis of Darwinism and genetics gained momentum in the 1930s and 1940s, the few scientists who still expressed an interest in the Lamarckian effect switched their efforts to postulating ways in which control of the genes could be modified indirectly, as with Conrad Hal Waddington's notion of 'genetic assimilation' (Peterson 2011).

3.4 Indirect Evidence for Lamarckism

In the late nineteenth century, much of the support for Lamarckism had come from field naturalists and palaeontologists who were convinced that the theory offered the most plausible explanation of the phenomena they observed. Field naturalists such as Joel A. Allen noticed variations within species over their geographical range that seemed to correlate with climatic factors. 'Allen's law' noted the tendency for mammals to have smaller extremities (ears, tails, etc.) at the northern edge of their range. Such correlations were seen as evidence for the direct effect of the environment on the individual organisms. Another phenomenon seen as susceptible of the same explanation was the disappearance of the eyes in species inhabiting dark caves, studied for instance by Alpheus Packard (1894). It was argued that the cumulative effects of disuse offered a better explanation for the complete elimination of the organ than the mere relaxation of natural selection.

Some palaeontologists were convinced that they could see trends in the fossil record of various groups which were far too regular to be the result of so haphazard a process as natural selection. The 'American school of neo-Lamarckism' led by Edward Drinker Cope and Alpheus Hyatt focused on the trends towards specialization they saw in many groups (Bowler 1983: chap. 6; Pfeiffer 1965). Darwinism could, of course, explain specialization, but these palaeontologists saw an element of linearity and directness in the trends that would not be expected if variation were 'random'. Instead, it was assumed that the power of a newly adopted habit could impose a direction on the group's later evolution by directly controlling the acquisition of a new character. Cope and Hyatt also claimed to see evidence of parallel evolution: several lineages within the group independently advanced along the same path, predetermined by the animals' habits. The element of parallelism would remain a key feature of twentieth-century opposition to Darwinism, converted into support for orthogenesis (see below). This approach subverted the Darwinians' vision of divergent, open-ended evolution and saw generic characters not as the product of the species' descent from a common ancestor, but as evidence that they had independently advanced to the same point on a predetermined scale of development. Cope (1896) provided a detailed survey of this neo-Lamarckian position.

By the turn of the century, younger palaeontologists such as Henry Fairfield Osborn were turning away from the Lamarckian explanation of parallel evolution. Along with James Mark Baldwin, Osborn was one of the proponents of the idea of 'organic selection' in which the animals' chosen habit generates characters which are not inherited directly but define the channel along which natural selection will operate (Richards 1987, chaps. 8 and 10). He would later turn more to orthogenesis.

Lamarckism survived more actively among the field naturalists. The myrmecologist William Morton Wheeler preferred the Lamarckian explanation of the origin of instincts in ants to the Darwinian view (Sleigh 2004). In Germany, Bernhard Rensch was one of many students of geographical distribution and speciation who retained the Lamarckian explanation of adaptive evolution. As he and Ernst Mayr later explained (Rensch 1980; Mayr 1980a, b), the field naturalists were still suspicious of Darwinism and preferred Lamarckism despite the lack of hard evidence that acquired characteristics were really inherited. Confusion over the term 'mutation' (originally used to demote discontinuous evolutionary steps or saltations) fuelled their distrust of genetics and thus held up their recognition of the emerging synthesis of that science with Darwinism. Only when Theodosius Dobzhansky's translated the new Darwinism into terms comprehensible to the field naturalists were they able to realize that there was no longer any point in retaining the Lamarckian alternative.

4 Orthogenesis

The model of evolution proposed by neo-Lamarckian palaeontologists such as Cope and Hyatt did not merely offer an alternative to natural selection as an explanation of adaptation. By focusing on habit as a driving force that could predetermine a rigid pattern of future development, they introduced the idea of evolutionary parallelism, in effect subverting the whole Darwinian vision of a constantly branching 'tree of life', at least within each group. Taking this rival model further led them to imagine that the parallel trends they envisaged might go on beyond the limit of adaptive benefit that could be gained from specialization. Structures might get overdeveloped to a point where they became positively harmful. The theory thus opened up the possibility of predetermined trends that had no relevance to the demands of the environment. The suggestion that evolution might be a non-adaptive process driven by rigid variation trends imposed by internal processes arising from ontogeny was also raised by some field naturalists. This was the foundation of what became known as the theory of orthogenesis.

The theory had emerged in the aftermath of the original Darwinian debates as perhaps the most extreme alternative to natural selection. It was extremely active within the German scientific community (Levit and Hossfeld 2017). The term 'orthogenesis' was popularized by Theodore Eimer, who worked with living species but whose trajectory of thought followed the same pattern as the American palaeontologists. He began as a Lamarckian but soon moved on to propose non-adaptive trends which he claimed to observe in the colour patterns of butterflies (Eimer 1898). Mimicry was dismissed as the result of two species independently affected by the same variation trend—it had no adaptive significance. The search for orthogenetic patterns in living species continued into the twentieth century. C. O. Whitman (1919) saw non-adaptive patterns in the colouration of pigeons, and Jepsen (1949) listed a number of similar studies (for details see Esposito 2017).

By far the most powerful line of support for orthogenesis came from palaeontology. A significant number of the specialists seeking to reconstruct phylogenies within the animal kingdom interpreted the patterns of development they saw as evidence of predetermined evolution. They included the next generation of the American school, led by Henry Fairfield Osborn, but also a number of European palaeontologists. Until this work was challenged by new fossil evidence and by a more critical evaluation of its claims, palaeontology stood as a bastion of resistance to the Darwinian viewpoint.

4.1 Mechanisms of Orthogenesis

Exactly how the orthogenetic patterns were imposed on evolution was a matter of some debate. Suggestions ranged from vaguely defined 'laws of development' to mechanistic processes imposing restrictions on the kinds of variation that could appear within a population (Bowler 1983, chap. 7; Ulett 2014). Some thought that variation might be controlled by forces intrinsic to the nature of living matter. This approach can be seen in D'Arcy Wentworth Thompson's classic *On Growth and Form* (1917). His demonstration that changing the coordinates of how a single form is represented could produce structures corresponding to a wide range of different fish species became well known. But he also insisted that this geometrical insight implied that the range of variation that had allowed the different species to evolve must be controlled by laws of growth determined by a simple system of forces (Thompson 1917: 727). Thompson wrote an introduction to the English translation of *Nomogenesis* by the Russian naturalist Leo S. Berg (1926). Berg too rejected any role for chance in evolution and saw the whole process as being directed by internally programmed laws of development. In principle, these laws were imposed by the basic chemical composition of protoplasm, but Berg also insisted that the patterns of development they imposed were purposeful and progressive, leading critics to dismiss his ideas as a relic of the old teleological approach.

Most supporters of orthogenesis insisted that their theory was not teleological—indeed, they often went out of their way to argue that the trends they saw were actually harmful and might eventually lead to extinction. C. O. Whitman (1919) quite explicitly rejected teleology and tried to render the theory more plausible by arguing that the patterns he saw represented not rigidly predetermined variation but merely restrictions on the possible range of variation. Some suggestions as to a possible mechanism still focused on the possibility that factors affecting individual ontogeny could somehow affect development in a cumulative manner. At a symposium on orthogenesis held by the American Society of Zoology, the biochemist Lawrence J. Henderson (1922) argued that a tendency to overproduce growth hormones might generate an evolutionary trend (see also MacCord and Maienschein 2017).

Henderson's approach was favoured by some palaeontologists (see below), but as the new science of Mendelian genetics began to throw light on the nature of variation, the supporters of orthogenesis needed to explain how such tendencies could be generated by mutation. Whitman's idea of restricted variation could easily be translated into terms compatible with genetics, and there were a number of efforts to demonstrate that there were limits to the kinds of new characters that were likely to be produced. Richard Goldschmidt (1933) endorsed work by V. Jollos claiming to show that mutations occurred preferentially in a particular direction. A. F. Shull (1936: 123–133) was less convinced by this work but also expressed the hope that directed mutation might turn out to be the long-sought explanation for orthogenetic trends. Most field naturalists and palaeontologists were not well versed in the new genetics and simply ignored the issue.

4.2 *Orthogenesis in Palaeontology*

Building on research traditions established in the late nineteenth century, a significant number of palaeontolgists in both Europe and America continued to insist that their work provided evidence of non-adaptive evolutionary trends (see also Turner 2017). In the United States, the legacy of the orthogenetic element in the thinking of the neo-Lamarckian school remained active. Among invertebrate palaeontologists, Charles E. Beecher and others continued the tradition established by Hyatt in which the recapitulation theory was used to provide a model for the evolutionary history of each group (Rainger 1981). Hyatt had argued that although unfavourable conditions might trigger non-adaptive variation, the trends that led each group to degenerate into a senile phase were also a sign that a once-progressive group exhausted its evolutionary energy. His followers continued to argue that their work revealed patterns of development driving each group towards racial senility and eventual extinction. The concept of racial old age thus remained in play and was often appealed to by pessimists commenting on modern culture (Bowler 1989).

Vertebrate palaeontologists too endorsed the model of evolution which saw degeneration as the almost inevitable fate of any phylum. But where Hyatt had emphasized the loss of complex characters, those studying vertebrate phylogeny claimed to see a process of overdevelopment by which once-useful structures eventually reached harmful proportions. Henry Fairfield Osborn was a follower of Cope and became one of the most influential American evolutionists of the early twentieth century (Rainger 1991). His work on extinct mammals led him to support the view that their evolution was governed by orthogenetic trends which he called 'rectigradations', leading to the overdevelopment of structures such as horns. Although now accepting that the Lamarckian effect did not work, he struggled to develop a theory in which the environment and the animals' behaviour could somehow influence the hereditary constitution of the species (Osborn 1917). The trends thus induced were not always adaptive and could in some cases lead to overdevelopment so severe it might play a role in the order's eventual extinction.

Osborn's fellow palaeontologist William Berryman Scott also supported the theory of orthogenesis, although he was doubtful that the trends ever went far enough to cause extinction (Scott 1929: 532). F. B. Loomis (1905) introduced an analogy which became widely used: he wrote of once-useful trends gaining a 'momentum' that carried them on beyond the limit of adaptive benefit. This model was applied to explain popular examples of what were claimed to be overdeveloped structures, including the enormous horns of the so-called 'Irish elk' (Gould 1974). Other American palaeontologists who endorsed the view that orthogenetic trends could lead to overdevelopment and extinction included Richard Swan Lull who thought that non-adaptive trends might be triggered by unfavourable conditions and later wrote of them leading to 'racial disease' and extinction (Lull 1917, 1924).

Among European palaeontologists, Arthur Dendy (1911) adopted the theory that an excess production of hormones led to the overdevelopment of various structures.

W. D. Lang produced a study of the fossil Bryozoa which claimed that the group was affected by an uncontrolled tendency to produce calcium carbonate. The trend was built into the group's constitution and was only secondarily used to construct a protective shell, which is why it eventually led to overelaborate structures that were positively harmful (Lang 1921). D. M. S. Watson (1925–1926) described trends in fossil amphibians that persisted despite changes in their environment and which must therefore originate in an internal limitation on what kinds of variations could be produced. In contrast, Francis Nopsca (1930) shared Lull's view that a changed environment had led to the emergence of predetermined trends in the fossil amphibians he studied, triggering modifications in the thyroid gland affecting growth.

Although some apparently non-adaptive trends were hard for the Darwinist to explain, the supporters of the new synthesis became increasingly suspicious of the palaeontologists' evidence. Julian Huxley's concept of 'allometry' explained the huge development of horns such as those of the 'Irish elk' as a by-product of selection for increased overall size (Huxley 1932: 214–221). More seriously for the supporters of orthogenesis, the fossil evidence itself was becoming increasingly less supportive of the claim that evolution exhibited rigid trends. Showing that the evidence for parallel non-adaptive evolution was actually illusory would play a significant role in preparing the way for the emergence of the modern Darwinian synthesis. From the early decades of the century, there were palaeontologists such as W. D. Matthew who argued that as more fossils were discovered the apparently regular patterns of development seen by earlier workers fragmented into complex branching trees—just as a Darwinian would expect (Bowler 1996). The work of George Gaylord Simpson would cement the palaeontologists' rejection of the non-Darwinian stance adopted by the previous generation (but see Turner 2017).

5 Saltationism

The first decade of the twentieth century saw a dramatic resurgence of a long-standing alternative to the Darwinian theory. There had always been some naturalists who were suspicious of Darwin's insistence that small individual differences were the raw material of evolution. Even Thomas Henry Huxley thought that new characters were more likely to appear suddenly via dramatic 'leaps' or saltations that would create a new variety if not a new species instantaneously. During the 'revolt against morphology' around 1900, the search for demonstrable evidence of evolution focused renewed attention on direct observations of the processes of variation and heredity. For some naturalists, this led to renewed studies of the distinct varieties that exist within many species and generated the assumption that the most obvious explanation of their formation was by saltation.

This assumption led to a rejection of the Darwinian focus on the power of the environment to determine which variations survive and breed—the new varieties formed by saltation came into existence without any involvement by natural selection and seemed able to perpetuate themselves alongside many other forms. William Bateson's *Materials for the Study of Variation* expressed this position, and although he admitted that varieties must be 'approximately' adapted to the environment (1894: 15), it was clear that he did not think natural selection to be a powerful limiting factor. Saltationism thus reflected a similar position to orthogenesis on the role of the environment but tended to assume that a wide variety of new forms could be produced (although there were occasional suggestions that saltations might occur in a cumulative direction).

The most prominent expression of saltationism in the new century was the 'mutation theory' proposed by Hugo De Vries. He observed the appearance of apparently new and distinct varieties in a controlled environment. Although it was later shown that these were not genuinely new forms, for some time his observations were taken as evidence in favour of what he called the process of mutation. The theory became immensely popular for a short time, and it was probably the emergence of this new alternative that led Julian Huxley to refer to this period as the 'eclipse of Darwinism'. As Lamarckism declined in influence, this new rival took its place. De Vries himself sought a reconciliation with Darwinism, claiming that in the long run only better-adapted mutated forms would survive. But most of his followers—including Thomas Hunt Morgan for a time—shared Bateson's suspicion of the power of selection. It was this anti-adaptationism that led field naturalists such as Ernst Mayr to remain suspicious of the concept of mutation even after it had begun to acquire a new role as a component of Mendelian genetics.

The fact that the term 'mutation' has gained a new meaning in the modern world suggests that there was a significant link between saltationism and genetics. It was no coincidence that three major figures associated with the development of genetics began their careers as saltationists: Bateson, De Vries and Morgan. The model of evolution that pictured it as a series of events producing discrete new characters that would breed true paved the way for acceptance of the laws of heredity proposed by Gregor Mendel. As the 'rediscovery' of Mendel's laws by De Vries and others in 1900 developed into the new science of genetics, it was soon realized that new characters are indeed produced by the 'mutation' of genes—although they functioned within the existing population rather than founding a discrete new variety.

With hindsight we know that mutations would eventually be recognized as the source of the individual variations, the Darwinians postulated as the raw material of natural selection. But at the time, the anti-adaptationism of the original mutation theory carried through into the new model of heredity. Geneticists studied breeding under laboratory conditions where there was little environmental pressure and no evidence of natural selection. They thus remained suspicious of the claim that new characters had to confer adaptive benefit in order to spread in a population. Bateson in any case remained deeply hostile to the biometrical version of Darwinism developed by Karl Pearson and others, and it was only in the 1930s that this hostility was overcome. Bateson and the geneticists were certainly hostile to Lamarckism, which would undermine their concept of the gene as a fixed entity breeding true over the generations. But in the short term, at least their theory was perceived as yet another alternative to Darwinism.

5.1 The Mutation Theory

Support for saltationism surged in the first decade of the century, largely in response to the work of the Dutch botanist Hugo De Vries (Allen 1969; Bowler 1983: chap. 8). During the 1890s, De Vries had noticed apparently new varieties appearing suddenly within cultivated populations of the evening primrose, *Oenothera lamarckiana*. He interpreted these new forms as saltations produced by a sudden transformation within the species' hereditary material and dubbed them 'mutations'. He then used his observations as the basis for a complete theory of evolution, arguing that selection acting on small individual differences was powerless and that saltations were the true source of new characters. Crucially, he believed that each new form appeared in multiple individuals, allowing the instantaneous creation of a distinct new variety or even species. De Vries imagined that all species go through occasional phases in which they throw off saltations—*Oenothera* was valuable because it was currently in such a phase. His 'mutation theory' was proposed in a book (translated as De Vries 1910) and in a series of lectures delivered at the University of California (De Vries 1904).

De Vries presented himself as a reformer who would put Darwinism on a more secure footing, not replace it. He implied that most mutations were adaptively neutral and would perpetuate themselves in a natural environment but accepted that some were harmful and would soon be eliminated. In the long run, there would be some beneficial mutations, and these would replace the older forms. Natural selection still operated and would control the establishment of new species, but its raw material was mutated varieties rather than trivial individual differences. For De Vries, this was an essential point that ensured the theory did not represent a re-emergence of teleology.

The mutation theory soon became popular, being seen as a modern, experimentally verifiable form of evolutionism (Endesby 2013). However, most of its supporters did not share De Vries' willingness to compromise with Darwinism. They extended his belief that many mutations were adaptively neutral into a wholesale rejection of the claim that adaptation played a significant role in evolution. Even in the long run, they assumed, any new character produced as a mutation would be able to persist. One of the most vociferous advocates of this interpretation was Thomas Hunt Morgan (Allen 1978). Before he came to accept Mendelism, he took up the mutation theory and extended it into a wholesale critique of Darwinism. His *Evolution and Adaptation* (Morgan 1903) argued that the majority of characters defining species have no adaptive significance, including the colour schemes that the Darwinians interpreted as camouflage or the products of sexual selection. There is no 'struggle for existence' and new forms simply appear by mutation and continue to breed independent of any environmental constraint. Since there was no pressure from selection, Morgan argued that the few complex structures that do benefit the organism must be the product of a directed sequence of mutations (in effect, of an orthogenetic trend). The same argument appeared in a study of the mutation theory by R. Ruggles Gates (1915).

5.2 Saltationism and Genetics

Despite the initial enthusiasm, suspicion soon emerged that the forms observed by De Vries were not genuinely new characters and it was eventually shown that *Oenothera* is a complex hybrid species. De Vries played a major role in the rediscovery of Mendel's laws but soon lost interest. It was Bateson and Morgan who went on to help create the new science of heredity that became known as genetics. Bateson did not accept the idea that mutations create genuinely new genetic characters and interpreted the appearance of new forms as the result of destructive saltations removing genes that had originally masked the character (Bateson 1922). Morgan eventually converted to Mendelism and founded the experimental school that demonstrated how the genes could be understood as material units strung along the chromosomes (Allen 1978). The definition of 'mutation' now morphed into the meaning we accept today: far from creating entirely distinct breeding populations, mutation modifies a gene so that it codes for a new character, and the new gene feeds into the existing population. Richard Goldschmidt (1940) was one of the few geneticists who continued to support the possibility of a 'hopeful monster' establishing a new species (see also Turner 2017).

Genetics provide an explanation of the individual differences existing within each population, with mutations being seen as the ultimate source of novelties. Yet, there were at first only limited efforts to explore the possibility that this new approach could be reconciled with Darwin's belief that the range of variation within the population served as the raw material of natural selection. The appearance of discrete new characters in the laboratory did not seem an appropriate model for natural variation. Bateson retained his suspicion of the selection mechanism and continued to regard the continuous range of variation within populations as a product of short-term environmental influences. He strenuously opposed the position of Karl Pearson and the biometrical school which studied variation in wild populations and sought to demonstrate the effect of selection. Morgan too remained suspicious of natural selection, although he gradually came to admit that maladapted characters would eventually be eliminated. He still found it difficult to accept the concept of the 'struggle for existence' and to imagine selection as a creative force.

Genetics was originally perceived as a minimized saltationism, offering a similar alternative to Darwinism and the adaptationist programme. Lamarckism was undermined without strengthening Darwinism. It would take several decades for the possibility that genetics might explain the range of variation in wild populations to be recognized, let alone that natural selection might act to change gene frequencies and produce new structures beneficial to the organism. The story of how genetics was eventually synthesized with Darwinism will be told throughout the rest of this volume.

6 Conclusion

Our understanding of the rise of modern Darwinism must take into account the fact that it has not been a simple or continuous process. The theory of natural selection was controversial from the start, partly for its moral and religious implications but also because many naturalists found it unsatisfactory as a scientific explanation of evolution. The critics sought alternatives that would resolve both the scientific and the non-scientific problems. Darwinism certainly gained notable adherents in the late nineteenth century, but supporters such as August Weismann had to battle with the critics who preferred the alternatives. Far from diminishing in the early twentieth century, the alternatives proliferated, prompting Julian Huxley's later claim that the period had witnessed an 'eclipse of Darwinism'.

This chapter has surveyed the alternatives to the selection theory: the Lamarckian theory of the inheritance of acquired characteristics, orthogenesis and saltationism. It has shown how they represented a complex of positions opposed to the Darwinian view of the roles played by heredity, adaptation and continuity. The emergence of Mendelian genetics after 1900 was at first seen as a new element supporting the concept of evolution by discontinuous steps. But its supporters were hostile to the belief that acquired characters could be inherited, so as their approach to the study of heredity gained ground, the Lamarckian alternative was discredited. When genetics was synthesized with the selection theory in the 1930s, the Darwinian theory at last began to gain enough momentum to displace support for the various forms of non-adaptive evolution. Since all of the non-Darwinian theories were themselves now eclipsed, the synthetic form of Darwinism was able to gain a dominant position in the biology of the mid-twentieth century. From the perspective of its supporters, the 'Modern Synthesis' allowed Darwinism to be seen as the main line in the development of evolutionism, all the rivals being dismissed as dead ends or blind alleys.

More recent studies have increasingly suggested that the gene-centred paradigm of the twentieth-century synthesis—especially as it was consolidated in the English-speaking world—had deflected attention away from valid concerns about the role played by individual development in the shaping of organic forms. By insisting that the gene provided a complete blueprint determining the form of the organism, a whole generation of Darwinians was persuaded to ignore the possibility that the processes translating genetic information into living structures might themselves play a role. Concern for this factor was one of the main influences that had persuaded so many naturalists in the late nineteenth and early twentieth century to search for alternatives to natural selection. Whether in responding to the environment or in shaping the possible consequences of genetic mutation, developmental factors were seen as crucial. We can now appreciate that this concern was not merely a distraction from the main business of evolutionary biology, as especially shown in Depew (2017) and MacCord and Maienschein (2017). The historians who look back at these early non-Darwinian theories can, perhaps, see evidence of ideas being explored that may once again come to play a role in evolution theory. More certainly, they can see an interest being displayed in issues that became marginalized for a period in the mid-twentieth century but which have now re-emerged as areas of real interest.

References

Allen GE (1969) Hugo De Vries and the reception of the mutation theory. J Hist Biol 2:55–87
Allen GE (1975) Life science in the twentieth century. Wiley, New York
Allen GE (1978) Thomas Hunt Morgan: the man and his science. Princeton University Press, Princeton
Bateson W (1894) Materials for the study of variation: treated with especial regard to discontinuity in the origin of species. Macmillan, London
Bateson W (1922) Evolutionary faith and modern doubts. Science 55:55–61
Berg LS (1926) Nomogenesis: or evolution determined by law (trans: I Rostovtsov, introduced by D'Arcy Wentworth Thomson). London (reprinted with a preface by Theodosius Dobzhansky. MIT Press, Cambridge 1969)
Bergson H (1911) Creative evolution (trans: Mitchell A). Henry Holt, New York
Blacher LI (1982) The problem of the inheritance of acquired characters (trans: Churchill FB). Amerind Publishing, Smithsonian Institution Libraries, New Delhi
Bowler PJ (1983) The eclipse of Darwinism: anti-Darwinian evolution theories in the decades around 1900. Johns Hopkins University Press, Baltimore
Bowler PJ (1984) E. W. MacBride's Lamarckian eugenics and its implications for the social construction of scientific knowledge. Ann Sci 41:245–260
Bowler PJ (1989) Holding your head up high: degeneration and orthogenesis in theories of human evolution. In: Moore JR (ed) History, humanity and evolution. Cambridge University Press, Cambridge, pp 329–353
Bowler PJ (1996) Life's splendid drama: evolutionary biology and the reconstruction of life's ancestry, 1860–1940. University of Chicago Press, Chicago
Bowler PJ (2001) Reconciling science and religion: the debates in early twentieth-century Britain. University of Chicago Press, Chicago
Bowler PJ (2015) Herbert Spencer and Lamarckism. In: Francis M, Taylor M (eds) Herbert Spencer: legacies. Routledge, London, pp 203–221
Burkhardt RW Jr (1980) Lamarckism in Britain and the United States. In: Mayr E, Provine WB (eds) The evolutionary synthesis. Harvard University Press, Cambridge, pp 343–351
Cook GM (1999) Neo-Lamarckian experimentalism in America: origins and consequences. Q Rev Biol 74:417–437
Cope E (1896) The primary factors of organic evolution. Open Court, Chicago
De Vries H (1904) Species and varieties: their origin by mutation. MacDougal DT. Open Court, Chicago, In
De Vries H (1910) The mutation theory: experiments and observations on the origin of species in the vegetable kingdom (trans: Farmer JB, Darbyshire AD). Kegan Paul, London, 2 vols
Delisle RG (2017) From Charles Darwin to the evolutionary synthesis: weak and diffused connections only. In: Delisle RG (ed) The Darwinian tradition in context: research programs in evolutionary biology. Springer, Cham, pp 133–168
Dendy A (1911) Momentum in evolution. Report of the British Association for the Advancement of Science, pp:277–280
Depew DJ (2017) Darwinism in the 20th century: productive encounters with saltation, acquired characteristics, and development. In: Delisle RG (ed) The Darwinian tradition in context: research programs in evolutionary biology. Springer, Cham, pp 61–88

Eimer GHT (1898) On orthogenesis and the impotence of natural selection in species formation (trans: McCormack JM). Open Court, Chicago
Endesby J (2013) Mutant utopias: evening primroses and imagined futures in early-twentieth-century America. Isis 104:471–503
Esposito M (2017) The organismal synthesis: holistic science and developmental evolution in the English-speaking world, 1915–1954. In: Delisle RG (ed) The Darwinian tradition in context: research programs in evolutionary biology. Springer, Cham, pp 219–242
Gates RR (1915) The mutation factor in evolution. Macmillan, London
Gissis SB, Jablonka E (eds) (2011) Transformations of Lamarckism: from subtle fluids to molecular biology. MIT Press, Cambridge
Gliboff S (2006) The case of Paul Kammerer: evolution and experimentation in the early twentieth-century. J Hist Biol 39:525–563
Gliboff S (2011) The golden age of Lamarckism, 1866–1926. In: Gissis J (ed) Transformations of Lamarckism. MIT Press, Cambridge, pp 44–55
Goldschmidt R (1933) Some aspects of evolution. Science 78:539–547
Goldschmidt R (1940) The material basis of evolution. Yale University Press, New Haven
Gould SJ (1974) The origin and function of 'bizarre' structures: antler size and skull size in the 'Irish elk,' Megaloceros giganteus. Evolution 28:191–220
Gould JS (1977) Ontogeny and phylogeny. Harvard University Press, Cambridge
Graham L (2016) Lysenko's ghost: epigenetics and Russia. Harvard University Press, Cambridge
Hale P (2006) The search for purpose in a post-Darwinian universe: George Bernard Shaw, 'creative evolution,' and Shavian eugenics. Hist Philos Life Sci 28:191–214
Henderson LJ (1922) Orthogenesis from the standpoint of the biochemist. Am Nat 56:97–104
Hodge MJS (1971) Lamarck's science of living bodies. Brit J Hist Sci 5:323–352
Hofstadter R (1959) Social Darwinism in American thought, revised edition. George Braziller, New York
Huxley JS (1932) Problems of relative growth. Methuen, London
Huxley JS (1942) Evolution: the modern synthesis. Allen and Unwin, London
Jepsen GL (1949) Selection, 'orthogenesis', and the fossil record. Proc Am Philos Soc 93:479–500
Jones FW (1942) Habit and heritage. Kegan Paul, London
Joravsky D (1970) The Lysenko affair. Harvard University Press, Cambridge
Kammerer P (1924) The inheritance of acquired characteristics. Boni and Liveright, New York
Kellogg VL (1907) Darwinism today: a discussion of present day scientific criticism of the Darwinian selection theories. Henry Holt, New York
Koestler A (1971) The case of the midwife toad. Hutchinson, London
Lang WD (1921) Catalogue of the fossil Bryozoa (Polyzoa) in the department of Geology. British Museum (Natural History): the Cretaceous Bryozoa. London, vol, Trustees of the British Museum, p 3
Laublichler MD, Maienschein J (eds) (2007) From embryology to evo-devo: a history of developmental evolution. MIT Press, Cambridge
Levit GS, Hossfeld U (2017) Major research traditions in 20th century evolutionary biology: the relations of Germany's Darwinism with them. In: Delisle RG (ed) The Darwinian tradition in context: research programs in evolutionary biology. Springer, Cham, pp 169–194
Loison L (2010) Qu'est ce que le néolamarckisme?: Les biologistes français et la question de l'évolution des espèces. Vuibert, Paris
Loison L (2011) French roots of French neo-Lamarckism, 1897–1985. J Hist Biol 44:713–744
Loison L, Herring E (2017) Lamarckian research programs in French biology (1900–1970). In: Delisle RG (ed) The Darwinian tradition in context: research programs in evolutionary biology. Springer, Cham, pp 243–270
Loomis FB (1905) Momentum in evolution. Am Nat 39:839–843
Lull RS (1917) Organic evolution. Macmillan, New York
Lull RS (1924) Dinosaurian climatic response. In: Thorpe MR (ed) Organic adaptation to the environment. Yale University Press, New Haven, pp 225–279

MacBride EW (1914) Texbook of embryology, Invertebrata, vol 1. Macmillan, London
MacBride EW (1924) An introduction to the study of heredity. Williams and Norgate, London
MacBride EW (1927) Evolution. Ernest Benn, London
MacCord K, Maienschein J (2017) Cells, development, and evolution: teeth studies at the intersection of fields. In: Delisle RG (ed) The Darwinian tradition in context: research programs in evolutionary biology. Springer, Cham, pp 289–308
Mayr E (1980a) Prologue: some thoughts on the history of the evolutionary synthesis. In: Mayr E, Provine WB (eds) The evolutionary synthesis. Harvard University Press, Cambridge, pp 1–48
Mayr E (1980b) How I became a Darwinian. In: Mayr E, Provine WB (eds) The evolutionary synthesis. Harvard University Press, Cambridge, pp 413–429
Mayr E, Provine WB (eds) (1980) The evolutionary synthesis: perspectives on the unification of biology. Harvard University Press, Cambridge
McDougall W (1927) An experiment for the testing of the hypothesis of Lamarck. Br J Psychol 17:267–304
Morgan TH (1903) Evolution and adaptation. Macmillan, New York
Nopsca F (1930) Notes on stegocephalia and amphibia. Proc Zool Soc Lond:979–985
Osborn HF (1917) The origin and evolution of life on the theory of action, reaction and interaction of energy. Scribners, New York
Packard A (1901) Lamarck, the founder of evolution: his life and work, with translations of his writings on organic evolution. Longmans, Green, New York
Packard A (1894) The origin of the subterranean fauna of North America. Am Nat 28:727–751
Peterson EC (2011) The excluded philosophy of evo-devo? Revisiting C.H. Waddington's failed attempt to embed A.N. Whitehead's 'organicism' in evolutionary biology. Hist Philos Life Sci 33:310–320
Pfeiffer EJ (1965) The genesis of American neo-Lamarckism. Isis 56:156–167
Rainger R (1981) The continuation of the morphological tradition in American paleontology, 1880–1910. J Hist Biol 14:129–158
Rainger R (1991) An agenda for antiquity: Henry Fairfield Osborn and vertebrate paleontology at the American museum of natural history. University of Alabama Press, Tuscaloosa
Raven CE (1927) The creator spirit. Martin Hopkinson, London
Raven CE (1962) Teilhard de Chardin: scientist and seer. Collins, London
Rensch B (1980) Historical development of the present neo-darwinian synthesis in Germany. In: Mayr E, Provine WB (eds) The evolutionary synthesis. Harvard University Press, Cambridge, pp 284–303
Richards RJ (1987) Darwin and the emergence of evolutionary theories of mind and behavior. University of Chicago Press, Chicago
Robson CC, Richards OW (1936) The variation of animals in nature. Oliver and Boyd, Edinburgh
Roll-Hansen N (1985) A new perspective on Lysenko? Ann Sci 42:261–276
Roll-Hansen N (2011) Lamarckism and Lysenkoism revisited. In: Gissis S, Jablonka E (eds) Transformations of Lamarckism. MIT Press, Cambridge, pp 77–88
Ruse M (1996) Monad to man: the concept of progress in evolutionary biology. Harvard University Press, Cambridge
Russell ES (1916) Form and function: a contribution to the history of animal morphology. John Murray, London
Sapp J (1987) Beyond the gene: cytoplasmic inheritance and the struggle for authority in genetics. Oxford University Press, Oxford
Scott WB (1929) A history of land mammals in the Western hemisphere. Macmillan, New York
Shaw GB (1921) Back to Methuselah: a metabiological Pentateuch. Constable, London
Shull AF (1936) Evolution. McGraw-Hill, New York
Sleigh C (2004) The ninth mortal sin: the Lamarckism of W.M. Wheeler. In: Abigail L, Richards RA, Ruse M (eds) Darwinian Heresies. Cambridge University Press, Cambridge, pp 151–172
Stockard CR (1923) Experimental modification of the germ plasm and its bearing on the inheritance of acquired characteristics. Proc Am Philos Soc 42:322–325

Thompson DW (1917) On growth and form. Cambridge University Press, Cambridge
Thomson JA (1907) Heredity. John Murray, London
Thomson JA (1934) Biology for everyman. Dent, London, 2 vols
Turner DD (2017) Paleobiology's uneasy relationship with the Darwinian tradition: stasis as data. In: Delisle RG (ed) The Darwinian tradition in context: research programs in evolutionary biology. Springer, Cham, pp 333–352
Ulett MA (2014) Making the case for orthogenesis: the popularization of definitely directed evolution. Stud Hist Philos Life Sci 45:124–132
Watson DMS (1925–1926) The evolution and origin of the amphibia. Philos Trans R Soc Lond 214B:189–256
Weber BH, Depew DJ (eds) (2003) Evolution and learning: the Baldwin effect reconsidered. MIT Press, Cambridge
Whitman CO (1919) The posthumous works of Charles Otis Whitman, Orthogenetic evolution in pigeons, vol 1. The Carnegie Institute, Washington

The Organismal Synthesis: Holistic Science and Developmental Evolution in the English-Speaking World, 1915–1954

Maurizio Esposito

Abstract In 1915, the German physiologist Jacques Loeb published a paper titled "Mechanistic Science and Metaphysical Romance." In that article, Loeb lamented that scientific research was still infected by a "romantic" approach. Despite the triumphal achievements of the sciences based on mechanistic precepts, romantic and mystical speculations abounded. Life science, Loeb added, was besieged by mysticism, vitalism, and irrationalism. "Romantic" evolutionists indulged in unsupported theories and untested conjectures. But who were these twentieth-century "romantics" really? In this chapter, it will be argued that, contrarily to Loeb's rhetoric, such a "romantic" community was not always constituted by irrational and mystical cranks. Rather, it was often composed of reflective scientists criticizing the overoptimism of the neo-Darwinian agenda and the unwarranted ambitions of the mechanistic (physicalist) approaches to biology. The chapter has three aims: First, to outline the main ideas of the early twentieth-century organicist agenda, with particular emphasis on evolutionary and developmental biology. Second, to briefly present the background and works of a few representative figures involved in the international community of organismal biology from the 1920s onward. Third, to show that aside from the neo-Darwinian synthesis, these scholars proposed an alternative synthesis between the 1920s and 1950s, a biological synthesis aiming to link studies on evolutionary and developmental biology within an organismal framework. The points of convergence and divergence between the two syntheses will be assessed. Then, the question of whether or not they were two incommensurable alternatives will be addressed.

Keywords Organicism • Evolution • Developmental biology • Heredity • Mechanism • Reductionism

M. Esposito (✉)
Department of Philosophy, University of Santiago, Santiago, Chile
e-mail: maurizio.esposito@usach.cl

R.G. Delisle (ed.), *The Darwinian Tradition in Context*,
https://doi.org/10.1007/978-3-319-69123-7_10

1 Introduction

"Since no discontinuity exists between the matter constituting living and non-living bodies, biology must also be mechanistic." (Loeb 1915: 771). With this simple sentence Loeb described his agenda, and the program of the biological sciences in the years to come. In the future, he averred, all living phenomena could be successfully reduced to "...the motion of electrons, atoms, or molecules" (772). However, as Loeb pointed out, there was strong resistance to such an agenda. Not everyone accepted that science had to be mechanistic and therefore based on experiments and quantification. All those who rejected mechanistic biology were, after all, metaphysicians or reactionary "romanticists" who preferred Bergson and Nietzsche to Helmholtz and other serious physicists. These "romanticists," Loeb added, had pretended to explain the "riddles of the universe," but they only produced speculations and fantasies. Even worse, they swayed the masses with ideological jargon repellent to serious scientific arguments. In 1912, Loeb had published a very controversial book which included ten lectures he had given since 1893. The book, titled *The Mechanistic Conception of Life*, offered a clear and effective example of how biologists should work: from morphology to physiology, from embryology to animal behavior, from phenomena such as tropism to fertilization, Loeb showed how mechanistic science could and should be performed. At the same time, he thought he had disproved a vitalistic interpretation of life phenomena and defused the arguments of his worse enemies, the "romantic evolutionists" (see Pauly 1987).

Seven years later, a young biologist from California published two large volumes directly addressing Loeb's provocations. The two volumes were edited under the same title: *The Unity of the Organism or the Organismal Conception of Life*. The author was William Emerson Ritter, a pupil of Joseph Leconte at Berkeley. With his synthesis of Hegel and Lamarck, Leconte would have easily been classified as a "romantic evolutionist."[1] In contrast to Loeb, and drawing on Leconte, Ritter believed that the organism could not be reduced to its simplest physicochemical components precisely because what essentially characterized the organism was its unity and integration acquired during evolution. Once it emerged from the depths of the geological past, living organization inaugurated a new, irreducible phase in the

[1]Leconte's evolutionary view, based on the ideas of spontaneity, creativity, and holism, was very distant from a mechanistic and determinist perspective. In his cosmic theory of transmutation, he saw the emergence of new complex unities moving from inorganic matter to human societies. The evolutionary process had to be seen as a form of embryonic development. Heterogeneity followed a state of homogeneity, as Herbert Spencer, inspired by Ernst von Baer, had argued. For Leconte, heterogeneity (diversification) was followed by a process of integration and coordination, which produced new organic unities (Stephens 1978). In other words, organisms became more complex insofar as novel instruments of organic coordination and integration appeared. Cephalization and socialization were two of these instruments that life had used to attain higher levels of integration. The unity of the organism was therefore the result of evolutionary strategies of coordination, which, once attained, produced new irreducible entities.

cosmic evolution. Life was something that the biologists had to assume as given and not try to grasp through artificial partitions and analysis. The essential nature of the organism did not lie in its inorganic components but in the way these components were articulated and functioned. When the organism was parceled out through experiment, life was irremediably gone. So, for Ritter, the Loebian mechanistic conception of life had to be replaced by an organismal conception of life, if a deeper understanding of living organization was to be sought.

This chapter aims to reconstruct the organicist tradition that Ritter, among others, articulated starting from previous and contemporary ideas. Indeed, Ritter's polemical position in relation to Loeb was not a novelty. He drew upon a vast array of old and new sources to support his view (as will be shown later on). However, Ritter's organismal conception of biology was not only a philosophical position; it also defined a community of biologists who, from different perspectives, disciplines, and places, agreed that the idea of a Loebian mechanistic biology was not a perspective worth adopting. In fact, behind what von Bertalanffy had dubbed in the early 1930s an "organismal revolution" (Esposito 2016), there were many of those Loeb disparagingly called "romanticists": a group of scientists that was active from the early twentieth century and that survived the Second World War. However, contrary to Loeb's despective view, it will be shown that this "romantic" community was not always constituted by irrational cranks or mystical eccentrics. Rather, it was a community that was often composed of critical scientists who questioned the overoptimism of the neo-Darwinian agenda and the unwarranted ambitions of the mechanistic (physicalist) approaches to biology. In short, what these biologists dubbed "organicism" was indeed a rational stance that aimed to provide a more sophisticated and realistic representation of living phenomena including evolution, development, and heredity.

Most of these "romantics" were also historically informed and philosophically knowledgeable. They explicitly connected their scientific learning with authors and doctrines of the eighteenth and nineteenth centuries: in particular, Kant and post-Kantian philosophies and many of the scientific extensions and applications springing from these (Esposito 2016). This neo-Kantian syncretic view included important tenets that Kant himself had established when reflecting on the nature of organisms and the epistemic limits to understanding them: for instance, the idea that organisms are self-organized entities that can only be understood as teleological wholes; the idea that organisms are active, creative, and purposive things, whereby causes and effects are deeply intertwined and self-directed toward their reproduction—and therefore their maintenance and adaptation (Lenoir 1982; Mensch 2013); and, as a consequence, the idea that a living organization has to be assumed and never reduced to physicochemical mechanisms, simply because, as Goethe had himself poetically observed in *Faust*: "...though fast your hand lie the parts one by one, the spirit that linked them, alas is gone..." (Goethe 1988). Indeed, the Kantian complex views were filtered and reframed through Goethe's morphological studies (see Levit and Hossfeld 2017), von Baer's embryological reflections, and many other major and lesser post-Kantian and romantic figures active throughout the nineteenth century (Lenoir 1982; Harrington 1999; Sloan 2007).

The chapter, however, does not focus on the "romantic" philosophy of the organism in general but rather on evolutionary and developmental biology—in other words, on the way twentieth-century "romanticists" connected development, heredity, and evolution. Of course, ideas about what an organism is were directly related to how organisms develop, transmit their characters, and evolve. So, the general philosophy of the organism—the latter seen as an irreducible and dynamic entity—was the fundamental premise for articulating the relations between these phenomena. However, although there is a growing quantity of literature that reconstructs organicist philosophies of biology (Harrington 1999; Nicholson and Gawne 2015; Esposito 2016), little work has been done in regard to revising the ideas and models of evolution that organismal "romanticists" scholars have proposed. The undertaking is historically interesting because the very same concept of evolution that many "romanticists" held in the twentieth century had many remarkable connections with the biocentric wide perspectives maintained by some eighteenth- and nineteenth-century romantic naturalists and philosophers: what the German philosopher Adolf Trendelenburg succinctly defined as "the organic view of the world" (Beiser 2014)—especially the idea that a cosmos in permanent evolution, crossed by material forces in constant opposition, produced unexpected alterations in which new irreducible entities emerged. The cosmos followed a teleological pattern akin to an organism in its development. If the world could be seen as a complex organism, rather than a sophisticated mechanism, then the study of the organic world acquired unprecedented relevance. Such a developmental perspective was increasingly streamlined in biology insofar as embryogenetic phenomena themselves harbored the most important secrets of organic matter, including its phylogenetical transmutations. Haeckel biogenetic law represented only one variation of this larger connection between macrocosm and microcosm, and therefore, between the developing embryo and the whole history of life on earth (Gould 1977).

Starting from this very general and vague scheme, many twentieth-century "romanticists" rarely questioned certain beliefs: evolution had to be conceived as a great cosmic process in which new entities and relations constantly emerged. New structural relationship had produced, in the course of geological epochs, functional units that were irreducible to their physicochemical components. These complex units became the principal agents of organic evolution insofar as they, and not a population or species, had to be considered as the source of evolutionary change. As self-organizing, self-directed, and creative entities in an open relationship with the environment, they were extremely plastic and creative beings that could not be properly explained supposing transcendent, immaterial, vitalistic forces. Finally, these organic agents exhibited a teleological nature that could not be easily dismissed with mechanical explanations. Of course, some of these beliefs inform the discussions over the nature of the organisms ever since the eighteenth century (Bertoletti 1990)—discussions that intensified in the nineteenth century—especially in France and Germany, and that led to the successful institutionalization of biology as a professional discipline (Gusdorf 1985).

Now, the training of many biologists in Germany or German institutions toward the end of the nineteenth century makes the conceptual connection between the ideas mentioned and their revival in early twentieth century comprehensible. Some of these "romanticists" had been students of Rudolf Leuckart, Eduard Strasburger, Hans Przibram, and several other figures generally linked to the organicist world. Those who did not travel abroad had formed themselves on books and articles that contained Kantian and post-Kantian bio-philosophical ideas. In England, successful textbooks such as Balfour's *Elements of Embryology* or Sedgwick's *Student's Textbook of Zoology* were impregnated with German organicist thought (see Esposito 2016). Most of these people Loeb would have deemed "romanticists" had read Aristotle, Kant, Schelling, Lotze, and Schopenhauer directly. Many were acquainted with primary and secondary literature on eighteenth- and nineteenth-century biology. They were aware that Buffon, Blumenbach, Cuvier, Geoffroy, Goethe, and von Baer had demonstrated the profound connection between epigenesis and organicism (see Mensch 2013).[2] Finally, these "romanticists" were familiar with the sophisticated debates between neo-Darwinians and neo-Lamarckians, associating the former with mechanistic, materialist, and conservative philosophies and the latter with organicist, systemic, and progressivist positions.

This chapter focuses on a small group of these "romantic" biologists from the Anglophone world (the UK and USA) during the first five decades of the twentieth century. The figures here considered all had important institutional ties and engaged each other in discussing various aspects of their contemporary life science. They all agreed that biology was an independent and irreducible discipline that studied systemic, complex, and creative entities. They also agreed that Loeb's proposal of a pure and unrestrained mechanistic biology was a threat to a discipline, biology, which had to think more in terms of dynamic processes than in terms of inert structures, and more in terms of wholes than in terms of parts. In other words, from the beginning of the twentieth century, there was a noisy and competent international community of "romantic" biologists opposed to neo-Darwinian biologies. Figures such as E. S. Russell, John S. Haldane, D'Arcy W. Thompson, Joseph H. Woodger, William E. Ritter, Frank and Ralph Lillie, Ernest Just, and Charles M. Child constituted a little—nonexclusive—Fleckian "Denkkollektiv" proposing an alternative biological synthesis that aimed to link studies of evolutionary and developmental biology within an organismal framework. In particular, the main idea that all these biologists shared was the conviction that ontogeny "produced" or "created" phylogeny, did not "recapitulated" it, as the British embryologist Walter Garstang concisely put it in 1922.

[2]The strong association between epigenesist and organicism, and the latter with biology, should not appear surprising today. After all, the notion of "organism" was one of the central concepts of romanticism, and "biology" itself, as the French philosopher Gusdorf recognized long ago, was largely a romantic word (Gusdorf 1985).

Finally, it is important to briefly clarify the reasons why I deem Loeb's category of "romanticists" to be historically useful and pregnant, in spite of its derogative intent. First, the adjective "romantic" allows us to see the historical continuity of an old, venerable tradition that was still active in the twentieth century. Indeed, to consider "organismal biology"—or, more generally, "organicism"—as a radical break from the past not only implies doing violence to the available historical evidence but also means denying the historical awareness of the protagonists of the organicist tradition (as will be shown in the next sections). The clash between Loebians and "romanticists" was not only a disagreement about two abstract, ahistorical, philosophical undertakings; it was also, and especially, a clash between two traditions that had a long, controversial history. Secondly, the adjective "romantic" also helps to comprehend, at least more generally, why and how important boundaries between different scientific sensibilities in biology emerged, at least in regard to the first half of the twentieth century. For instance, the contrast between anti-reductionist and systemic "thought style" and a more technocratic, reductionist, and pragmatic approach. After all, especially in evolutionary biology, the clash between developmentalist views of evolution and neo-Darwinian perspectives can be seen as a part of a larger epistemological trend that involved the scientific enterprise as a whole, particularly in the context of the political and social transformations following World War II. However, it must be clear that the use of the adjective "romanticist" should not be intended as an essentialist label that includes and excludes, neatly and sharply, genuine "romanticists" from "non-romanticists." I use the notion "romantic biology" as a signpost that refers to a movement of thought, as an analytical concept denoting a particular way of interpreting life phenomena, as an instrumental label referring to a concentration of ideas articulated and disarticulated according to different contexts and conveniences, and as a name standing for a set of convictions—not always consistent—that oriented scholars of diverse places and generations toward specific views and results. In sum, in order to understand the clashes that we perceive between different ways of understanding biology during the first decades of the twentieth century, we need some tentative categories that can be used as guides or markers. Once these labels have provided us with a better view of particular historical trends, they can be dismissed in favor of more fine-grained categories.

2 From the UK to the USA

One year before his death, the Scottish biologist Edward Stuart Russell published one short essay on Schopenhauer's contribution to biology (1953). The essay situated Schopenhauer's ideas within a larger history of organicism, from the nineteenth century until Ralph Lillie's *General Biology and Philosophy of Organism* (1945), and included a sharp critique of contemporary biology. In particular, Russell used Schopenhauer's philosophy as an effective weapon against mechanistic approaches in developmental and evolutionary biology (i.e., neo-Darwinism).

Indeed, recovering Schopenhauer, Russell observed, was important because "...with the gradual spread of the holistic conception of the organism, the integral view has been coming back into favour" (206). In revising Schopenhauer's short essay *Über den Willen in der Natur*, Russell underscored...three main fundamental points: the law of adaptive specialization, the unity of plan, and the purposiveness of organic nature. The first refers to the correlation of organs in the overall morphological plan; i.e. all organs and functions are necessarily connected in a whole organized unity. Although this idea could be traced back to Cuvier, with his principle of the correlation of parts, Russell noted that the novelty of Schopenhauer's position lied in his interpretation of this principle in terms of the "will": the Kantian thing-in-itself, which is behind all phenomena. The second point, the unity of plan, could be linked to Geoffrey Saint-Hilaire and explains that all structures can be understood as functional adaptations. Finally, the third point refers to the teleological nature of organisms, which can be interpreted neither as an external intelligent goal nor as the result of undirected selection of variations: the organic purposive phenomena are, rather, the result of an internal drive, Schopenhauer's will, which directs and shapes the living matter. For Russell, Schopenhauer's philosophy of the organism was directly opposed to post-Darwinian insights and vitalistic tendencies: "In post Darwinian speculation on evolution—he wrote—too much stress was laid on the separate 'characters' of the organism, especially those which vary inside the species, and the primary fact of the fundamental wholeness and integral adaptive specialization of the living things was lost from sight" (1953: 206). And again, Schopenhauer would "...have rejected also any theory of dualistic vitalism, such as Driesch's theory of entelechy" (208).

Russell's historical reconstruction of Schopenhauer's bio-philosophy was therefore a polemic effort directed against mechanist and Neo-Darwinist evolutionary hypotheses and, at the same time, a strategic move aiming to demonstrate that there was a viable alternative between mechanism and vitalism—an alternative that not only found support in a venerable history but also in twentieth century biology and physiology. Such an alternative would temper overenthusiasm for genetical explanations of evolutionary diversification and, at the same time, would emphasize the importance of individual development in the processes of speciation. Such an alternative, Russell added, might also find support in the philosophical intuitions of Henry Bergson and the more recent scientific ideas of Ralph Lillie (on Bergson, see Loison and Herring 2017).

Russell was not the only scientist interested in Schopenhauer's philosophy. In Britain, the first translation of *Die Welt als Wille und Vorstellung* was published in 1883 and the translator had been the brother of the physiologist John Scott Haldane, Richard Buldon. A student of the neo-Kantian philosopher Hermann Lotze in Germany, Richard would cooperate with his brother John in diffusing Kant and post-Kantian philosophies in the UK. At a very young age, John Scott and Richard would make clear that Kantian and post-Kantian speculations were relevant to biological thinking. As they argued, stressing one of the main themes of romantic thinkers, the organism had to be comprehended through the category of *reciprocity*, where: "...every part of the organism must be conceived as actually or potentially acting on and being acted on by other parts of the environment, so as to form with them a self-conserving system" (1883: 45). Almost three decades later, John Scott introduced, in the Anglophone world, the word "organicism" as a short label for this idea[3]—a notion that, as he noted, could be found in the earlier teachings of Xavier Bichat, Von Baer, Claude Bernard, and Yves Delage—and referred to a third position between mechanistic hypotheses and vitalistic views. This position implied a whole philosophy of the organism that can be briefly summarized as following: life is a category that, in virtue of its visible properties, is irreducible to physico-chemical analyses. The organic parts are shaped and constituted in a dynamic interaction with the whole organism and its environment. These dynamic wholes have to be conceived as teleological and self-sustaining entities, able to adapt and change their form and behavior according to the external circumstances. Precisely because the organism has to be regarded as a whole self-contained unit, organic evolution has to be related to an organism's complex capacity of responding to environmental pressures.

Like Russell later on, Haldane spent his life criticizing what he saw as an over-reductionist neo-Darwinism based on hasty speculations about material inheritance (i.e., Mendelian genetics and then population genetics). And, like Russell, Haldane was heavily indebted to Kantian, Romantic, and idealist philosophies, who denied a mechanist and purely material world. Indeed, British idealists such as T. H. Green and F. H. Bradley, Haldane acknowledged, had been very important for his formation. Although he did not specify what he took from these neo-idealistic philosophies, we could advance the hypothesis that Bradley had taught him to be wary of atomistic worldviews in science (from physics to psychology), which conduced to mere abstractions taken from an absolute, organic, and dynamic whole. Green, instead, an arch-critic of Spencer's individualism, had probably taught him to be very skeptical about social Darwinism and, at the same time, to think of human society as a complex organic whole (Simhony 1991). What Green, Bradley, Kantian, and post-Kantian philosophy had certainly taught him was the old Aristotelian saying that, in the organic realm, the whole was more than the parts. Although Haldane applied such a maxim to his physiological and medical practices, he also believed that evolutionary biology and heredity had to be conceived accordingly.

Haldane argued that whereas heredity could not be reduced to independent and discrete particles, which biologists increasingly frequently named genes, evolution could not be reduced to a statistical distribution of characters fixed through a process of directionless selection. In other words, the production of new characters in evolution could not be focused on mere intracellular processes whereby the genetic material was statistically reshuffled in sexual reproduction. Rather, an evolutionary novelty had to be seen as an "...active adaptation of pre-existing life, and its transmission to descendants is a sign that in the adaptation the life of the organism itself is expressing itself" (Haldane 1935: 74). Heredity belonged to the whole striving organism, which expressed itself through efficient adaptations: "the chromosomes...of a living germ-cell are an expression of its whole life, and its further life in essential

[3]See Haldane (1917: 3).

connexion with embryonic and adult environment furnishes the only key to an understanding of their real nature" (74). While the organism could not be severed from its environment, heredity could not be severed from the whole life cycle of the organism itself. Changes in the life cycle of the individual, therefore, from the earliest stages of development to the adult life, could induce changes in the whole species. New characters arise from this individual process of adaptation during development. Haldane believed that the organism could not be understood as a passive subject of evolution, but rather must be understood as an active entity able to adjust itself according to its needs in a variable environment. Although Haldane did not mention Lamarck directly, his vision was much more akin to a Lamarckian understanding of the evolutionary process than it was to the blind mechanism of Darwinian selection.

Unlike his ambitious cogitations on physiology, Haldane never developed a full-fledged vision of evolutionary biology, of the kind his son J. B. S. developed later. However, John Scott's thinking on evolution was both informed and extended by one of his oldest friends: D'Arcy W. Thompson. Both born in Edinburgh in 1860, they shared most of educational background and interests. Both leaned toward German philosophy and science, and both polemicized, frequently between themselves, regarding materialism, vitalism, and reductionism in biology. While Thompson believed that mechanistic hypotheses could be heuristically fruitful in explaining life phenomena, Haldane remained deeply skeptical about any kind of mechanist approach. However, what they shared was a general distaste toward neo-Darwinism and Mendelian genetics. Even though Thompson had a deep respect for Darwin, he could not accept the overenthusiastic conclusion that many of his later disciples suggested: that natural selection could fully explain adaptation and morphology. And yet, although he seriously considered the works and results of Mendelian geneticists, his view about how cells worked and reproduced was strongly opposed to Mendelian models. Indeed, Thompson believed that the cell had to be understood as a dynamic "sphere of action," rather than as an entity with a static structure containing small powerful entities able to direct and form living organization: "The things which we see in the cell are less important than the actions which we recognise in the cell" (Thompson 1942: 289). Picturing the inner structure of the chromosomes would not reveal the complex function of the whole cell. As a relatively self-contained entity, the cell was a system of forces in relative equilibrium. Speaking about "hereditary substance" in cell nucleus was equivalent to saying that "...a particular portion of matter is the essential vehicle of a particular charge or distribution of energy, in which is involved the capability of producing motion, or doing the work" (1942: 288). As a consequence, for Thompson, the idea that during reproduction organisms transmitted discrete "factors" uniquely responsible for the emergence of phenotypes was a deceptive figure of speech. What was really transmitted during reproduction was a whole system of forces that would produce phenotypes according to the environmental context.

Thompson's cytological view was directly related to his evolutionary view. With Haldane, Thompson maintained that the organism could not be thought of without considering the crucial contribution of the environment. From bacteria to jellyfish, from amphibians to mammals, all organic forms were the response of physical forces and stresses coming from the environment and, more generally, from the material conditions of animal life. Depending on the size of the body, organisms had to face diverse challenges: i.e., surface tension for small organisms and gravity for the largest ones. These physical forces, in concert with what Thompson called "morphological heredity"—i.e., systems of force which engender growth rates during development—produced the different adult forms in the general course of evolution:

> The deep-seated rhythms of growth, which, as I venture to think, are the chief basis of morphological heredity, bring about similarities of form which endure in the absence of conflicting forces; but a new system of forces, introduced by altered environment and habits, impinging on those particular parts of the fabric which lie within this particular field of force, will assuredly not be long of manifesting itself in notable and inevitable modification of form (1942: 1025).

While the organism had to be conceived of as an "integral and indivisible whole," as Aristotle, Kant, Goethe, and Cuvier had taught, such integral wholes were also historical entities, as Lamarck and Darwin had showed. The organicist view, together with the awareness that life had history, conducted Thompson toward the idea that evolution was a process of holistic diachronic transformations of related *bauplans* that could be geometrically described through Cartesian grids. The transformation of whole integrated morphologies could neither be simply ascribed to natural selection nor to animal needs or vital forces. Rather, the transformations were the outcome of physical, chemical, and electrical forces that acted in concert with internal and external stimuli. Altogether, the forces influenced the growth rates of the morphological parts and therefore the overall organic form. As the title of Thompson's famous book indicates, form was the result of differential growth rates.

The relation between growth and form, and therefore between development and morphology, fostered Thompson's interest in developmental phenomena. The American physiologist Charles Manning Child and the Canadian embryologist Frank Lillie were two of his important references for understanding the relationship between growth rates and morphology. While Child had conceived of the organism as a reaction system crossed by specific patterns of metabolic activities that he dubbed "metabolic gradients," Lillie had stressed the power of the environment in relation to morphological evolution. Both had severely criticized Mendelian genetics and both had offered an alternative view according to which evolution could not be understood as change in gene frequencies but rather as changes in developmental paths related to gradients or physiological mechanisms. As I have shown elsewhere, the evolutionary ideas of Thompson were not as bizarre as many recent interpreters have claimed (Esposito 2014): rather, they were an original variation of a general view held by many embryologists and physiologists during the first half of the twentieth century—a view which can be best resumed through Walter Garstang's lucid remark: "Ontogeny does not recapitulate phylogeny: it creates it" (1922: 87). If evolution was the result of developmental changes triggered by internal or external (environmental) stimuli, then natural selection had a very secondary relevance in explaining speciation.

The biologist and philosopher H. Woodger offered another variation on Garstang's theme. Born in 1894, Woodger is one of the most fascinating theoretical biologists of the...first half of the twentieth century (See Nicholson and Gawne 2014). As an admirer of Haldane's organicism, Woodger believed that such a philosophical position had to be related to the view that the living entities had to be understood through three general concepts: functionality (organisms are entities that are always functioning), morphology (this functioning is always related to a particular structure), and environment (function and structure work, directly or indirectly, always in reference to what surrounds the organism). As he explained: "Just as the organism is more than the sum of its material parts, so it is more than a bundle of functions. Just as its parts are organized and unified, so are it functions in reality not separable but all interconnected and integrated, as to result, in the living animal, in one great function—the behaviour of the animal as a whole" (1924: 457). On D'Arcy Thompson's advice (Esposito 2016), Woodger undertook a short period of research leave at the Vienna Vivarium in Austria. This institution, headed by Hans Przibram, was, as Pouvreau recorded, "characterized by the opposition to Darwinian and neo-Darwinian theories, accepting, of course, evolution, but denying the idea that natural selection alone could explain evolution... the dominant conception was that the organism had to be conceived as a system in an active relationship with its environment, and that the organic morphogenesis had to be seen as the result of epigenetic processes..." (Pouvreau 2006: 13). An enthusiast of Thompson's morphological ideas, Przibram had been a pupil of the physiologist Rudolf Leuckart—a deep admirer, in turn, of Kantian philosophy (Lenoir 1982). In the vibrant environment of Vienna, Woodger would learn about systemic thinking through the scientific and philosophical discussion that took place between Paul Weiss and Ludwig von Bertalanffy (Drack and Apfalter 2007). However, one of the concepts that most attracted Woodger's interest was the notion of "hierarchy" in an organic system and how it relates to heredity, development, and, eventually, evolution. Organisms, after all, could be conceptualized as hierarchal systems whereby causes and effects, intra and extra level, were deeply connected. Woodger believed that a solid concept of biological hierarchy could solve different conceptual issues related to dichotomies such as inborn and acquired characters and, therefore, could provide a larger, more sophisticated context in which the process of speciation could be framed. Indeed, Woodger believed that the term "heredity" had to be eliminated, in favor of a different conceptualization of the transmission and manifestation of characters. In order to do that, Woodger distinguished between intrinsic (or immanent) and relational properties. When we think about a hierarchical organization, we assume that the parts have specific relations with other parts, which may or not be static or dynamic. Thus, while intrinsic properties were those characters (phenotypes) which were transmitted with relative constancy generation after generation, the relational properties had to be considered as new characters generated during development, representing the engine of novelty in evolution. In other words, the intrinsic properties were none other than previous acquired relational properties:

> If two cells, which are assumed to have "equal" nuclei behave differently in the same environment, we should say that they differed intrinsically in their cytoplasm, since their relations are supposed to be the same. But that intrinsic cytoplasmatic difference may have been acquired in consequence of relational differences during development, and would therefore be an acquired relational property. But since it now persists in spite of changed relations (since by hypothesis both cells are in the same environment now) we should have to call it an acquired intrinsic property (Woodger 1930: 15).

The dialectic between relational and intrinsic properties could explain how novelty can be generated over time, supposing that the acquired relational properties were consequence of developmental changes. The unit of evolution, therefore, could not be the population, but rather the individual organism, which, during development, reproduced patterns and structures typical to its own species and, at the same time, small or large variations triggered by environmental stresses. Of course, Woodger never offered a full-fledged evolutionary theory. He was mainly concerned with defining what an organism was. Nevertheless, he reached the conclusion that it was within an organized, hierarchical, and dynamic entity that new relations between parts in a whole could be acquired during development. After all, Woodger argued: "Development is a process in which with temporal passage new spatial parts come into being all with the same genetic endowment." (Woodger 1929: 376). More specifically, he maintained that both embryogenesis and evolution were two forms of development: "We speak of individual development and evolutionary development as two examples of development" (Woodger, 1930:391), and yet, "Developmental theories, whether individual or racial, do not deal with characters but with processes, i.e. with organisms as events in relation to the events constituting their enviroment" (Woodger, 1930: 423). As a consequence, evolution, as a specific form of development, was characterized by the temporal emergence of new modes of living organization. Although Woodger admitted that he could not find any convincing hypothesis about how these new modes of organization actually occurred, he concluded that evolution itself had to be eventually conceived as a dynamic, not "uniform", epigenetic event (Woodger, 1030:427).

While for Russell the French philosopher Henri Bergson, with his conception of elan and creative evolution, could make sense of the essential productivity of the organic matter, Woodger found in Whitehead's process philosophy the correct inspiration for a new philosophy of the organism. In the Romantic spirit of Whitehead, who in 1925 wrote that: "nature is a process of expansive development... nature is a structure of evolving processes" (Whitehead 1925: 135), Woodger felt that evolution could be seen as a process in which new relations and hierarchies come into being ("concretions" in Whitehead's terminology). The dialectic between relational and intrinsic properties involved a Whiteheadian evolutionary process, going from the structuring of novelty to the fixation of the characters. When Russell noticed in his *The Interpretation of Development and Heredity* (1930) that evolution could not be explained as being the result of blind mutation and selection, but rather as systemic alterations of developmental paths, he was only stressing one central tenet advanced by what I call "romantic biologists": the largely shared belief that ontogeny produced phylogeny—which is quite different from the more famous, and probably more controversial, argument that ontogeny recapitulates phylogeny. Indeed, while in the latter phylogeny produced ontogeny, in the former it was ontogeny which produced evolution. For Woodger after all, the origins of evolutionary novelties depended on hierarchical changes along the ontogenetical processes. Evolution was the overall drama in which the creativity of each individual organisms was staged.

In 1954, Russell felt that new findings supported this idea, and many of the new proposals came from the USA (in particular, he mentions Joseph Lillie and William Ritter). The future of organicism could be glimpsed, in its newest developments, in the USA.

3 From the USA to the UK

In the chapter on Schopenhauer's biology, Russell had concluded that the history of organicism could count on one important recent contribution from the USA: Ralph Lillie's *General Biology and Philosophy of Organism* (1945). Lillie was a physiologist based at the University of Chicago from 1924. Until his retirement, he had often published in philosophy journals. Indeed, *General Biology* was a compilation of different papers Lillie had published from the 1920s onward, for a general public. The influence of Whitehead and Bergson, as well as the references to D'Arcy Thompson, Russell, and Ritter and many organicists, makes Lillie's monograph an excellent instance of what Loeb understood with "romantic evolutionism." The book tackled what Lillie considered the most important and controversial topic of general biology: the difference between living and non-living systems—the former deemed to be the evolutionary consequence of the latter. Indeed, life was a particular kind of emergent organization from matter—an irreducible organization which implied directionality. In general, Lille believed that there were two principal tendencies in the organic nature: a conservative and an active side. Conservation and transformation are two sides of the same coin. For instance, Lillie maintained that there could be no creativity in evolution without a general framework of regularity and permanence. Living entities were very conservative systems. Complex mechanisms of reproduction guaranteed the continuity of the species. However, the regularity of biological systems did not imply their complete predictability and determinism. The process of embryogenesis, although highly stable, could also be highly creative. Using Whitehead's terminology, Lillie defined development as a process of concrescence "...in which a variety of materials come together to form a closely unified whole" (Lillie 1945: 23). Life, Lillie added, quoting Bernard, is creation: such intrinsic creativity of the living systems had to be linked to the capacity of life to transcend regularity under environmental stresses: "This dependence of local development (i.e. of special parts and organs) on local physical conditions is an intimate one, and experimental alterations of these conditions (by excision or displacement of parts, chemical influence, transplantation, etc.) produces correspondingly constant changes in development" (Lillie 1945: 203). Although environmental pressures could trigger changes in the organism's developmental path, the real issue was to understand the origin of novelty in evolution: "What constitutes the real biological problem is... the tendency toward novelty, synthesis and higher organization, as seen both in individual development and in evolution" (Lillie 1945: 204). Lillie did not have a clear answer to this problem. He believed, however, that speciation had to be somehow related to a psychical directive factor:

> The factors of evolutionary diversification still remain essentially unknown, but the selection of purely fortuitous variations does not seem to be a sufficient explanation for the origin of the more complex adaptive characters. The entrance of some directive or integrative factor seems to be required (Lillie 1945: 47–48).

Like Russell, and Richard Semon, or Richard Hering before him, Lillie played with the mnemonic theory of heredity (Esposito 2013). Although there were different versions of such theory, the common denominator was that the characters of the organisms could be analogically explained as "engrams": mnemonic traces engraved in the germinal matter. These traces were produced through an organism's interaction with the environment and were reproduced during development, just as a phonograph reproduced the traces of a vinyl record. However, Lillie recognized that the "engrams" could only be used as analogical or metaphorical concepts for explaining heredity and could not really explain the emergence of novelties in evolution. Even though the causes of evolutionary change were far from being understood, apart from the general descriptive dialectic between conservation and novelty that organisms exhibit over geological eras, what the biologist could surely maintain was that evolution referred to the historical process of the emergence of new organized entities. For Lillie, the evolutionary process that Lloyd Morgan had proposed in his influential *Emergent Evolution* (1923) described very clearly the generation of irreducible novelties—of new levels of systemic organization—which could be simply the result of accidental combinations in the physical world or teleological assortments in the organic world.

Not surprisingly, Loeb despised Lillie's speculations, which he considered as too mystical or "Bergsonian." But Loeb also scorned the work of Lillie's elder brother, Frank Lillie (Manning 1985). A pupil of Charles Otis Whitman at the University of Chicago,[4] Frank replaced Whitman as director at the Marine Biological Laboratory at Woods Hole in 1908. Like Ralph, Frank was deeply critical of mechanistic and reductionist approaches in the life sciences. And, like Haldane, Thompson, Russell, or Woodger, Frank shared the same uneasiness toward particulate theories of heredity. Indeed, he distinguished between the latter and physiological theories of heredity: the first pertaining to the study of the transmission of characters within populations and the other concerning the development of those characters in individual organisms. As he clarified, for example, the observed differences in the color of mammals' fur did not simply result from the presence or absence of some supposed determinants in the germ cells but from physiological and developmental mechanisms: "the development or inheritance of color... can certainly not be due to the presence of black or brown or red or yellow determinants in the germ, assumed for theoretical purposes by some students of heredity, but to a specific power of oxidation of the protoplasm" (Lillie 1914: 248). The visible characters, after all, could not be merely

[4]Whitman, in turn, had been a student of Leuckart in Leipzig.

related to some postulated elements but had to be conceived of as the unique result of different physiological processes happening in different times and contexts during the whole process of embryogenesis.

The studies and experiments Lillie started to perform from the 1930s onward on feather development in Brown Leghorn fowl—the very same studies that had fascinated D'Arcy Thompson—showed that the inheritance of morphology and pigments (feather patterns) were related to hormones and growth rates in different regions. Lillie injected hormones and showed that this could produce particular pigment patterns, depending on the growth rates and the quantities administered: "As rate of growth is a fixed property of the different feathers tracts...it is possible to produce birds by means of suitable administration of female hormone, with female feathers in the slowly growing tracts, and feathers of male characters in the more rapidly growing tracts" (Lillie and Juhn 1932: 177). In observing variations in, and the transmission of, plumage patterns, Lillie was following to some degree the example of his mentor, Charles Whitman, who had previously studied the evolution of pigeons through the observation of plumage patterns (Gould 2002). However, while Whitman used his observations as evidence to support an orthogenetic theory of evolution, Lillie saw plumage patterns as a window onto physiological heredity. But in spite of their differences, both believed that development was "...the more general problem of biology" (Whitman 1919). The nature of development, with its law-like processes and teleological patterns, mirrored evolution. Ontogeny and phylogeny, after all, had not to be considered as two distinct phenomena insofar as the nomological, linear, and seemingly teleological direction of evolution mimicked the nomological, linear, and teleological direction of individual development. Indeed, Lillie maintained that development was the proper window through which to observe evolution (Gilbert 2003).

Although Lillie did not develop any theory about how these changes could be transmitted, one of his students, the zoologist Ernest Just, did. Just followed a similar career path of many American biologists during the late nineteenth and early twentieth century. After the PhD at the University of Chicago, he complemented his formation at Dohrn Zoological Station in Naples and at Kaiser-Wilhelm-Institut für Biologie in Berlin. In 1939, Just published the *Biology of the Cell Surface*, an ambitious book showing the importance of cytoplasm in controlling development in interaction with the cell nucleus. The book begins with Goethe's short poem "on the contrary (to the physicists)" which anticipates the whole monograph's content: "Natur hat weder Kern, Noch Schale, Alles ist sie mit einem Male" (that can be translated as "Nature has neither core nor shell. But is all at once"). To Just, the poem expressed the intuition that organisms are constituted of parts, but the parts, in turn, are shaped and articulated into the whole, so that neither parts nor the whole could be thought independently from each other. Just also applied the poem's idea to his cytological theory: nucleus and cytoplasm were both responsible for the expression of hereditary characters and the generation of novelties in evolution. In the cell, there was neither core nor shell, but all is at once.[5]

[5]For a recent revisitation of some of Just's biological ideas, see Byrnes and Newman (2014).

In particular, Just speculated that the crucial differences among species had to be related to differences in the structure and function of the cell surface, rather than structural changes in the chromosomes. Thus, environmental stimuli could produce changes in the ectoplasm (the outer region of the cytoplasm) that, in turn, could modify the nucleic substance and therefore could be transmitted to the following generations. As he explained: "...species arose through changes in the structure and behaviour of the ectoplasm. In the differentiation of ectoplasm from ground-substance we thus seek the cause of evolution" (Just 1939: 361). If heredity manifested itself during the embryogenetic process, then both heredity and evolution had to be understood through the analysis of embryogenesis—and the latter, in turn, revealed that the developing organism and the environment were "...one reacting system" (Just 1933: 23). In other words, the proper engine of evolution was the dynamic dialectic between environmental pressures and cellular alterations.

Just had also been a student of the physiologist Charles Manning Child, a colleague of Lillie's brothers at the University of Chicago. As one of the last students of the organicist physiologist Rudolf Leuckart in Leipzig, Child maintained his criticism of any reductionist theory of heredity until the end of his life. He believed that the phenomena of animal regeneration could reveal the essence of development and, therefore, the origins of variation in evolution. In order to explain the origin of order in embryogenesis, Child introduced the concept of "metabolic gradient." He observed that cellular organization followed a specific path along the axis of the organic bodies (whether radial, bilateral, or spherical morphological symmetries), and this path could be understood as a physiological gradient orienting cellular differentiation and specialization. Thus, organisms were crossed by metabolic gradients which directed and dictated growth rates in different regions. His experiments performed on *Planarias* and *Tubularias* showed that gradients established axes of activity that were directly related to metabolic rates. Reproduction, for instance, was interpreted by Child as the reestablishment, in sexual or asexual reproduction, of a new metabolic gradient. If the organism was a collection of integrated metabolic gradients forming dynamic tensions, heredity had to be conceived of as the capacity of an individual organism to produce a new whole. In other words, the organism had to be seen as a "system of reaction" capable of transmitting specific potentialities which could trigger, in the right context, new waves of metabolic gradients. Child therefore concluded, as against Weismannians and Mendelians, that the unity of heredity had to be the whole developing organism: "the original specific reaction system in which the gradient arises is the fundamental reaction system of the species, the basis of inheritance and development" (Child 1911: 152).

If the whole developing organism had to be considered as a unity of inheritance, it followed that morphological changes in evolution had to be ascribed to changes in the whole reaction system. As Child argued: "Evolution is not directly concerned with morphological characters, but with the physico-chemical constitution of the reaction system, and so with the rate and character of its reactions and the conditions under which they occur" (Child 1915: 205). The very notion of "acquired characters" had to be reconceptualized as the structural changes experienced by the reaction system due to the continuous pressure of external stimuli. Despite the evidence, Child added, "...it is difficult to understand how biologists can continue to maintain the distinction between soma and germ plasm, and to content themselves with the assertion that natural selection is adequate to account for adaptation in the organic world" (Child 1915: 205). Evolution, therefore, was to him all that followed from structural and functional changes in the developing embryos.

Child's view was taken further by one of his friends in southern California, William Emerson Ritter, the same scholar who, as already mentioned, had challenged Loeb's mechanistic views in 1919. In 1903, Ritter had founded a small marine zoological station on the shores of San Diego. Child spent his research summers there, where he could count on a rich marine flora and fauna upon which he could perform his experiments relating to animal regeneration and development. Like Child, Ritter spent an important part of his formative years in Europe. He had befriended Child at the Dohrn Marine zoological station before concluding his research trip in Berlin. With Child, Ritter was preoccupied with the fact that an important tradition in biology, which he dubbed "organismal biology," was at risk of disappearing due to the extraordinary appeal of reductionist and mechanistic approaches in the biosciences. The proliferation of particulate theories of heredity, as well as neo-Darwinian hypotheses, overshadowed more traditional ideas and methods that could provide a deeper understanding of life phenomena. As we have seen, in order to contrast this dangerous tendency that Loeb had clearly expressed in his 1912 *Mechanistic Conception of Life*, Ritter published the two volumes titled the *Organismal Conception of Life* (1919). The books included the history, philosophy, and scientific content of organismal biology, and an informed criticism of what Ritter called "elementalism"; i.e., the idea that life processes are best understood when reduced to their simplest components. To "elementalism" Ritter opposed "organismalism", which regarded living organisms as active and reactive irreducible systems that could never be understood as a collection of physicochemical properties. Ritter's historical narrative, in emphasizing this philosophical intuition, started with Aristotle, passed through Cuvier, and the French comparative anatomists, and concluded with the American embryological school, which principally included C. Whitman, F. Lillie, E. Wilson, and C. Child.

The volumes also included philosophical reflections. As previously mentioned, Ritter had been influenced by his mentor, Leconte, at Berkeley. However, he had also been influenced by the neo-Kantian philosopher Josiah Royce, while studying at Harvard. Ritter believed that Royce's idealism was a good candidate for explaining how organic unity was the manifestation and, at the same time, the result of psychic integration (Ritter 1919, vol. II). In an article published in 1928, Ritter complemented his biological synthesis with the British philosopher Alfred Whitehead, who, while at Harvard, had developed a new philosophy that he had dubbed "process or organic philosophy": a metaphysical stance which Whitehead compared to the nineteenth-century romantic critiques against eighteenth-century materialism (Whitehead 1925). Ritter's article aimed to offer a new updated and synthetic view on organismal biology. He recognized the general contributions of Haldane and Russell to organicism and argued that the organismal approach had to be completed with Jan Smuts' and Lloyd Morgan's conception of emergent evolution. Synthesizing the intuitions of the "emergentists," Ritter restated the idea that evolution was a creative and productive process that could not be understood as a mere mechanical process.

While Ritter took seriously Whitehead's process philosophy, he was also engaged in a strenuous conceptual struggle against Mendelian biology, which he considered to be at odds with the evidence provided by organismal biologists. As Ritter explained, in criticizing Bateson's Mendelism and its relation to evolution: "My main reason for believing the enterprise will never be carried through, seriously, is that the organismal standpoint has already advanced so far on secure observational and experimental and inductive foundations, that the scientific uselessness if not folly of such elementalistic systems will deter working biologists from spending their time on them" (Ritter 1919, vol. 2: 22). In fact, with Lillie and Child, he believed that the developing organism, as a reaction system, had to be thought of as the unit of inheritance and, consequently, as the privileged site of investigation for understanding evolution. Ritter believed that the organism could not be separated into heredity and development because heredity was equivalent with the developing reaction system. In an unpublished manuscript titled "Biology Greater than Evolution" (undated, but probably dating from the 1920s; see Esposito 2016), he complained that neo-Darwinians pretended to explain variations only through the modification of the whole "race" (population), overlooking the individual, developing organism, which was the veritable source of novelty in evolution. Thus, Ritter maintained, although Darwinian biology had been essential for our understanding of life on earth, it needed to be complemented with the essential tenet that the individual organism was the veritable source of variation and, therefore, speciation. While heredity studies had to focus on what Ritter called "descriptive ontogenesis"—i.e., observing the way morphological characters emerge during development—evolutionary biology had to rely on empirical and experimental investigations of the whole reaction system. In short, Ritter underscored Child's intuition that evolution was the consequence of functional and structural changes in the reaction systems during development.

In 1931, Ritter traveled to London to chair a session at the Second International Congress of the History of Science. The session included, among others, most of the British organicists: Russell, Woodger, Haldane, Joseph Needham, and D'Arcy Thompson attended and talked about the relationships between physics and biology. This was one of the last occasions on which most of the Anglophone organismal biologists were gathered together. Indeed, 20 years later, while Russell was writing his chapter on Schopenhauer's bio-philosophy, many of the most important organismal biologists from both sides of the Atlantic Ocean had died and the reductionist and mechanist approaches in biology had largely overshadowed the organismal conception of life. The organism itself, as a self-contained, self-organized, dynamic, and creative entity, was reframed in terms of molecules and their interactions and, later, in terms of information encoded within macromolecules. The functional and teleological phenomena could be understood as epiphenomena of purely adaptive mechanisms. Evolution could be simply explained in terms of changes in gene frequencies within a population. A biology based on processes and activities was translated into a biology rooted in structures. In the increasingly technocratic life sciences emerging during the Cold War, the Loebian dream of a biology freed of "romanticists" was generally realized (although never totally achieved).

4 Conclusion

Mayr's famous distinction between typological and population thinking can help us in introducing the opposition between developmental and neo-Darwinian views of evolution (Mayr 1982). In the first case, the individual is the most significant source and unit of evolutionary change, while in the latter the population is the main site of evolutionary novelties. The "romanticists" I have introduced in this chapter definitely uphold the first alternative and criticize the second. We can also contextualize Mayr's dichotomy within Bowler's narrative about non-Darwinian revolution. Bowler showed how nineteenth-century developmental traditions survived well into the twentieth century, challenging and often even overwhelming neo-Darwinian hypotheses. Orthogenesis and other transformist theories proliferated at the end the nineteenth century and continued to be accepted and updated until a certain consensus was reached with the modern synthesis after the 1950s—at least in the Anglophone world (Bowler 1988, 1994). Indeed, throughout the first decades of the twentieth century development continued to be a powerful metaphor for understanding phylogeny. The orderly and teleological processes of embryogenic development worked as a powerful metaphor for evolutionary thinking. However, as we have seen, the developmentalist perspective advanced by these "romanticists" was neither equivalent with Ernst Haeckel's biogenetic law nor represented a variant of orthogenetic evolution, but rather endorsed Walter Garstang's idea according to which ontogeny does not recapitulate phylogeny, it creates it. While in the biogenetic law ontogeny is a mirror of phylogeny, and in many cases its outcome, in the latter phylogeny was the product of ontogeny. In other words, the creative site of evolutionary diversification was the individual developing organism. In that sense, this perspective is closer to contemporary evo-devo than nineteenth-century recapitulationist theories.

But Mayr's philosophical dichotomy and Bowler's historical thesis needs to be linked to even larger epistemological trends in biology (and beyond) during the late nineteenth century and the first half of the twentieth. As shown, accepting or rejecting a theory (neo-Darwinism or developmentalism) means accepting or discarding a certain kind of biology and therefore a specific way to explore, understand, or explain life phenomena. The philosophy of the organism that many "romanticists" defended was definitely at odds with the idea that organisms were the outcome of relatively independent characters. If the organisms had to be conceived of as integrated, systemic, and creative wholes, evolution could only be the consequence of the dynamic activity of these wholes taken individually. The clash between alternative understandings of evolution was much deeper than an empirical or theoretical conflict between rival hypotheses. The real issue behind developmentalists and neo-Darwinians lay in the different tendencies regarding how to perceive the organism and what biology as a discipline should be. And this, in turn, was linked to a host of social and political concerns involving mechanist and materialist world views, often connected with conservative scientific or anti-scientific ventures (eugenics, social Darwinism, racial biology, etc.). Indeed, social progress and democracy, Manning Child and Edwin Conklin averred, were directly rooted in biological principles of organic integration and coordination (Child 1924; Conklin 1938). And, as Ralph Lillie lamented: "What the pure mechanist contends is that past conditions determine present conditions completely and unconditionally, and he extends this kind of determinism to human behaviour." However, as Lillie added, "the present is part of the general creative advance of nature" (Lillie 1945: 169). In short, to some biologists, organicism guaranteed democracy and free will. Mechanism only involved fatalism and despotism. For others, the opposite was true. In the twentieth-century discussions over the nature of biological phenomena, vague notions such as mechanism, determinism, reductionism, physicalism, etc., were not mere philosophical options supporting neutral views about development and evolution; rather, they also worked, frequently, as premises for political positioning (Esposito 2016).

Now, if after the 1940s, the "romanticists'" views gradually fell in the background, in favor of neo-Darwinian hypotheses (supported by new disciplines such as population genetics, increasingly paralleled by a deeper understanding of the hereditary mechanism through molecular biology), then the question is: why did that happen? The question is particularly relevant because, in the last few decades, the consensus about the neo-Darwinian synthesis has been eroded in favor of a new form of developmental evolutionism (evo-devo), and novel versions of organicist philosophies have again entered onto the stage. As Meloni (2016) has recently stressed, new versions of soft heredity, in concert with notions such as the reactive genome, have made the classic paradigm of the modern synthesis problematic at least. The overreaching perspective offered by the developmental system theory, or by the complex epigenetical models of gene expression, is certainly closer to the systemic view of the "romanticists" than the adaptationist models of the modern synthesizers. The organismic system approach defended by Callebaut et al. (2007) or evo-devo's agenda based on the intuition that the processes of growth in development offer a privileged window for observing evolutionary changes is perfectly in line with the visions of the individuals I have here presented. Of course, differences and similarities between early twentieth-century organicisms and more recent organicist proposals need to be handled with care. We should avoid "Whiggish" historical interpretations, suggesting, even implicitly, that past reconstructions fit, or support, contemporary research programs, as Pigliucci seems to argue with his critical remarks of this chapter (Pigliucci 2017). Indeed, the chapter does not have the ambition to foster a contemporary revival of what I have called "romantic biology." It would be an anachronistic and purely nostalgic attempt without any chance to succeed. The meaning of this historical exercise of recovering different twentieth-century traditions in biology, at least within the scopes of this volume, is to figure out how Darwinism itself developed and changed through, and against, his critical detractors.

Although a comprehensive answer to the question about why "romanticists" became a small minority after the Second World War goes beyond the aims of this chapter, there are at least two related hypotheses which could be profitably explored. First, the postwar anti-Lysenko campaign in the Western bloc has certainly prevented the development and diffusion of hypotheses that smelled of Lamarckism (Esposito 2015). It is not very difficult to realize how organicist conceptions, such as those of Ritter or Russell, were seen to be dangerously closer to Lysenko than neo-Darwinism. Secondly, the pessimistic epistemology of the organicists, who dismissed the possibility that the behavior of creative and unpredictable entities such as organisms could be successfully forecasted and controlled. In an increasingly technocratic environment, in which science has had to play a decisive role in managing and controlling reality, the holistic philosophies that animated romantic views of life and evolution were perceived as useless conjectures or indulgently aristocratic speculations in a world broken by socialist threats and capitalist excesses. In short, anti-Lamarckian models, based on unilineal and simpler representations of heredity and evolution, matched better with the more utilitarian expectations of the postwar biomedical establishment, epistemically and politically. In the 1950s, while Russell was recovering Schopenhauer's bio-philosophy, hoping to dismiss mechanistic ideals and neo-Darwinian views, he did not realize that the world around him had radically changed. In this new scientific world, where controlling was more important than understanding, there was increasingly little space and patience for "romanticist" speculations on the real nature of life.

References

Beiser F (2014) After Hegel: German philosophy, 1840–1900. Princeton University Press, Princeton

Bertoletti S (1990) Impulso, Formazione e Organismo. Leo Olschki, Firenze

Bowler P (1988) The non-Darwinian revolution. Reinterpreting a historical myth. John Hopkins University Press, Baltimore

Bowler P (1994) Darwinism and modernism: genetics, palaeontology and the challenge to progressionism, 1880–1930. In: Ross D (ed) Modernist impulses in the human sciences—1870–1930. John Hopkins University Press, Baltimore, pp 236–254

Byrnes MW, Newman SA (2014) Ernest Everett just: egg and embryo as excitable systems. J Exp Zool B 322(4):191–201

Callebaut WG, Muller G, Newman S (2007) The organismic systemic approach. In: Sansom R, Brandon R (eds) Integrating evolution and development. MIT Press, Cambridge, pp 25–92

Child CM (1911) Studies on the dynamics of morphogenesis and inheritance in experimental reproduction. J Exp Zool 13(1):265–320

Child CM (1915) Individuality in organism. Chicago University Press, Chicago

Child CM (1924) Physiological foundation of behaviour. Henry Holt, New York

Conklin EG (1938) The biological bases of democracy. Barnwell Address 62:1–10

Drack M, Apfalter W (2007) Is Paul Weiss' and Ludwig von Bertalanffy's system thinking still valid today? Syst Res Behav Sci 24(5):537–546

Esposito M (2013) Heredity, development and evolution: the unmodern synthesis of E.S. Russell. Theor Biosci 132(2):165–180

Esposito M (2014) Problematic "idiosyncrasies": rediscovering the historical context of D'Arcy Wentworth Thompson's science of form. Sci Context 27(1):79–107

Esposito M (2015) More than the parts: W. E. Ritter, the Scripps Marine Association, and the organismal conception of life. Hist Stud Nat Sci 45(2):273–302

Esposito M (2016) Romantic biology—1890–1945. Routledge, London

Garstang W (1922) The theory of recapitulation: a critical re-statement of the biogenetic law. Linn J Zool 35(232):81–101

Gilbert SF (2003) Edmund Beecher Wilson and Frank R. Lillie and the relationship between evolution and development. Developmental Biology. Sinauer

Goethe JW (1988) Faust. Act I–IV. Bantam Classics, New York

Gould SJ (1977) Ontogeny and phylogeny. Harvard University Press, Cambridge

Gould SJ (2002) The structure of evolutionary theory. Harvard University Press, Cambridge

Gusdorf G (1985) Le Savoir Romantique de la Nature. Les Éditions Payot, Paris

Haldane JS (1883) The relation of philosophy to science. In: Seth A, Haldane R (eds) Essays in philosophical criticism. Burt Franklin, New York

Haldane JS (1917) Organism and environment as illustrated by the physiology of breathing. Yale University Press, New Haven

Haldane JS (1935) The philosophy of a biologist. Clarendon Press, Oxford

Harrington A (1999) Reenchanted science: holism in German culture from Wilhelm II to Hitler. Princeton, Princeton University Press

Just EE (1933) Cortical cytoplasm and evolution. Am Nat 66:61–74

Just EE (1939) The biology of cell surface. Technical Press, London

Lenoir T (1982) The strategy of life. University of Chicago Press, Chicago

Levit GS, Hossfeld U (2017) Major research traditions in 20th century evolutionary biology: the relations of Germany's Darwinism with them. In: Delisle RG (ed) The Darwinian tradition in context: research programs in evolutionary biology. Springer, Cham, pp 169–194

Lillie FR (1914) The theory of individual development. Pop Sci Mon 75(14):239–252

Lillie RS (1945) General biology and philosophy of organism. Chicago University Press, Chicago

Lillie FR, Juhn R (1932) The physiology of development of feathers. Physiol Zool 5(1):124–184

Loeb J (1912) The mechanistic conception of life. University of Chicago Press, Chicago

Loeb J (1915) Mechanistic science and metaphysical romance. Yale Rev 4:766–785

Loison L, Herring E (2017) Lamarckian research programs in French biology (1900–1970). In: Delisle RG (ed) The Darwinian tradition in context: research programs in evolutionary biology. Springer, Cham, pp 243–270

Manning K (1985) Black Apollo of science. The life of Ernest Everett just. Oxford University Press, Oxford

Mayr E (1982) The growth of biological thought. Harvard University Press, Cambridge

Meloni M (2016) Political biology. Science and social values from eugenics to epigenetics. Palgrave, London

Mensch J (2013) Kant's organicism: epigenesis and the development of critical philosophy. Chicago University Press, Chicago

Nicholson D, Gawne R (2014) Rethinking Woodger's legacy in the philosophy of biology. J Hist Biol 47(2):243–292

Nicholson D, Gawne R (2015) Neither logical empiricism nor vitalism, but organicism: what the philosophy of biology was. Hist Philos Life Sci 37(4):345–381

Pauly P (1987) Controlling life. Jacques Loeb and the engineering ideal in biology. Oxford University Press, Oxford

Pigliucci M (2017) Darwinism after the modern synthesis. In: Delisle RG (ed) The Darwinian tradition in context: research programs in evolutionary biology. Springer, Cham, pp 89–104

Pouvreau D (2006) Une Biographie non Officielle de Ludwig von Bertalanffy (1901–1972). Bertalanffy Centre for the Study of System Science, Wien
Ritter WE (1919) The unity of the organism, or, the organismal conception of life, vol 2. Gorham Press, Boston
Ritter WE (1928) The organismal conception, its place in science and its bearings on philosophy. Univ Calif Publ Zool 31(14):307–358
Russell ES (1930) The interpretation of heredity and development. Clarendon Press, Oxford
Russell ES (1954) Schopenhauer's contributions to biological theory. In: Underwood EA (ed) History, philosophy and sociology of science II. Arno Press, New York, pp 203–211
Simhony A (1991) Idealist organicism: beyond holism and individualism. Hist Polit Thought 12 (3):515–535
Sloan PR (2007) Kant and the British bioscience. In: Huneman P (ed) Understanding purpose: Kant and the philosophy of biology. University of Rochester Press, Rochester, pp 149–170
Stephens L (1978) Joseph Leconte's evolutional idealism: a Lamarckian view of cultural history. J Hist Ideas 39:465–480
Thompson DW (1942) On growth and form. Cambridge University Press, Cambridge
Whitehead A (1925) Science and the modern world. Macmillan, New York
Whitman CO (1919) Orthogenetic evolution in pigeons. Carnegie Inst. Washington Publication N.257, Washington
Woodger JH (1924) Elementary morphology and physiology for medical students. Oxford University Press, Oxford
Woodger JH (1929) Biological principles: a critical study. K. Paul, Trench, Trubner, London
Woodger JH (1930) The concept of organism and the relation between embryology and genetics—Part I. Q Rev Biol 5(1):1–22

Lamarckian Research Programs in French Biology (1900–1970)

Laurent Loison and Emily Herring

Abstract The situation of biology in France in the twentieth century has always been considered something of an oddity. The theories of the Darwinian Modern Synthesis and of population genetics were not included in standardized university curricula and the main research programs until the 1970s. Against the Darwinian picture that was developing abroad, French life scientists promoted various forms of Lamarckism. The aim of this chapter is to produce a general picture of these different twentieth century Lamarckian research programs which deeply structured various fields of the French life sciences, like morphology, zoology, paleontology but also microbiology and virology. We first recall the failure of the first Lamarckian program, based on a mechanistic understanding of life, and which aimed at explaining evolution in terms of cumulative adaptation through the inheritance of acquired characters. We show that during the interwar period, French Lamarckians were no longer unified in their understanding of the evolutionary process but instead defended a heterogeneous array of concepts. In particular, we examine philosopher Henri Bergson's legacy, which was pivotal in the setting up of a second Lamarckian program that started to develop in the 1940s with the work of zoologists Albert Vandel and Pierre-Paul Grassé. While it is true that the various forms of Lamarckism delayed the reception of Darwinism and, to a lesser extent, genetics, we assess their impact on the way the Modern Synthesis and molecular biology were conceived and developed in France by non-Lamarckian biologists like Georges Teissier, Philippe L'Héritier, André Lwoff, or Jacques Monod.

Keywords French biology • Lamarckism • Inheritance of acquired characters • Henri Bergson

L. Loison (✉)
CNRS, IHPST (UMR 8590, CNRS, Paris 1, ENS), 13, rue du Four, 75006 Paris, France
e-mail: laurentloison@yahoo.fr

E. Herring
University of Leeds Woodhouse Lane, Leeds LS2 9JT, UK
e-mail: preh@leeds.ac.uk

R.G. Delisle (ed.), *The Darwinian Tradition in Context*,
https://doi.org/10.1007/978-3-319-69123-7_11

1 Introduction

No other scientific nation opposed Mendelian genetics, the chromosomal theory and the Evolutionary Synthesis as strongly, and during such a significant portion of the twentieth century, as France. During the same period, Lamarckian-oriented research programs were flourishing. For nearly a century, French biological thought was almost completely dominated by various forms of Lamarckism, forms that often presented substantial differences and were sometimes even theoretically incompatible. Despite these differences, all posited that the Darwinian approach was neither a satisfactory nor fruitful way of theorizing about organic evolution. Natural selection provided an explanation for the survival of the fittest but could by no means account for the *origination*—i.e., the causal/physiological formation—of the fittest. In other words, the creativity of natural selection was either denied or misunderstood.

The aim of this chapter is to provide a balanced overview of these various Lamarckian programs as well as their relationship with the programs they opposed, namely genetics and Darwinism. However, our main focus will not be to describe how these Lamarckian programs contributed—among other factors—to prevent the positive reception of these theories (this aspect will be raised briefly in the conclusion). Illuminating studies already provide detailed explanations of how and why Darwinism and genetics were ignored or opposed in France during the nineteenth and part of the twentieth century (Conry 1974; Boesiger 1998; Burian et al. 1988; Gayon 2013a, b). In this chapter, we limit our study of the Lamarckian reception and rejection of genetics and the Evolutionary Synthesis to those elements allowing us to clarify the internal logic of these Neo-Lamarckisms. We attempt to summarize the main conclusions of some of our previous works (especially Herring 2016; Loison 2010, 2011, 2012; Loison et al. 2017), by contrasting the various ways in which French life scientists theorized along Lamarckian lines and by providing a general picture of their research programs.

The terms "Lamarckism," "Neo-Lamarckism," and "Lamarckian" are not historiographical categories we use to retrospectively label past theories: most of the life scientists we reference abundantly made use of this vocabulary and explicitly identified themselves as Lamarckians or Neo-Lamarckians. The epistemological adequacy or inadequacy of these terms is an issue that will not be addressed here (Loison 2011: 737–741), and for the sake of clarity and concision, we will use them indiscriminately as convenient labels. The same does not apply, on the other hand, to the "research program" concept, which is a historiographical category that was not necessarily employed by the life scientists in our study. We use it to emphasize that these Lamarckisms were not merely vague and audacious speculations: most of them were designed as frameworks for concrete research intended to guide everyday work in the laboratory or in the field. In addition, it is important to note that although our characterization of the "research program" concept is similar and compatible with the concept developed at length by Imre Lakatos, we do not wish to import the whole Lakatosian philosophical apparatus into our study (Lakatos 1986).

We are utilizing a less restrictive and more common usage of the concept as a basis to examine how metaphysics, scientific theory, and practice were intertwined.

The first two sections of our study (Sects. 2 and 3) contrast two very different forms of Lamarckism. In the Sect. 2, we give a brief account of the failure of "experimental transformism," the (relatively) unified view about evolutionary dynamics shared by most French Neo-Lamarckians at the end of the nineteenth century. This first research program which focused on the issue of adaptation was unable to produce any conclusive evidence in support of the inheritance of acquired characters. Therefore, from the 1920s onwards, with no experimental evidence to back it up, all that was left of this program were doctrinal arguments against genetics and Darwinism. Section 3 is devoted to the delayed influence of Bergson's philosophy on French biological thought. It starts with a short summary of Bergson's famous metaphysical take on biological evolution, *L'Evolution créatrice* (1907), and describes how some of his ideas gradually became appealing to certain French biologists. At first quite elusive in the 1930s, references to Bergson and his "élan vital" started being used authoritatively in the 1940s and by the 1950s, zoologists Albert Vandel and Pierre-Paul Grassé were developing a vitalist and spiritualist Lamarckian program which drew heavily on some of Bergson's views. This second main form of French Lamarckism is partly responsible for the prolonged negative reception of the Modern Synthesis in France after the Second World War.

Sections 4 and 5 deal with the influence of the Lamarckian intellectual environment within French biology on research programs which did not explicitly position themselves within the Lamarckian traditions. In Sect. 4, we analyze how Vandel and especially Grassé opposed the Evolutionary Synthesis and how they interacted with some of its founders like Theodosius Dobzhansky. In France, the Modern Synthesis was at first embraced and developed solely by Georges Teissier and Philippe L'Hérétier, whose works were pivotal in the setting up of experimental population genetics (Gayon and Veuille 2001). We describe how Teissier and L'Héritier's interests, ideas, and conjectures, despite their Darwinian inclinations, were influenced by the Lamarckian atmosphere of French biology. This example perfectly shows how non-Darwinian ideas influenced the development of the Modern Synthesis, a core hypothesis of Depew's (2017).

Our fifth and last section does not concern evolution per se but instead what would come to be known as molecular biology in the 1950s. We argue that the Lamarckism of the first kind (i.e., the concept of the inheritance of acquired characters) was instrumental in the birth of two research programs at the Pasteur Institute: the one on enzymatic adaptation and the other on bacteriophagy and lysogeny (Loison et al. 2017). During the period 1890–1940, these phenomena were interpreted in terms of Lamarckian heredity. It was only in the late 1940s and 1950s, when Lysenkoism became a central bone of contention in the French community, that André Lwoff, Jacques Monod, François Jacob, and their colleagues decided to break with this tradition and produce a strictly genetic and molecular account of bacterial adaptation, namely the operon model.

2 The End of an Era: Experimental Transformism's Lack of Experimental Support

2.1 Extending "Experimental" Science to Evolution

French Neo-Lamarckism first originated as an attempt to introduce experimentation in the study of organic evolution. One of the reasons that French biologists had opposed Darwin's *Origin of Species* since the 1860s was because Darwin did not mention a single instance of the transformation of one species into another based on rigorous experimentations (Conry 1974; Loison 2010, 2011). At the time, French biology was almost completely dominated by the figures of Claude Bernard and Louis Pasteur, and thus positive scientific knowledge was believed to be obtained only through rigorous laboratory experimentation (Burian et al. 1988; Loison 2010; Gayon 2013a).

That is why "transformism,"[1] the science of evolution, was to be based on a materialist understanding of living matter and of its various forces and had to be developed in accordance with the guiding principles of the experimental method. During the period 1880–1910, these methodological and theoretical ambitions were emphasized in several articles and books written by the most prominent biologists of the time, like Edmond Perrier (1844–1921), Gaston Bonnier (1853–1922), Julien Costantin (1857–1936), and Yves Delage (1854–1920).

Botanists were the very first to try to apply this methodology, both in the field and in the laboratory. The Bernardo-Pasteurian project of an "experimental transformism" (De Varigny 1891) developed simultaneously in several branches of biology, such as microbiology and teratology, but it was in botany that the results obtained were by far the most conclusive. In the early 1880s, Gaston Bonnier, professor of botany at the Sorbonne, launched a vast experimental program with his students and collaborators in order to establish that the morphology, anatomy and physiology of plants were dominated by abiotic parameters such as luminosity, temperature, and humidity (Bonnier 1890). Part of this research program consisted in comparative cultures. The standard protocol could be summarized as follows: cuttings of the same seedling were planted at various stations in the French Alps and Pyrenees (at altitudes of 1060–2030 m) and in his laboratory near Paris (Fontainebleau), and differences between individuals were periodically measured. After only a few weeks, he and his colleagues observed that many features of the plants had changed, and that these induced modifications could, at least in some cases, completely transform the individual so that the original species was no longer recognizable. These types of results seemed to show that natural selection was an unnecessary mechanism: by changing

[1]The anthropologist and biologist Armand de Quatrefages (1810–1892) popularized the term "transformism" [*transformisme*] during the debates surrounding the reception of Darwin's *Origin of Species*. To avoid the problems resulting from the polysemy of "evolution," he proposed that "transformism" should be preferred to designate what would later be called the evolutionary theory (see de Quatrefages 1870: 14–15). From the 1870s, "transformism" was frequently used by French scientists for almost a century.

their conditions of growth, individual organisms were capable of physiologically conforming to the new requirements and these internal modifications could in the end—it was supposed—lead to morphological evolution. The mechanism of adjustment was never clearly articulated by scientists like Bonnier and others. In general, they believed that the intimate relationship of an organism with its environment was a sufficient explanation: the physiological/nutritive working of the protoplasm was the way in which the environment affected morphology (Loison 2010).

On some occasions, Bonnier directly referenced Claude Bernard and explained that he was trying to extend Bernard's ideas and methodology to anatomy and morphology (Bonnier 1893). Bernard strictly separated morphology and anatomy from physiology (Bernard 1966 [1878]; Loison 2013b) claiming that only physiology had the potential to become a true experimental science. Physiology studied the present functioning of living beings whereas anatomy and morphology studied organisms' form and structure which were mainly the consequence of heredity, meaning they were not the result of present "determinism" and could therefore not be studied experimentally (Bernard 1966 [1878]: 341; Gayon 2013c). This led Bonnier and others to argue that Bernard's account of experimentation was too restrictive and that morphology could also be included in the realm of experimental science. Most of the French Neo-Lamarckians of this period understood "experimental" the way Bonnier did, as a category which needed to be extended in order to include transformism (Loison 2012). But in the French context, "experimental" also referred to another, substantially different, practice that also challenged Bernard's restrictive characterization. Bernard's account valued experimentation as a practice where the scientist controlled the phenomenon at stake, whereas observation was reduced to a strictly passive activity. Many zoologists felt that Bernard's intransigent and often arrogant claims amounted to an unfair characterization of their practice (Paul 1985: 98–103). In the 1860s and 1870s, zoologists were especially averse to their work being characterized as the mere "contemplation" of nature. The most influential of them, Henri Lacaze-Duthiers (1821–1901) extensively opposed Bernard's demarcation between experimentation and observation. Inspired by Ernest Chevreul's ideas, Lacaze-Duthiers argued that regardless of the type of scientific practice, the scientist's mode of reasoning would always be "experimental" (i.e., hypothetico-deductive). In other words, both nature and the laboratory presented cases requiring more than just passive observation, and therefore, zoology could be said to be experimental too.

One of Lacaze-Duthiers' students, Alfred Giard (1846–1908), was also one of the most prominent French Neo-Lamarckians around 1900 (note that Lacaze-Duthiers himself, like many of his generation, always remained skeptical about the seemingly adventurous hypothesis of organic evolution). Despite being opposed to his previous mentor on almost every possible subject, Giard nevertheless continued to support Lacaze-Duthiers' understanding of what it meant to be "experimental" (Loison 2013b). Founder of the marine station of Wimereux, in the North of France, he trained dozens of zoologists and taught them what he considered to be the principles of experimental transformism. This sort of transformism was not based on laboratory experimentation but on the careful observation of living beings in their specific "milieu" (De Bont 2010). The significant adaptation of their anatomy and behavior to the requirements of their environment was interpreted as the cumulative ongoing result of individual morpho-physiological accommodations (Loison 2010).

Hence, experimental transformism, in the context of French biology, designated two different practices. Despite their methodological disagreements, these research programs pictured the evolutionary process identically: evolution was completely reduced to changes at the level of the individual, and thus the organism was the only relevant level for studying the operations of evolutionary mechanisms. This made the inheritance of acquired characters a necessary postulate; otherwise, these induced individual modifications would have been evolutionarily irrelevant. At the beginning, this hypothesis seemed rational, as it appeared to be supported by various observations and susceptible to further experimental tests. But, as time went on, the power of these empirical arguments weakened, especially because of August Weismann's sharp criticisms (part of Weismann's texts were translated into French as soon as 1892). During the 1910s, it became clear, even for the most radical French Neo-Lamarckians, that no conclusive experimental results asserted the reality of a general process of the inheritance of acquired characters. And, because natural selection was still seen as a secondary evolutionary cause, the scientificity of transformism was once again in question.

2.2 *Renouncing the Experimental Method: The Theoretical Agnosticism of the Interwar Period*

Camille Limoges correctly highlighted that because of the failure of the experimental transformism research program, most French biologists of the interwar period were "led to an attitude of theoretical agnosticism" (Limoges 1998: 323). These second generation Lamarckians, in contrast to their predecessors, could no longer take the reality of the inheritance of acquired characters as given and were forced to renounce studying the mechanisms of evolution through what they conceived as the standard experimental method. This epistemological renunciation gave birth to various theoretical attitudes.

Some biologists proposed that the mechanisms of evolution could no longer be studied experimentally because they were no longer operative. This view was supported by the zoologist Maurice Caullery (1868–1958), the successor of Alfred Giard in the chair of "evolution of organized beings" at the Sorbonne (1909–1939). In his main book *Le problème de l'évolution* (1931), he argued that Lamarckian mechanisms had once been operational in the distant past but this was no longer the case and that this could explain the failure of current attempts to demonstrate the reality of Lamarckian processes. Because natural selection was still understood as of secondary importance, Caullery and others (like Jean Rostand) maintained that the main steps of phyletic adaptation were the consequence of Lamarckian mechanisms. But to do so, they had to renounce the uniformitarianist stance the previous generation of Neo-Lamarckians had adopted: the new form of Neo-Lamarckism could not be tested in the present day and had thus become a strictly ideological doctrine. In the early 1960s, Georges Teissier (1900–1972), one of the founders of experimental population genetics, recalled the pessimism that characterized French transformism during this period. According to him, during the 1920s, the issue of the mechanism of organic evolution "had become a desert that one only crossed as a tourist in a hurry because it didn't deserve much attention" (Teissier 1962: 362). Because of this apparent lack of scientificity, scientists "had become convinced that this problem did not concern them any longer and should be left to the philosophers" (Teissier 1962: 362).

But the idea that evolution had come to a halt thus rendering its experimental study impossible was not the only option explored by French biologists to account for the experimental failure of the inheritance of acquired characters. Another student of Alfred Giard's, Etienne Rabaud (1868–1956), professor of experimental biology at the Sorbonne, developed an alternative solution to this problem. Rather than renouncing the principle of uniformitarianism, he challenged the basic idea that morphological adaptation was the cornerstone of evolution. He went as far as to say that the adaptationist vision of evolution was an expression of a fundamental bias in favor of teleological explanations in the living world. Since morphological adaptations were mostly projections of the naturalist's teleological mind, it thus became irrelevant to construct a hypothesis—like the inheritance of acquired characters—to explain them (Rabaud 1922).

As it has been previously emphasized, these various forms of renunciation reduced Neo-Lamarckism to a state of explanatory impotence during the interwar period (Loison 2011). In contrast to the project of its founders, this evolutionary doctrine could no longer explain adaptation, at least in its classical morphological sense; in fact, it failed to produce any heuristic research programs whatsoever. The lack of positive arguments for Lamarckism meant that during this period most Neo-Lamarckians concentrated their efforts on producing negative arguments directed against their opponents which contributed to the specific French resistance against Mendelian genetics and the chromosomal theory.

2.3 *Demoting Genetics to Secondary Knowledge*

Richard Burian and Jean Gayon have given masterly accounts of the fate of genetics in the history of French biology to which we refer readers seeking a more detailed exposition of this reception (Burian et al. 1988; Burian and Gayon 1999, 2004; Gayon and Burian 2000). Our aim here is simply to give a broad picture of the way French Neo-Lamarckians opposed, or at least undermined, genetics.

The aforementioned scientists shared a common understanding of biological heredity. In accordance with their Bernardian framework, French biologists pictured the entire mechanism of heredity as the continuation of the physiological state of the protoplasm (Loison et al. 2017). Reproduction was reduced to the division of a mass of protoplasm in which each of the new parts preserved the nutritional dynamics of the initial entity. This developmental account of inheritance implied that acquired characters were automatically heritable and opposed all forms of particulate inheritance. By particulate inheritance, we mean all the hypothetical and sometimes speculative explanations of inheritance, from Darwin's pangenesis to Mendelian genetics, based on the existence and transmission of discrete representative particles.

From 1900 onwards, French biologists did not reject the results of Mendelism, but they argued that the phenotypic characters at stake were of secondary importance, i.e., only superficial or ornamental. Important physiological and anatomical features were not the consequence of hypothetical genes (or of any other sort of hereditary particles) but of the workings of the entire protoplasm conceived as an integrated whole. If genes existed, they should be seen as analogous to microbes (in a typical Pasteurian sense), i.e., microscopic discrete entities able to disturb the normal regulation of physiological processes and thus bring abnormal traits into existence. Félix Le Dantec (1869–1917), another student of Giard, was from the outset one of the strongest opponents to genetics and did not hesitate to write, as early as 1904, that "Mendelian heredity does not concern heredity, properly conceived, but rather a kind of *contagion* affecting the gametes" (Le Dantec 1904: 515).

Alongside Rabaud, Le Dantec was no doubt the fiercest critic of genetics, and most French biologists shared at least part of his suspicion against what they considered to be a secondary science. For instance, Caullery, who was one of the few who taught the principles of genetics during the interwar period (Burian et al. 1988), nevertheless always remained convinced that a true developmental theory of biological inheritance still needed to be conceived and that genetics was at best nothing more than a practical model (Caullery 1916: 424, 1931: 336).

This almost systematic opposition to Mendelian genetics and to the chromosomal theory—with the remarkable exception of Lucien Cuénot (Burian et al. 1988)—strengthened the received view that the problem of the mechanism of organic evolution was not to be found within the boundaries of experimental and positivist science. Therefore, the explanation for evolution was a problem for metaphysicians and philosophers, rather than for true scientists. The success of Bergson's *L'Evolution créatrice* (1907) strongly reinforced this view and contributed to the reluctance, on the biologists' part, to engage in any form of theorization in the fields of evolution and heredity.

3 Bergson's Legacy

3.1 L'Evolution créatrice

In the first decades of the twentieth century, French philosopher Henri Bergson (1859–1941) was one of the most famous intellectual figures in the world. After teaching secondary school philosophy for just under two decades, in 1898 he obtained a professorship at the *Ecole Normale Supérieure*, where he himself had studied philosophy. Two years later, he accepted the chair of Greek and Latin philosophy at the Collège de France. At this time, Bergson was already famous, renowned for his theories of time, in his own words *durée* (duration), mind, and matter (Bergson 1889, 1896) and later, the nature of laughter (Bergson 1900). His lectures were so crowded that people sometimes resorted to climbing up the walls of the prestigious Parisian institution for a chance to listen in (Soulez and Worms 1997). It was not until 1907 that Bergson reached the height of his fame after the publication, and rapid translation into several languages, of his metaphysical take on biological evolution, *L'Evolution créatrice*. The book was a huge commercial success, and Bergson's ideas were discussed in and outside of philosophical circles propelling the philosopher to international stardom (Azouvi 2007). Although many biologists dismissed Bergson's ideas as nonscientific, others from both Lamarckian (Herring 2016) and Darwinian (Gayon 2008) traditions found some of Bergson's ideas appealing, for instance, his criticism of Laplacian determinism, his idea of evolution as the progressive pursuit of certain tendencies without predetermined telos, and the coextensiveness of life and consciousness.

In his previous books, Bergson had discussed and challenged existing theories of mind, memory, and brain. In *L'Evolution créatrice*, he focused on the different biological theories of his time. Bergson was well versed in the life sciences and had read and mastered biological classics including Cuvier, Lamarck, Darwin, and Weismann as well as more contemporary works by his compatriots such as Cuénot, Delage, Giard, Le Dantec, and Perrier and by life scientists outside of France like Bateson, Cope, De Vries, Driesch, and T. H. Morgan. Bergson used a metaphor, the *élan vital*, to describe evolution as an unpredictable, unitary, and creative force, turning inert matter to its advantage and striving to free itself from material constraints. He was writing in the midst of the period Julian Huxley later called the "eclipse of Darwinism" (Huxley 1942: 22–28) during which various evolutionary theories were proposed. In the first chapter, Bergson discussed his controversial notion of the *élan vital* in relation with the four main rival theories of the time (Darwinism, mutationism, orthogenesis, and Lamarckism).

Bergson was often accused of defending a vitalistic agenda. However, he was not attempting to propose a fifth rival scientific theory, "the theory of the élan vital." Rather, the *élan vital* was a metaphor (Bergson 1907: 258) serving philosophical purposes. In fact, it was, according to Bergson, the metaphor that best expressed his metaphysical picture of life because the main characteristics of life could be subsumed under this one image. Firstly, the image of an original impetus common to all of life allowed Bergson to consider evolutionary history as a coherent whole consisting of successive divisions stemming from the same origin. He also stressed the inherently unpredictable (therefore non-teleological and nondeterministic) and creative nature of evolution. Finally, adaptation was not for Bergson—in contrast to the views of French Neo-Lamarckians of the first generation—the driving force of evolution. Changes in environment were merely contingencies, obstacles to the development of the main tendencies of evolution. Even though these environmental changes required organic innovations that some might call adaptations, these did not constitute the driving force of evolution. Evolution was not driven by external environmental causes; it was internally driven and engaged in the direction of the liberation of mind and the complexification of mental phenomena.

Because of the strong psychological component of his evolutionary philosophy, Bergson was sympathetic, although not committed, to American Neo-Lamarckism represented by people like paleontologist Edward Drinker Cope (1840–1897) who characterized evolutionary change as internally motivated by some kind of force analogous to a psychological effort (Bergson 1907: 77–78). However, Bergson was highly critical of the French tradition of Neo-Lamarckism. He aligned himself with the life scientists who, at the time, were expressing strong doubts about the heredity of acquired characters. He concluded that even if one day it became apparent that, in some instances, acquired characters were inherited, these phenomena would be too marginal to account for the level of coordination required for the creation of complex organic structures. In addition, he was strongly opposed to the mechanistic component of the French Neo-Lamarckian philosophical framework. This did not, however, mean that he defended a teleological vision of evolution instead. The creative nature of evolution meant for Bergson that the outcome of evolution could not be predicted by deterministic mechanical laws nor by a predetermined telos, and he insisted that evolution could be directional without being teleological. He believed that the different branches of the evolutionary tree followed the same tendencies, mainly the development of higher forms of consciousness (some more successfully than others), not because of a shared goal but because of a shared origin since all of life was the result of the same unique initial impulse. This special conception of teleology involving notions of creativity and inventivity eventually made its way into French biology, but not until the first generation of French Neo-Lamarckism had definitively died out (both literally and metaphorically).

3.2 Emile Racovitza and René Jeannel: Rethinking the Lamarckian Issue of Adaptation

The failure of the experimental transformism research program broadly undermined the explanatory power of the concept of the inheritance of acquired characters during the interwar period. However, this did not mean that the ideas Bergson put forward in 1907 were positively received by French biologists. While in Britain some biologists like Arthur D. Darbishire (Wood 2015), J. A. Thomson (Bowler 1996), D'Arcy Thompson (Esposito 2013), and Darwinian zoologist Julian Huxley (Gayon 2008) showed immediate and explicit enthusiasm for some of Bergson's ideas; in France during the 1910s and the 1920s, biology was still mostly mechanistic and therefore Bergson's "élan vital" was dismissed as vitalistic and nonscientific. For instance, in his scathing review of *L'Evolution créatrice*, Félix Le Dantec described the book as a good piece of poetry but denied it any scientific value (Le Dantec 1907: 232). During the late 1930s and early 1940s, however, the attitude towards Bergson's philosophy substantially changed and his notions of "creative evolution" and of the "*élan vital*" started receiving approval from several French biologists. In 1941, Lucien Cuénot published a book that rapidly became a classic, entitled *Invention et finalité en biologie* (Cuénot 1941). Cuénot expressed his disaffection regarding the traditional mechanistic account of life and made extensive use of finalist vocabulary of inventive and creative evolution to describe living matter's puzzling ability to create purposive structures over the course of organic evolution. Cuénot's tone is often very Bergsonian, and the philosopher's name appears 14 times in the book.

Research carried out during this period on the evolution of underground creatures provides a particularly interesting case study for these changes that were gradually affecting French biological thought. Firstly, cavernicolous animals still provided what appeared to be evidence supporting adaptation via the inheritance of acquired characters, for instance, the progressive loss of functional eyes because of the Lamarckian law of use and disuse (see Bowler 2017). Secondly, biologists involved in the study of underground life gradually made use of increasingly finalist and vitalistic vocabulary in their writings. In France, "biospeology" (*biospéologie*), the science of cavernicolous life, was founded by Emile Racovitza (1868–1947) and his colleagues at the beginning of the twentieth century. Racovitza was a Romanian biologist trained in Paris and in the marine stations of Banyuls and Roscoff under the supervision of Lacaze-Duthiers. Before 1920, he spent most of his career in France where he was recognized as an exceptional naturalist. Racovitza started studying cavernicolous animals in 1904, and was quickly joined by René Jeannel (1879–1965), an entomologist who became director of the Museum of Natural History in Paris after the Second World War.

At first, Racovitza and Jeannel explained morphological adaptations to the underground environment in typical Lamarckian terms: as a result of the disuse of certain organs in the specific cavernicolous milieu, these structures had most probably gradually become atrophied through the cumulative effect of the inheritance of acquired characters (Racovitza 1907: 418, 453). However, both naturalists eventually changed their minds and became more and more interested in the idea of orthogenesis because underground evolution appeared to result from linear and oriented evolution rather than simply from adaptations. The two ideas were not seen as incompatible: because of the remarkable invariability of the underground milieu, the adaptive processes would always advance in the same direction and thus create orthogenetic lines of evolution (Racovitza 1929).[2]

Because of his involvement in the development of a speleological institute in Cluj, Racovitza's scientific activity decreased in the 1930s. In contrast, Jeannel, who had by then become a renowned entomologist, proposed a vitalist, finalist, and orthogenetic

[2]We would like to thank our friend and colleague Cristiana Oghiva-Pavie (Angers University, France) who translated for us part of Racovitza's (1929) book that was published in Romanian.

vision of evolution over the course of several significant publications (Jeannel 1950: 51–52). His 1950 book with an explicitly teleological title, *La marche de l'évolution* (*The march of evolution*), marks the definitive transition from the period of mechanistic adaptive Lamarckism (1880–1910) to a new Lamarckian program which incorporated some major aspects of Bergson's philosophy of life. In the first pages, Jeannel argued that teleological thinking had acquired new respectability mainly because of Bergson's influence. Nevertheless, Jeannel did not renounce the classical Lamarckian explanation of adaptation and even developed it a little further: he believed that the induced modification of a somatic character lead to the synthesis of specific hormones able to reach the gametes and then modify germinal genes in the same direction as the phenotypic character (Jeannel 1950: 100). He labeled this hypothetical explanation of the inheritance of acquired characters "somatic induction" (Jeannel 1950: 8). It is interesting to note that this explanation was based on Mendelian genes, a significant difference from explanations supported previously by the first Neo-Lamarckians.

Orthogenesis was still viewed as the necessary consequence of an ongoing process of adaptation, and Jeannel also recuperated the distinction made by Racovitza (1929) between two types of adaptations: specialized adaptations that significantly constrained the subsequent evolutionary potential and "séclusions" that were adaptations which reinforced the autonomy of animals (for instance, homoiothermy). The former necessarily drove evolution towards a dead-end because of excessive specialization, whereas the latter opened the door to the development of new lineages, new organic types. Jeannel used the term "relais" to label this shift from one organizational body plan to another (Jeannel 1950: 52), a concept and a term that would be pivotal in the zoologist Albert Vandel's Lamarckism.

3.3 Putting Adaptation Aside: Albert Vandel, Pierre-Paul Grassé, and the Project of a Bergsonian Theory of Evolution

Vandel (1894–1980) held the chair of Zoology at the University of Toulouse which he occupied until the end of his career. He was a specialist of terrestrial isopods and published over 150 papers on the matter covering subjects ranging from genetics to systematics and evolutionary biology, but he was mainly interested in their sexuality and geographical distribution. In 1948 Vandel became the head of a CNRS funded underground laboratory situated in a cave in the French Pyrenees (Moulis in the Arriège region) and created at the initiative of Jeannel. In his studies of cave fauna, Vandel concluded that the animals' regressed ocular organs were neither the result of the direct impact of the environment transmitted via the inheritance of acquired characters nor the result of the selection of fortuitous mutations. In fact, said Vandel, the loss of eyesight was a form of evolutionary regression, not an adaption at all (Vandel 1964: 563–564). As such Vandel was in direct opposition to Racovitza's interpretation and a lot closer to Cuénot's notion of preadaptation. Vandel belonged to a generation of life scientists beginning their career in the 1920s at the time when the first generation of French Neo-Lamarckism was declining, failing to secure new followers to carry out their experimental program. Zoologist Pierre-Paul Grassé (1895–1985) was another prominent representative of this generation. They both proposed similar yet nonequivalent versions of Neo-Lamarckism, discarding soft inheritance and focusing on internally driven and progressive evolution instead. They positioned their version of Lamarckism within a tradition inspired partly by Bergson and Jesuit paleontologist Pierre Teilhard de Chardin and motivated by conservative and religious views. Both men were respected members of the *Académie des Sciences de Paris* and had highly successful careers. Grassé comprehensively studied the anatomy, systematics, and behavior of termites and published a lifetime's worth of research on the matter in the early 1980s. In 1941, he obtained the chair of "evolution of organized beings" after Maurice Caullery's retirement. Until his death in 1985, he worked on his monumental zoological encyclopedia which ended up being published in 48 volumes and was the absolute reference for French biology students who would, up till quite recently, refer to it as "Le Grassé."

Contrary to their Lamarckian predecessors, Grassé and Vandel did not believe that the study of evolution would be possible through experimentation. For them, evolution was neither a matter of speciation nor adaptation, or rather, speciation and adaptation were not representative of the true forces at work. Both believed that the true nature of evolution—namely that it was progressive, creative, and directional—could only be discovered through the erudite study of palaeontology and comparative anatomy. True progressive evolution only arose through the creation of new organic types and therefore could only be observed at the level of the phylum. Therefore, they viewed themselves as being part of the intellectual lineage of French naturalists of the past, like Lamarck, Buffon, or Cuvier, rather than the more recent strands of Neo-Lamarckism.

For Vandel, to explain evolution was to explain the creation of absolutely novel organic types at the level of the phylum. He was explicit about the Bergsonian undertones of this claim and insisted on the unpredictability and irreversibility of evolution (Vandel 1942, 1958). He analyzed the history of life into different key stages which he called "paliers" (levels), each new level being irreducible to preceding ones. From each new level, only a few truly progressive routes appeared from which new irreducible characteristics emerged. All phyla went through the same cycle (a progressive stage, an expansive stage and gradual death) but only a few managed to break their cycle and open the path for new forms while the others gradually lost all creative potential before dying out. Human evolution, said Vandel, is currently engaged on the most advanced and progressive route (i.e., with the most creative potential). Like Jeannel, Vandel used the term "relais" to designate the progressive succession of evolutionary cycles (Vandel 1942: 82–84).[3] Vandel didn't propose any explanatory mechanism for the passage from one level to the next. He considered himself an "organicist" which he took to mean that the mechanism of evolution was to be found in "the organization of the living being itself" (Vandel 1958: 12). He speculated that processes occurring at the embryonic level may be able to account for these structural changes.

Grassé, on the other hand, emphasized the oriented nature of evolution. He conceived evolution as "progress towards a certain form which takes place, within a phylum, by adding up similarly oriented variations, completing each other for millions and millions of years" (Grassé 1973: 173). He called the form or type towards which evolution progressed, the *idiomorphon*. Grassé never clearly defined this notion; however, in his view the *idiomorphon* was first created[4] without being actualized and then, over the course of evolution, was realized through different forms within various different lineages in a manner analogous to musical variations on a same theme. Evolution progressed in the direction of the realization of the *idiomorphon* until the emergence of a new form which would become evolution's new goal. Grassé believed that to explain the advent of new organic forms, biologists would have to explain the creation of new genes and their associated enzymes, rather than mutations which were, in Grassé's opinion, merely the rearrangement or destruction of existing genes.

With no concrete evolutionary mechanism, the explanations for Vandel's and Grassé's evolutionary pictures ultimately rested on metaphysical theories. In Vandel's case, evolution rested upon Bergson's metaphor, the *élan vital*, depicting evolution as an unpredictable, irreversible creative movement; whereas Grassé's theory relied on a supernatural or spiritual force responsible for both the creation of new types and the orientation of evolution (much like in Teilhard de Chardin's theological theory of orthogenesis). Therefore, in order to establish themselves as legitimate alternatives to the Modern Synthesis despite not having a rival mechanism to propose, they launched a series of attacks against their Darwinian opponents based not only on scientific issues but also on general epistemological questions concerning the correct way to conduct a scientific investigation in evolutionary studies. These in turn raised philosophical problems about the meaning of evolution for humankind.

---

[3] Vandel claimed that he was inspired by Jesuit paleontologist Pierre Teilhard de Chardin's description of the law of "relais."

[4] Grassé heavily implies on several occasions that the creation of the *idiomorphon* involves, in some way or another, supernatural forces. Grassé became increasingly vocal about his catholic faith as he got older.

4 Dialogues with the Modern Synthesis: Teissier and L'Héritier in a Lamarckian milieu

4.1 A "True" Synthesis Against the Deficient Darwinian Synthesis

Vandel and Grassé repeatedly attacked their Neo-Darwinian adversaries on two main theoretical fronts which they viewed as the two fundamental components of the Modern Synthesis: adaptation via natural selection and random genetic mutations as the selected "material." Because they viewed evolution as internally driven, they rejected the idea of organisms being passively adapted by natural selection (which they took to be the Darwinian stance). They insisted that it was extremely unlikely that random mutations could give rise to complex structures such as the eye or the human brain (Vandel 1969: 269; Grassé 1973: 176–178). In addition, the chances of the highly coordinated and synchronized evolution of systems such as the eye and nervous systems resulting from random mutations were even slimmer. Therefore, either the Neo-Darwinians of the Modern Synthesis secretly regarded natural selection as a supernatural teleological force or they were unaware that this was a direct consequence of their theoretical stance. Either way, the Darwinian picture did not hold.

Vandel and Grassé claimed that these complex structures, which could not be explained within a Darwinian framework, indicated that evolution involved internal and directed processes. However, as we have already highlighted, they proposed no explanatory mechanism for evolution. Therefore, as one of us has previously argued (Herring 2016), they self-identified as Lamarckians as part of a strategy to constitute their theories as legitimate alternatives to the increasingly dominant Darwinian Synthesis. On several occasions, they wrote their own histories of evolutionary biology with Lamarck systematically cited as the father of evolutionism and the founder of biology. They made a point of identifying with a French tradition of thought in the life sciences and, on an intellectual level, they genealogically linked themselves to Lamarck, Bergson, and Pierre Teilhard de Chardin.

In addition to attacking the architects of the Modern Synthesis for the content of their theories, Vandel and Grassé claimed that the general enterprise of the Modern Synthesis represented a defective manner of conducting scientific research. The two naturalists criticized the Modern Synthesis on the grounds that it was a superficial juxtaposition of different specialist areas of knowledge with no internal logic or harmony rather than a true synthesis. True understanding of evolution required decades of erudite study of all aspects of nature including philosophical reflections. Therefore, the true synthesis was the one taking place within the minds of erudite naturalists such as themselves. However, Vandel and Grassé's depiction of the architects of the Modern Synthesis as pure specialists was not quite accurate. Neo-Darwinians like Theodosius Dobzhansky and Julian Huxley also proposed views of evolution incorporating data from a whole range of biological disciplines while being thoughtful about philosophical and metaphysical questions (for a detailed analysis of Neo-Darwinian philosophy see Delisle 2009).

The Neo-Darwinians did not feel threatened by these repeated attacks that spanned decades (from Vandel and Grassé's early careers in the 1940s until the late 1970s in Vandel's case and as late as the mid-1980s for Grassé). Therefore, interactions between both parties were quite rare, with a few notable exceptions. For example, in 1947, two architects of the Synthesis, geneticist John B. S. Haldane and paleontologist George G. Simpson, attended a colloquium in Paris on the relations between paleontology and transformism, organized by the Sorbonne's professor of paleontology, Jean Piveteau. The two Darwinians were up against, among others, Cuénot, Grassé, and Teilhard de Chardin who all claimed in one way or the other that evolution was directional and that the picture proposed by the Modern Synthesis failed to account for this directionality. While Simpson attempted to debunk once and for all the orthogenetic view of evolution (Simpson 1950), Haldane admitted that Darwinism did not yet manage to explain all evolutionary phenomena which did not mean that it should be abandoned (Haldane 1950). Despite this genuine attempt to create a dialogue between both parties, neither side convinced the other. Another example is the interaction between Dobzhansky and Grassé. In 1975, Dobzhansky wrote a scathing albeit respectful review (Dobzhansky 1975) of Grassé's main book on evolution *L'Evolution du vivant* (Grassé 1973) later translated into English (Grasse 1977). Dobzhansky admitted that Grassé counted among the "great modern biologists" (Dobzhansky and Boesiger 1983: 17) of his time and that *L'Evolution du vivant* demonstrated undeniable brilliance but maintained that Grassé's poor understanding of the relationship between genotype and phenotype meant that the main theses presented in the book needed to be dismissed (Dobzhansky and Boesiger 1983: 158–159).

These critiques did not deter Grassé and his anti-Darwinism seems to have grown stronger as he grew older: by 1980 he was comparing Darwinism to an incurable disease (Grassé 1980: 150). Vandel's and Grassé's hostility towards Darwinism was tied with their philosophical and ideological views. They believed that humans represented the highest point and most progressive route of organic evolution. As such, humans summarized all the organic levels below them. Therefore, true erudition, true knowledge of as many aspects of evolution as possible, is the only way to grasp the complexity of human evolution and this would not be achieved through an enterprise like the Modern Synthesis. They also proposed that humans represented the only route for evolution because evolution currently took place on an exclusively spiritual level. Therefore, their erudite synthesis could provide the means for evolution on a spiritual level to progress. An important consequence of Vandel's and Grassé's theories of human evolution was that humans were responsible for their own evolution. This was incompatible with the Darwinian idea that evolution originated in blind mutations: in other words chance. Organisms could not be passive and evolution had to be progressive, creative, and directional for humans to be able to take evolution into their own hands.

4.2 *"Ne dédarwinisons pas": Teissier's and L'Héritier's Commitment to Darwinism*

In the mid-twentieth century, French biology was thus deeply committed to a renewed form of Bergsonian-oriented Lamarckism; Grassé and Vandel being its main representatives. However, despite the dominance of Lamarckian ideas, not all French biologists were opposed to the core concepts of the Modern Synthesis. Two in particular stood out through their work and teachings from the 1930s to the 1970s: Georges Teissier and Philippe L'Héritier (1906–1994).

When Mayr and Provine asked Ernest Boesiger, a Swiss population geneticist and a former student of Teissier, to recount what had happened in France at the time of the Synthesis, Boesiger could think of only two biologists involved in a genuinely Darwinian research program: Teissier and L'Héritier (Boesiger 1998). Teissier and L'Héritier are still remembered today because of their collaborative work in population genetics (Mayr 1982: 574). In the 1930s, they provided the first experimental evidence for natural selection. Based on the new population cage technique applied to the species *Drosophila melanogaster*, they were able to precisely measure the various parameters controlling competitive interactions among individuals (Burian and Gayon 1999; Gayon and Veuille 2001).

Jean Gayon and Michel Veuille's in-depth study of Teissier's and L'Héritier's joint work reconstructs in great detail the origins of this French school of population genetics. In particular, Gayon and Veuille explain how these two French geneticists were able to show that selection is frequency dependent: fitness coefficients are not constant but depend on the frequency of the alleles in the studied population. They were also the very first to demonstrate heterozygote advantage, which was a purely mathematical hypothesis before then (Gayon and Veuille 2001: 86–88).

In a recent paper, one of us opposed the pervasive idea, first put forward by Mayr (Mayr 1998: 321), that Teissier was not affected by the Lamarckian atmosphere of French biology because he started out as a mathematician (Loison 2013a). On the contrary, the evidence demonstrates that Teissier was from the outset a true zoologist who was very much aware of what was at stake with the different evolutionary explanations available during the interwar period (Lamarckism, mutationism, orthogenesis, etc.). For example, in personal notebooks he kept between the ages of 17 and 18, Teissier summarized biological articles published in French academic journals and discussed at length issues such as adaptation, evolution, "experimental transformism," etc.

Here, we would like to put the emphasis on another aspect of Teissier's and L'Héritier's commitment to Darwinism: their involvement in popularizing Darwinian thought. Teissier and L'Héritier were of course aware of the generalized hostility towards Mendelian genetics and Darwinism (broadly speaking) within French biology, and they knew that publishing experimental data in the nascent field of population genetics would not be enough to change their colleagues' minds, especially because of the complex mathematical component of this abstract discipline (most French biologists were still at this time "naturalists," i.e., field biologists with erudite knowledge of systematics, but with no specific interest in abstraction and theorization).

As a result, Teissier and L'Héritier gave several talks in a wide set of circumstances which were sometimes explicitly devoted to reaching a wide audience of biologists beyond disciplinary boundaries. Both of them also made a point of publishing papers in non-specialized journals, for instance *L'Année biologique*, the *Revue de l'Encyclopédie française*, or *La Pensée*. They believed their colleagues' rejection of Darwinism originated, above all, from a misunderstanding of the central concepts of population genetics and thus of the emerging Synthesis. To counter this trend, Teissier published two articles in 1945 (Teissier 1945a, b) in which he proposed a more comprehensible account of these concepts. The central claim of these papers was that evolution could only be understood as a strictly gradual process. In other words, there was no need for any kind of "internal" or "vitalist" principle or force driving evolution: macroevolution was conceived as nothing more than the long-lasting consequences of microevolutionary mechanisms.

Teissier's and L'Heritier's pedagogical concerns were also apparent in their experimental work. The experiments they devised in order to demonstrate the selective advantage of the loss of wings for insects living in windy environments aimed to confirm Darwin's speculations on the matter (Gayon 2014) and thus convince their peers of the efficiency of natural selection.

In 1962, *L'Année biologique* published a series of conferences on organic evolution. Most of the contributors—including Grassé—were at least skeptical about the possibility of using nothing but spontaneous mutations and natural selection to explain long-term evolutionary transformations. In his talk, Jean Rostand agreed with Vandel about the inadequacy of the Modern Synthesis and he concluded saying that a "dedarwinization" (*dédarwinisation*) of the evolutionary theory was in order (Rostand 1962: 356). Despite the overwhelming, almost universal, hostility towards the Synthesis from his French peers, their reluctance to accept it, and their unwillingness to rigorously study its principles, Teissier continued to push his own Darwinian agenda and programmatically concluded his own contribution: "Oh, and I almost forgot: we must not dedarwinize" (Teissier 1962: 374; our translation).

4.3 Beyond (Population) Genetics: Plasmagenes, Non-Mendelian Inheritance, and the Issue of Macroevolution

Despite their indisputable commitment to Darwinism, Teissier and L'Héritier also showed interest in certain aspects of inheritance and evolution that did not belong to the classical Mendelian–Darwinian account of evolution. Here, we would like to briefly sketch these unorthodox dimensions of their work and emphasize their connections with the predominantly Lamarckian atmosphere of French zoology during the 1930s and 1940s.

Jan Sapp has documented in detail how, after the setting up of the standard Mendelian account, some heterodox geneticists deliberately decided to focus on non-Mendelian aspects of inheritance (Sapp 1987). Among the various scientists in his study (Sonneborn, Goldschmidt, Spemann, Wettstein, Ephrussi, Nanney, etc.), Sapp includes L'Héritier and his work on cytoplasmic maternal inheritance in *Drosophila melanogaster*. Beyond their foundational work in experimental population genetics, Teissier, and especially L'Héritier, were also key figures of the debate surrounding the concept of the plasmagene and cytoplasmic inheritance during the period 1940–1960.

In 1937, while they were carrying out their experimental study of natural selection, L'Héritier and Teissier discovered that in certain strains of flies, light doses of carbon dioxide were fatal (L'Héritier and Teissier 1937). They coauthored a series of papers on the patterns of transmission of this unusual sensitivity. After the end of the war, when Teissier started a genuine research school focused on the maintenance of genetic polymorphism (Gayon and Veuille 2001), L'Héritier chose to devote all his time to developing research on what appeared to be the first case of a cytoplasmically inherited character ever documented in animals (Burian et al. 1988: 377–378).

Our aim here is not to reconstruct the history of L'Héritier's research program in the field of cytoplasmic and/or nonchromosomal heredity, but to emphasize that his results directly impacted on his (and Teissier's) understanding of the evolutionary process. As we have just seen, during the 1930s and 1940s, Teissier was strongly committed to a gradualism typical among the synthesists: in several publications, he repeatedly stated that the distinction between micro- and macroevolution should not be seen as qualitative (Teissier 1938: 11, 1945a: 5). He firmly believed that population genetics was the only relevant basis for the Modern Synthesis (Teissier 1945a: 3), i.e., that the causal forces of the evolutionary process could be analyzed quantitatively and experimentally.

In the early 1950s, things started to shift. The rapid emergence of what would later be called epigenetics[5] forced Teissier and L'Héritier to rethink, at least in part, the issue of gradualism and the relationship between micro- and macroevolution. Epigenetics was not understood as a new field, but rather as a mere addition or extension to classic genetics (Teissier 1952: 40). Since they saw genetics as a pivotal foundation of the Synthesis, a significant change in genetics meant that a change in the structure of the Evolutionary Theory would be needed (Loison 2013a). Both of them were also perfectly aware of the critical judgements of most embryologists against genetics who viewed Mendelian nuclear genes as being involved in

[5]As early as in the late 1950s and early 1960s, L'Héritier used the term "epigenetics" [*épigéné tique*] to denote this emerging field. Not only did he use the term, but he also proposed a definition which already fitted with our modern understanding of the concept (and despite the fact that, of course, molecular processes like DNA methylation were still completely unknown at the time): "The second [hereditary mechanism] [...] only modifies the modes of expression of encoded structures [...]. To designate this second type of hereditary mechanism, the term epigenetics has been proposed and seems well chosen." (L'Héritier 1962: 16, our translation).

nonessential characters, whereas the main organizational structures of the body were controlled by the cytoplasm.

As Jan Sapp (1987: 141–142) has rightly noted, around 1950, L'Héritier came to oppose the standard Darwinian interpretation of inheritance which, to him, was far too narrow because it did not take into account cytoplasmic heredity (L'Héritier 1955: 494). L'Héritier saw cytoplasmic heredity as a major component in the control of the body plan, and, as such, a plausible key factor in the process of macroevolution (thereby partially decoupled from microevolution).

In 1952, Teissier—and this is even more surprising—also published a paper in which he detailed a speculative account of the working of plasmagenes which could potentially explain major morphological transitions (Teissier 1952). On the basis of the knowledge available at the time, he thought that it was plausible that no essential differences existed between nuclear genes and cytoplasmic plasmagenes. In his view, plasmagenes controlled the physiological functioning of living beings. Teissier proposed that during the event of major changes in environmental conditions, plasmagenes temporarily became nuclear genes: as such, they would be able to evolve rather quickly in an adaptive way because "they entered the selective competition" (Teissier 1952: 43; our translation). In other words, Teissier here reintroduced an ontological demarcation between micro- and macroevolution, a position far from the classical uniformitarianism of the Synthesis.

L'Hérititier's and Teissier's heterodox position underlines the specificity of the French context: at the time of the Synthesis, French biology was under the domination of Lamarckian–Bergsonian thought which prioritized the separation between adaptation and true evolution and which tended to favor non-Mendelian modes of heredity: these two main characteristics were central to L'Héritier's and Teissier's rethinking of the structure of the Evolutionary Synthesis.

5 Lamarckism and the French School of Molecular Biology

In the twentieth century, most of the various forms of Lamarckism which developed in French biology had not been very productive: they contributed to the delayed reception of genetics and Darwinism and largely pushed French research into intellectual dead ends. In Lakatos' terms, these Lamarckian research programs were "regressive" (Lakatos 1986): whatever their starting points, they could not be made to relate meaningfully to empirical data. In the field of biology (i.e., if we exclude disciplines such as psychology and the like), only one Lamarckian program could be viewed as "progressive": the one that paved the way to what is usually called "the French school of molecular biology" (Morange 1998). Our aim here is to present how the first Lamarckian accounts of the phenomena of enzymatic adaptation and lysogeny were indeed fruitful starting points in the history of Pasteurian molecular biology.

This French school included people like François Jacob, André Lwoff, and Jacques Monod and its main contribution to the birth of molecular biology consisted in the setting up of the first model of genetic regulation based on experimental evidence, namely the operon model (Loison and Morange 2017). It is acknowledged that this model was formed at the intersection between two lines of research in microbiology: enzymatic adaptation (Monod) and lysogeny (Jacob, Lwoff). A recent study seeks to document how certain forms of Lamarckian explanations of variation and heredity contributed to the shaping of early debates on enzymatic adaptation and lysogeny in the Pasteur Institute (Loison et al. 2017). We would like to present here a summary of our most important findings.

5.1 *Lamarckian Explanations of Enzymatic Adaptation and Lysogeny at the Pasteur Institute: An Overview*

Since the late 1890s, it was known that some cells were able to produce specific enzymes if the appropriate substrate was present in the culture medium (Loison 2013c). In 1930, the Finnish microbiologist Henning Karström distinguished between this type of enzyme, which he called "adaptive enzymes," and the "constitutive enzymes," which were continuously synthesized by cells whatever the composition of the medium. With the work of Emile Duclaux (1840–1904) and Frédéric Diénert, the Pasteur Institute was at the forefront of research on what would become enzymatic adaptation during the interwar period.

In the late 1890s, Duclaux was working on two types of fungi: *Aspergillus glaucus* and *Penicillium glaucum* (Duclaux 1899: 83–93). He observed that some enzymes (like "saccharase") would only be produced and secreted in the culture medium in the presence of certain substrates (like saccharose). His student, Diénert, started working on yeast that would later become the most commonly used organism in experiments in the emerging field of enzymatic adaptation research. Diénert's results suggested that the substrate could directly and adaptively transform the enzymes that were already present in the cytoplasm (Diénert 1900: 68). He thought of enzymatic adaptation in terms of "physiological acclimatization," i.e., within a global Lamarckian framework (Diénert 1900: 71). Duclaux himself was convinced of the efficiency of the inheritance of acquired characters in the adaptation of microbes to varying conditions. He was in particular very interested in the plasticity of the protoplasm. Enzymatic adaptation was one key component of the adaptive ability of cells and was seen as the first step in a process of Lamarckian transformation (Duclaux 1898: 605).

The phenomenon of lysogeny was extensively studied in the Pasteur Institute, much more than that of enzymatic adaptation. Lysogeny was a puzzling phenomenon—after exposure to bacteriophages (bacterial viruses), some strains of bacteria seemed to be able to hereditarily transmit the ability to produce these viruses to their progeny—and during the interwar period, several microbiologists proposed hypothetical mechanisms to explain it. Physician and microbiologist, Eugène Wollman (1883–1943) played a prominent role in the debate about lysogeny and bacteriophagy. In the 1920s and

1930s, in close collaboration with his wife Elisabeth, he developed the idea that lysogeny could be thought of in terms of the theory of pangenesis proposed by Darwin (Gayon and Burian 2017). He later came to the conclusion that the virus was integrated in the genetic material of the bacterium in a latent nonpathogenic form: as such, lysogeny was explicitly viewed as an indisputable example of the inheritance of acquired characters (Loison et al. 2017).

It is obvious that Duclaux or Wollman were not committed to Lamarckism in the same way as Le Dantec or Perrier were. However, it remains the case that their Lamarckian inclinations were explicit, and that this framework helped put the phenomena of enzymatic adaptation and lysogeny on the Parisian–Pasteurian agenda.

5.2 Breaking with Lamarckism, Fighting Against Lysenkoism: The Rise of the Operon Model in Context

The Lamarckian connotations surrounding enzymatic adaptation and lysogeny were of course well known by Monod, Lwoff, and Jacob when they took up these research programs. For example, André Lwoff (1902–1994) was a close friend of the Wollmans and as a young researcher, during the interwar period, he often discussed the problem of the mechanism of lysogeny with Eugène Wollman (Loison et al. 2017).

After the Second World War and the death of the Wollmans,[6] Lwoff decided to take part in the debate on lysogeny. Lysogeny represented a challenge for the Pasteurian group because: (1) it seemed to escape any kind of experimental determinism (to such an extent that Max Delbrück, the head of the famous "phage group," contested the very existence of the phenomenon); (2) it seemed to offer strong evidence supporting the typical Lamarckian account of heredity. Quite the same was true of enzymatic adaptation.

The challenge was even greater because of the beginning of the "Lysenko affair" at the end of 1948 (Loison 2014: 15–19). Many French intellectuals who at the time shared affinities with the communist party tried to support Lysenko's claims about heredity, despite knowing next to nothing on the subject. The young Monod described Lysenkoism as nonscientific, (Monod 1948) and this led to a series of vehement attacks from one of Lysenko's main supporters, the famous poet, novelist, and editor Louis Aragon (born Louis Andrieux).

The Pasteurians were deeply affected by what they perceived as the devastating eruption of irrationality in science. Years later, Jacob claimed that one of the reasons he chose to become involved in genetics was his determination to fight Lysenkoism. In his own words, "to do genetics was [...] to insist on substituting

[6] Because of their Jewish origins, they were arrested by the French police and deported to Auschwitz in December 1943 where they died (Gayon and Burian 2017).

reason for intolerance and fanaticism" (Jacob 1998: 32). It was, therefore, essential that enzymatic adaptation on the one hand and lysogeny on the other be entirely explicable in strictly genetic terms.

As a consequence, Lwoff and later Jacob—who entered the Pasteur Institute at the end of 1950—were careful never to use the term "inheritance of acquired characters" when they described, and later explained, lysogeny (Loison et al. 2017). Monod was also deeply committed to putting aside any teleological component in his account of enzymatic adaptation. For this reason, he imposed a shift in the vocabulary. He convinced the main participants of the debate to use "enzymatic induction" rather than "enzymatic adaptation" in order to avoid any confusion about the molecular role of the substrate: it acts as an inducer of a preexisting genetic machinery, not as a template directly modeling enzymes (Monod et al. 1953).

The operon model was the main result of the close collaboration between Monod and Jacob which started in 1957 (Judson 1996; Morange 1998). It provided a unified explanation of lysogeny and enzymatic adaptation in which both phenomena are understood exclusively in molecular and genetic terms (operator, repression, structural gene, regulator gene, transcription, translation, etc.). Indisputably, it marked the defeat in microbiology of both traditional adaptive Lamarckism and Lysenkoism. Nevertheless, as historians we are duty-bound to note that the outcome of history should not obscure its origins. If enzymatic adaptation and lysogeny became central topics in the hands of Monod, Jacob, and Lwoff, it is because they had previously been developed by people like Duclaux, Diénert, and especially the Wollmans. Therefore, the Lamarckian dimension of these first works has to be valued as a positive and perhaps necessary contribution in the tortuous course of the works which culminated in the joint article published by Jacob and Monod in 1961 (Jacob and Monod 1961).

6 Conclusion

French biology was traversed by several forms of Lamarckism during the twentieth century. Our aim here was to provide a synthetic description of the main Lamarckian programs and to contrast them. We have seen that these Lamarckisms exhibited substantial differences: what does the spiritualist–teleological Lamarckism of Grassé have in common with the materialist–mechanist one of Le Dantec or Rabaud? Almost nothing, except their shared belief that Darwinism—i.e., natural selection—proposed a completely unsatisfactory evolutionary mechanism, incapable of accounting for the different trajectories of evolution over millions of years. French Lamarckisms were consistently anti-Darwinian. Why such a rejection of Darwin?

Jean Gayon and Richard Burian have provided the key insights to properly tackle this difficult question (Burian et al. 1988; Gayon 2013a). Firstly, when *The Origin of Species* was translated into French (1862), the names of Claude Bernard and Louis Pasteur already stood for the kind of *excellence* valued by their colleagues. Biology was seen as a science that took place in the laboratory, by means of the experimental method. Darwin did not propose anything resembling an experimental case of transformation of one species into another. Therefore, "transformism" was developed against Darwinism in accordance with the French epistemological requirement of an experimental basis (Loison 2010). Secondly, French biology was greatly influenced by Auguste Comte's positivism for more than a century (1840–1940). Therefore, empirical facts were seen as the true material of science and theoretical speculation was always considered suspicious. Thirdly, the excessive centralization of the French academic system during this period must be taken into account. A handful of Parisian bosses were responsible for virtually all major decisions affecting the orientation of French research, thus putting the whole scientific community at the mercy of their decisions and personal inclinations.

This specific situation was reinforced by the relative isolation of part of the French community. After the end of the golden age of French zoology and anatomy (i.e., after the death of Georges Cuvier), French biologists did not show much interest in foreign science. For instance, because they had their own marine laboratories (like Roscoff or Banyuls), they did not feel the need to go to the marine station in Naples which attracted many international scientists. This prevented the development of fruitful and lasting relations between French scientists and their European contemporaries for decades. This isolation also resulted from their lack of familiarity with foreign languages, especially German and English. When English became the international scientific language, French biologists resisted as strongly as possible (even defender of the Modern Synthesis Georges Teissier, for example, was unable to speak or write in English).

In the late 1940s, when biologists like Vandel or Grassé reached the top of the academic system, French isolation had increased further because of the Second World War. It took time before French biology fully reconnected with the international science scene. In 1967, Charles Bocquet was appointed professor at the Sorbonne in the chair of "evolution of organized beings." This position was created in 1888 for Alfred Giard. After his death, Giard was succeeded by Caullery, and Caullery by Grassé. Bocquet, a former colleague of Teissier and himself an eminent population geneticist (Carton 2014), was thereby the very first Darwinian to hold this position. Although in the late 1970s, and until his death in 1985, Grassé was still publishing vehement anti-Darwinian pamphlets (Grassé 1971, 1980),[7] he was increasingly isolated and could be described as the last veteran of radically anti-Darwinian Lamarckism in France. Indeed, the 1970s were the period during which disciplines concerned with evolution finally got to grips with the Modern Synthesis in France, and Lamarckian research programs were no longer seen as promising fields.

[7]In 1985, year of his death, Grassé was working on some new material, a book he would have entitled *La Face cachée de l'évolution* (The hidden side of evolution).

References

Azouvi F (2007) La gloire de Bergson Essai sur le magistère philosophique. Gallimard, Paris

Bergson H (1889) Essai sur les données immédiates de la conscience. Félix Alcan, Paris

Bergson H (1896) Matière et Mémoire. Félix Alcan, Paris

Bergson H (1900) Le Rire. Félix Alcan, Paris

Bergson H (1907) L'Evolution créatrice. Alcan, Paris

Bernard C (1966 [1878]) Leçons sur les phénomènes de la vie communs aux animaux et aux végétaux. Vrin, Paris

Boesiger E (1998) Evolutionary biology in France at the time of the evolutionary synthesis. In: Mayr E, Provine WB (eds) The evolutionary synthesis, perspectives in the unification of biology. Harvard University Press, Cambridge, pp 322–328

Bonnier G (1890) Cultures expérimentales dans les Alpes et les Pyrénées. Rev Gén Bot 2:513–546

Bonnier G (1893) L'anatomie expérimentale. Rev Sci 31:225–231

Bowler P (1996) Reconciling science and religion. The debate in early-twentieth-century Britain. The University of Chicago Press, Chicago

Bowler PJ (2017) Alternatives to Darwinism in the early twentieth century. In: Delisle RG (ed) The Darwinian tradition in context: research programs in evolutionary biology. Springer, Cham, pp 195–218

Burian RM, Gayon J (1999) The French school of genetics: from physiological and population genetics to regulatory molecular genetics. Annu Rev Genet 33:313–349

Burian RM, Gayon J (2004) National traditions and the emergence of genetics: the French example. Nat Rev Genet 5:150–156

Burian RM, Gayon J, Zallen DT (1988) The singular fate of genetics in the history of French biology, 1900–1940. J Hist Biol 21:357–402

Carton Y (2014) L'œuvre de Charles Bocquet (1918–1977), Directeur du LGEB du CNRS de 1965 à 1977. In: Loison L (ed) Le laboratoire CNRS de génétique évolutive de Gif. De part et d'autre de l'oeuvre de Georges Teissier, Paris, pp 81–109

Caullery M (1916) L'état présent du problème de l'évolution. Rev Sci 54:417–425

Caullery M (1931) Le problème de l'évolution. Payot, Paris

Conry Y (1974) L'introduction du darwinisme en France au XIXe siècle. Vrin, Paris

Cuénot L (1941) Invention et finalité en biologie. Flammarion, Paris

De Bont R (2010) Organisms and their Milieu. Alfred Giard, his pupils and the history of early ethology, 1870–1930. Isis 101:1–29

de Quatrefages A (1870) Charles Darwin et ses précurseurs français, Etude sur le transformisme. Germer Baillière, Paris

de Varigny H (1891) Le transformisme expérimental. Rev Sci 29:769–777

Delisle R (2009) Les Philosophies du néo-darwinisme. PUF, Paris

Depew DJ (2017) Darwinism in the 20th century: productive encounters with saltation, acquired characteristics, and development. In: Delisle RG (ed) The Darwinian tradition in context: research programs in evolutionary biology. Springer, Cham, pp 61–88

Diénert F (1900) Sur la fermentation du galactose et sur l'accoutumance des levures à ce sucre. Doctoral dissertation, Sceaux

Dobzhansky T (1975) Darwinian or 'oriented' evolution? Evolution 29(2):376–378

Dobzhansky T, Boesiger E (1983) Human culture, a moment in evolution. Columbia University Press, New York

Duclaux E (1898) Traité de microbiologie, Microbiologie générale, vol 1. Masson, Paris

Duclaux E (1899) Traité de microbiologie, Diastases, Toxines et Venins, vol 2. Masson, Paris

Esposito M (2013) Romantic biology 1890–1945. Pickering and Chatto, London

Gayon J (2008) L'Evolution créatrice lue par les fondateurs de la théorie synthétique de l'évolution. In: Fagot-Largeot A, Worms F (eds) Annales bergsoniennes IV L'Evolution créatrice 1907–2007: épistémologie et métaphysique. Presses universitaires de France, Paris, pp 59–84

Gayon J (2013a) Darwin and Darwinism in France before 1900. In: Ruse M (ed) The Cambridge Encyclopedia of Darwin and evolutionary thought. Cambridge University Press, Cambridge, pp 243–249

Gayon J (2013b) Darwin and Darwinism in France after 1900. In: Ruse M (ed) The Cambridge Encyclopedia of Darwin and evolutionary thought. Cambridge University Press, Cambridge, pp 300–312

Gayon J (2013c) Claude Bernard et l'hérédité. In: Duchesneau F, Kupiec JJ, Morange M (eds) Claude Bernard, La méthode de la physiologie. Editions Rue d'Ulm, Paris, pp 115–132

Gayon J (2014) Une expérience de sélection naturelle en conditions naturelles: L'Héritier, Neefs et Teissier, 1937. In: Loison L (ed) Le laboratoire CNRS de génétique évolutive de Gif. De part et d'autre de l'oeuvre de Georges Teissier, Paris, pp 65–80

Gayon J, Burian RM (2000) France in the Era of Mendelism (1900–1930). Comptes rendus de l'Académie des Sciences, Paris. Life Sci 323:1097–1106

Gayon J, Burian RM (2017) Eugène et Elisabeth Wollman: la question de la lysogénie. In: Loison L, Morange M (eds) L'invention de la régulation génétique. Les Nobels 1965 (Jacob, Lwoff, Monod) et le modèle de l'opéron dans l'histoire de la biologie. Editions Rue d'Ulm, Paris, pp 31–65

Gayon J, Veuille M (2001) The genetics of experimental populations: L'Héritier and Teissier's populations cages. In: Singh R, Krimbas C, Paul D, Beatty J (eds) Thinking about evolution: historical, philosophical, and political perspectives. Cambridge University Press, Cambridge, pp 77–102

Grassé P-P (1971) Toi, ce petit Dieu ! Essai sur l'histoire naturelle de l'homme. Albin Michel, Paris

Grassé P-P (1973) L'évolution du vivant: matériaux pour une nouvelle théorie transformiste. Albin Michel, Paris

Grasse P-P (1977) Evolution of living organisms: evidence for a new theory of transformation. Academic, New York

Grassé P-P (1980) L'Homme en accusation: De la biologie à la politique. Albin Michel, Paris

Haldane JBS (1950) Les mécanismes de l'évolution. In: Arambourg C (ed) Paléontologie et transformisme. Albin Michel, Paris

Herring E (2016) Des évolutionnismes sans mécanisme : les néolamarckismes métaphysiques d'Albert Vandel (1894–1980) et Pierre-Paul Grassé (1895–1985). Rev Hist Sci 69(2):369–398

Huxley J (1942) Evolution. The modern synthesis. Allen and Unwin, London

Jacob F (1998) Of flies, mice, and men. Harvard University Press, Harvard

Jacob F, Monod J (1961) Genetic regulatory mechanisms in the synthesis of proteins. J Mol Biol 3:318–356

Jeannel R (1950) La marche de l'évolution. Editions du Muséum, Paris

Judson HF (1996) The eight day of creation, makers of the revolution in biology. Cold Spring Harbor Laboratory Press, New York

L'Héritier P (1955) Les virus intégrés et l'unité cellulaire. Année Biol 31:481–496

L'Héritier P (1962) Le problème de l'hérédité non chromosomique. Année Biol 4(1):3–34

L'Héritier P, Teissier G (1937) Une anomalie physiologique héréditaire chez la Drosophile. C R Acad Sci Paris 205:1099–1101

Lakatos I (1986) The methodology of scientific research programmes. Cambridge University Press, Cambridge

Le Dantec F (1904) L'hérédité des diathèses ou hérédité mendélienne. Rev Sci 17:513–517

Le Dantec F (1907) La Biologie de M. Bergson. Rev Mois 4:230–241

Limoges C (1998) A second glance at evolutionary biology in France. In: Mayr E, Provine WB (eds) The evolutionary synthesis, perspectives in the unification of biology. Harvard University Press, Cambridge, pp 322–328

Loison L (2010) Qu'est-ce que le néolamarckisme? Les biologistes français et la question de l'évolution des espèces. Vuibert, Paris

Loison L (2011) French roots of French Neo-Lamarckisms, 1879–1985. J Hist Biol 44:713–744

Loison L (2012) Le projet du néolamarckisme français (1880–1910). Rev Hist Sci 65:61–79

Loison L (2013a) Georges Teissier (1900–1972) and the modern synthesis in France. Genetics 195:295–302
Loison L (2013b) Controverses sur la méthode dans les sciences du vivant: physiologie, zoologie, botanique (1865–1931). In: Duchesneau F, Kupiec JJ, Morange M (eds) Claude Bernard, La méthode de la physiologie. Editions Rue d'Ulm, Paris, pp 63–82
Loison L (2013c) Monod before Monod. Enzymatic adaptation, Lwoff, and the legacy of general biology. Hist Philos Life Sci 35:167–192
Loison L (2014) Introduction. In: Loison L (ed) Le laboratoire CNRS de génétique évolutive de Gif. De part et d'autre de l'oeuvre de Georges Teissier, Paris, pp 5–21
Loison L, Morange M (eds) (2017) L'invention de la régulation génétique. Les Nobels 1965 (Jacob, Lwoff, Monod) et le modèle de l'opéron dans l'histoire de la biologie. Editions Rue d'Ulm, Paris
Loison L, Gayon J, Burian RM (2017) The contribution – and collapse – of Lamarckian heredity in Pastorian molecular biology: 1. Lysogeny, 1900–1960. J Hist Biol 50:5–52
Mayr E (1982) The growth of biological thought: diversity, evolution and inheritance. Harvard University Press, Cambridge
Mayr E (1998) The arrival of Neo-Darwinism in France. In: Mayr E, Provine WB (eds) The evolutionary synthesis. Harvard University Press, Cambridge, pp 321–322
Monod J (1948) La victoire de Lyssenko n'a aucun caractère scientifique. Combat, 15 Sept 1948
Monod J, Cohn M, Pollock MR, Spiegelman S, Stanier RY (1953) Terminology of enzyme formation. Nature 172:1096
Morange M (1998) A history of molecular biology. Harvard University Press, Cambridge
Paul HW (1985) From knowledge to power: the rise of science empire in France, 1860–1939. Cambridge University Press, Cambridge
Rabaud E (1922) L'adaptation et l'évolution. Etienne Chiron Editeur, Paris
Racovitza E (1907) Essai sur les problèmes biospéologiques. Arch Zool Exp Gén 6:371–488
Racovitza E (1929) Evolutia si problemele ei. Astrei, Cluj
Rostand J (1962) Le passé du transformisme. Année Biol 4(1):339–357
Sapp J (1987) Beyond the gene, cytoplasmic inheritance and the struggle for authority in genetics. Oxford University Press, Oxford
Simpson G (1950) L'orthogenèse et la théorie synthétique de l'évolution. In: Arambourg C (ed) Paléontologie et transformisme. Albin Michel, Paris
Soulez P, Worms F (1997) Bergson. Flammarion, Paris
Teissier G (1938) Une controverse sur l'évolution. Rev Encyclopédie Française, June 11–14
Teissier G (1945a) Mécanisme de l'évolution. La Pensée 2:3–19
Teissier G (1945b) Mécanisme de l'évolution. La Pensée 3:15–31
Teissier G (1952) Dynamique des populations et taxonomie. Ann Soc R Zool Belgique 83:23–44
Teissier G (1962) Transformisme d'aujourd'hui. Année Biol 4(1):359–374
Vandel A (1942) L'évolution du monde animal et l'avenir de l'espèce humaine. Sci Vie IXII (300):81–88
Vandel A (1958) L'Homme et l'évolution, édition revue et augmentée. Gallimard, Paris
Vandel A (1964) Biospéléologie. La biologie des animaux cavernicoles. Gauthier-Villars, Paris
Vandel A (1969) Pages choisies: Textes réunis par les amis de A. Vandel à l'occasion de son jubilé scientifique. Éditions-Impressions, Toulouse
Weismannn A (1892) Essais sur l'hérédité et la sélection naturelle. C. Reinwald, Paris
Wood P (2015) Darbishire expands his vision of heredity from Mendelian genetics to inherited memory. Stud Hist Philos Biol Biomed Sci 53:16–39

Molecularizing Evolutionary Biology

Michel Morange

Abstract The encounter in the 1960s between molecular biology and evolutionary biology had short- and long-term consequences. Comparison of protein sequences suggested that evolution proceeded at a regular pace, obeying a molecular clock. It rapidly led evolutionary biologists to give neutral variations a larger role in their models. The development of genetic engineering technologies opened the door to progressive replacement of the abstract notions of gene and gene mutation hitherto used by evolutionary biologists by precise molecular descriptions. The precise structural and functional characterization of mutations assumed an increasing role and supported the introduction of a hierarchy between genes and between gene mutations that is clearly visible in evolutionary developmental biology. I will examine how far the accumulation of molecular data has challenged the Modern Synthesis established in the 1940s. In particular, different molecular mechanisms have been successively proposed to support a Lamarckian form of evolution. My conclusion will be that molecularization of evolutionary biology is still in its infancy, and that the Modern Synthesis will be replaced by a functional synthesis in which models of evolutionary biology and a description of molecular mechanisms will be intimately dovetailed.

Keywords Evo-Devo • Functional synthesis • Lamarckism • Molecular biology • Molecular clock • Neutralism

1 Introduction

Is extant evolutionary biology "molecularized"? The answer is not obvious. On the one hand, molecular phylogenies are widely used to estimate the evolutionary distances between different species and their place in the evolutionary tree. Research often aims to describe evolutionary steps at the level of the genetic modifications that were involved and of the mechanisms by which these genetic mutations affected the

M. Morange (✉)
Centre Cavaillès, République des Savoirs, USR 3608, Ecole normale supérieure, 29 rue d'Ulm, 75230 Paris Cedex 05, France
e-mail: michel.morange@ens.fr

R.G. Delisle (ed.), *The Darwinian Tradition in Context*,
https://doi.org/10.1007/978-3-319-69123-7_12

phenotype. On the other hand, the description of abstract—since not related to precise mechanisms—strategies and scenarios remains hugely important in the work of evolutionary biologists. Sometimes, the molecular events behind these scenarios can be described, at least partially, and sometimes not. The latter situation does not limit the explanatory value of these evolutionary models.

The problem of molecularization is not limited to evolutionary biology. It has concerned all biological disciplines since the 1960s: genetics, embryology, ecology, even medicine, and the answer to the previous question would be probably similar for all of them, although different in the details and in the tempo of this molecularization.

What exactly does molecularization mean? The answer is not obvious, no more so than are attempts to define molecular biology! The most visible signs of molecularization are the introduction of the techniques of molecular biology, more precisely of the genetic engineering techniques developed in the 1970s. But to reduce molecular biology to an ensemble of techniques is clearly insufficient. A technique is never neutral, but brings with it a certain ontology, a conception of the objects present in the world, and of their relations. To manipulate DNA, and to transfer it from one organism to another, is more than a tool to transform organisms. It tells us something essential about the role of genetic information in determining the characteristics of organisms. What is unique in molecular biology is the privilege given to one level of the organization of matter: the macromolecular. The reason is not that this level had been unexplored before the molecular revolution. The diversity and complexity of macromolecules is a characteristic of organisms, and their description is a favored way to understand the functions of these organisms. One emblematic example of the explanatory power of molecular descriptions was the characterization of the double helix molecule of DNA, the constituent of the gene. Its structure explained how this molecule could be the bearer of genetic information.

This emphasis on molecular structures explains why those working in other biological disciplines saw the molecularization of biology as an attempt at reductionism. To say that the description of macromolecules is important in understanding biological phenomena does not mean, as some molecular biologists have asserted, that explanations need only be sought at the macromolecular level, nor even that the macromolecular is to be preferred. A description of the macromolecular level may be considered necessary, but this does not mean that it is sufficient!

Molecularization of the biological disciplines was considered a threat by practitioners of these disciplines and generated strong opposition. Evolutionary biology was no exception. In 1961, one of the fathers of the Modern Synthesis and eminent evolutionary biologist Ernst Mayr published an influential article entitled "Cause and effect in biology," which explored an ensemble of strategies to defend evolutionary biology against the molecular biology invasion (Mayr 1961). The first strategy was to borrow a term, program, already used by the small community of molecular biologists, and explicitly defined by François Jacob and Jacques Monod the same year as the central mechanisms by which the genetic information is controlled (Jacob and Monod 1961), but to give it a slightly different meaning by which the genetic program is the result of the complex evolutionary history of organisms (Peluffo 2015). The second strategy was to distinguish two types of explanations of biological phenomena, the proximate and the ultimate, the first being produced by physiologists and molecular biologists and the second by evolutionary biologists. This was an elegant way of asserting that, fashionable as they may be, explanations of molecular biology will never replace those of evolutionary biologists. The third strategy adopted by Mayr was more problematic. It was to consider that what is characteristic of biology is the action of natural selection, i.e., of ultimate causes and explanations. Proximate explanations are physical–chemical ones, not biological explanations. The "true" biological explanations are those produced by evolutionary biology. Mayr never renounced this idea. Two years before his death, he still considered that evolutionary biology had been only superficially affected by the tremendous developments in molecular biology.

I believe that this assessment is totally wrong. Evolutionary biology has not disappeared, but it has been deeply transformed by its interactions with molecular biology. It is the steps and nature of these transformations that I want to describe in this contribution. The Modern Synthesis was built on an abstract conception of the structure and function of genes for the simple reason that no other conception was available in the 1930s and 1940s, when it was elaborated. Now that the precise determination of the nature of genes and of their mutations has been achieved, and that a description of the mechanisms relating the genes (and their modifications) to the phenotype is constantly progressing, we must consider the consequences for the Modern Synthesis and for evolutionary biology in general. As we will see, a decisive event in the transformation was the development of genetic engineering tools in the 1970s.

I will first briefly remind the reader that the Modern Synthesis and molecular biology were born at the same time and share a number of assumptions regarding the nature of organisms. The 1960s were the years of the first confrontation, with the acquisition of molecular data questioning the dominant role of natural selection and leading to the emergence of the neutralist theory of evolution. The second wave of molecularization was the emergence of evolutionary developmental biology (Evo-Devo) in the 1980s. Before describing the complex nature of Evo-Devo, I will first examine its roots, which were clearly distinguishable since the beginning of the 1960s, and then attempt to go beyond Evo-Devo. What does an extension of the Modern Synthesis mean exactly? I will argue that the richness of Evo-Devo research should not mask another more silent transformation, a progressive general dovetailing of molecular and evolutionary biology, the emergence of what Antony Dean and Joseph Thornton have called "a functional synthesis" (Dean and Thornton 2007).

2 What Is Common Between Molecular Biology and the Modern Synthesis?

It would probably be going too far to speak of a common origin for these two disciplines. Nonetheless, many of the ideas and convictions that supported the rise of molecular biology and the construction of the Modern Synthesis were shared by researchers who participated in the two transformations. The first molecular biologists and evolutionary biologists alike harbored the same hope of unifying biology

(Smocovitis 1992). Biochemists made a decisive contribution to this process of unification in the 1920s–1930s by their demonstration of the conservation of metabolites and metabolic pathways between bacteria and elephants: "From elephant to butyric acid bacterium—it is all the same," in the words of Albert Kluyver (Kamp et al. 1959). This desire for unification also pushed geneticists to search for the presence of genes, and evidence of mutations, in microorganisms, and in particular in bacteria. Molecular biologists and evolutionary biologists both believed in the existence of rules in the realm of life. They were both convinced that the work of biologists was no longer that of "stamp collectors," but consisted of the search for these rules. Both groups condidered that biological phenomena had yet to be fully naturalized. Vitalism ought to be definitively banished from biology. Many evolutionary biologists also wanted to eliminate teleological interpretations from their field of research and any reference to orthogenesis, internal forces, or trends. The first molecular biologists such as Max Delbrück were Darwinians. The Luria–Delbrück experiment of 1943 demonstrated that the adaptation of bacteria to their environment—in this case to bacteriophages—did not result from a direct action of bacteriophages inducing resistance in bacteria, but from the selection of preexisting rare variants of bacteria resistant to bacteriophages. This experiment is often considered as the last decisive blow to Lamarckism in a field—microbiology—where it had flourished since the time of Pasteur (Luria and Delbrück 1943).

This attachment of molecular biologists to Darwinism was also apparent in their day-to-day practice. The isolation of mutants, in bacteria and in bacteriophages, was the preferred way to have access to and distinguish the different functions of gene products. This does not mean that molecular biologists were doing in vitro evolution: mutations were selected by experimenters in totally artificial conditions. But their practice utterly convinced molecular biologists of one of the credos of evolutionary biologists, the creative power of natural selection: whatever characteristics of the mutants the experimenters sought, they always found. The two communities were not impermeable to one another. Theodosius Dobzhansky, one of the architects of the Modern Synthesis, observed with attention the results obtained by molecular biologists. He and Hermann Muller were responsible after 1944 for the dissemination among geneticists of the results of Oswald Avery, showing that the transforming principle of Pneumococcus was DNA. This commonality of convictions and practices between molecular biologists and evolutionary biologists does not mean, however, that the two groups looked for the same types of explanations.

3 The First Encounter: Molecular Clocks and the Neutralist Theory

I will be brief on this first episode of the confrontation between molecular biology and evolutionary biology, which has already been studied extensively by historians (Dietrich 1994, 1998, 2007; Suarez and Barahona 1996; Morgan 1998). The starting point was the attempts by Emile Zuckerkandl and Linus Pauling to use protein sequences to measure evolutionary distances between organisms (Zuckerkandl and Pauling 1965). Francis Crick had argued years before that the comparison of sequences would provide rich material for evolutionary biologists. The most surprising result obtained by Zuckerkandl and Pauling was that the percentage of variations was proportional to the time of divergence of the two species compared from their last common ancestor. The interpretation provided by the two authors was that the rate of amino acid substitution was constant: the molecular clock of evolution ticked regularly.

This observation generated a huge debate that is considered by most historians as the first clash between molecular biology and evolutionary biology. Richard Lewontin locates the origin of the debate further back in time, in discussions among population geneticists on the frequency of heterozygosity that took place years before. Even if Lewontin is right, it does not change the issue: it was also the new molecular data—in this case, the search for heterozygosity by an electrophoretic analysis of enzymes—that initiated the debate by revealing a natural high frequency of heterozygosity. This was not the first encounter between molecular data and evolutionary studies. Before molecular biologists, biochemists had already tried to extract information on evolution from the comparative analysis of proteins or metabolic pathways. Under the leadership of Marcel Florkin, biochemical evolution was an active field of research in the 1940s and 1950s (Florkin 1949). The techniques used, immunological for instance, permitted quantitative comparisons less easily than protein sequences.

The debate and the interpretation of the molecular data were far from simple. The first answer of evolutionary biologists was to consider that molecular data—protein sequences—were not appropriate for evolutionary studies because they did not occur at the level at which natural selection acts, the level of the organism.

The molecular/morphological paradox—the absence of correlations between the regular molecular variations and the discontinuous modifications of the form—was succeeded by another debate on the role of natural selection in evolution. Was the regular replacement of amino acids in proteins compatible with the sieving action of natural selection? Motoo Kimura (1968) and Jack King and Thomas Jukes (1969) suggested that the simplest explanation of the observations was that most of the variations at the molecular level are neutral and thus invisible to natural selection. This prompted Kimura to propose some years later the "neutral theory of molecular evolution" (1983). The debate on the limits of natural selection was not totally new. Sewall Wright had emphasized the importance of genetic drift—the random transmission of genetic variants during reproduction—in the 1930s, but his suggestions had not been fully integrated into the Modern Synthesis.

The neutralist model was not proven: a regular replacement of amino acids could also be the result of the action of natural selection. In addition, the molecular clock was shown by Kimura himself not to tick at the same rate for different proteins. Some additional confusion was introduced when amino acid sequences were replaced by polynucleotide (DNA) sequences. At the level of DNA, many variations are obviously neutral: if they occur in sequences of the genome that have no coding role, nor a regulatory one, or if they occur in a coding sequence at the third position of a codon and that, due to the degeneracy of the code, they do not change the nature of the amino acid that is encoded. These observations made on DNA sequences cannot be interpreted in the same way as those of Zuckerkandl and Pauling on protein sequences. Neutral mutations now have a place in the evolutionary synthesis, and the importance of genetic drift is no longer denied. As we mentioned previously, molecular phylogenies have acquired a preeminent place. Was the debate important for evolutionary biology? The introduction of neutral mutations did not shake the foundations of this discipline. Nevertheless, it was a first blow to the dominant position of natural selection in the Modern Synthesis.

4 1960–1980: The Roots of Evo-Devo

As alluded to above, molecular biologists have always been interested by evolution. They regularly proposed "mechanisms of evolution" that differed greatly from the mechanisms discussed by evolutionary biologists (Morange 2009). I will give three examples of such mechanisms that will illustrate the vision of evolution by (some) molecular biologists. In 1944, in a book that was largely ignored, because it was written in French and published during the war, André Lwoff proposed a mechanism of evolution by loss of function (Lwoff 1944). Organisms evolve by losing some of their biochemical functions. This evolution is somehow a consequence of the second principle of thermodynamics according to which the Universe evolves from order to disorder. Organisms compensate for this loss of function by finding the missing biochemical components in their food. In contrast, loss of function permits organisms to acquire new, more complex functions: it liberates them from useless tasks, which are replaced by new ones. This theory of evolution by loss of function was not new: it had already been proposed by William Bateson. It has not disappeared: it is regularly mentioned in the case of human evolution and, more recently, has been used to guess the characteristics of the last universal common ancestor (LUCA) of bacteria, archaea, and eukaryota.

A second model, the opposite, is evolution by an increase in the total number of genes. This is an idea that naturally comes to mind when one compares the sizes of genomes. This idea faded at the end of the 1990s, when the sequence of the human genome revealed its limited number of genes. It is interesting to note that, when Jacob compared in 1977 the action of evolution to that of a tinkerer permanently recombining the same pieces, he excluded from this comparison the big evolutionary steps: they were due to an increase in genetic information (Jacob 1977). A third mechanism, a precise variant of the second, was the growth of genomes by gene duplication. Pseudoalleles, an object of predilection for geneticists since the 1920s, were considered as an intermediary step in this process, following gene duplication and preceding the specialization of the new genetic copies. This mechanism of evolution was proposed again by Susumu Ohno in the 1960s and was largely accepted (Ohno 1970).

This rather brief description of three examples of mechanisms of evolution hypothesized by molecular biologists permits us to outline three characteristics of these models. The first, obvious and unsurprising, is the importance attributed to genes and their modifications in evolution. In contrast with the models proposed by evolutionary biologists, the type of mutation (gene inactivation, gene duplication, addition of new genetic material) is described and considered important. The second characteristic is the conviction of their proponents that these mechanisms are important in understanding evolution, as or even more important than the action of natural selection. The third characteristic is that these mechanisms have the capacity to account for a certain form of progress, the enrichment of the genetic material, and the diversification of the functions encoded by it.

To understand these characteristics, and more generally to appreciate the efforts molecular biologists made to explain evolutionary phenomena, we must consider a historical and philosophical dimension of the debate. Molecular biologists were Darwinians, but the form of Darwinism they espoused was that of the founders of the Modern Synthesis. They were unaware of the new ways evolutionary biologists reasoned, in terms of evolutionary strategies. More significantly, they had adopted a progressive vision of evolution that was still held by some founders of the Modern Synthesis, such as Julian Huxley and Theodosius Dobzhansky, but that was progressively abandoned by evolutionary biologists. This explains in part why the discussions between molecular biologists and evolutionary biologists were often difficult.

The philosophical dimension concerns the nature of explanations. Ernst Mayr was right when he noted that molecular biologists look for mechanistic explanations and evolutionary biologists for ultimate ones. More precisely, molecular biologists aim to describe the precise nature of the mutations and the mechanisms by which they modify the phenotype. "Naturally" also, they prefer the mutations that have a huge effect, provoke a leap in evolution, and concern a limited set of genes: the experimental work is simpler!

In contrast, evolutionary biologists are interested by the increase in fitness that the mutations provide. They are convinced that this increase might have been reached in different ways, and for this reason they are not interested in a precise description of the mechanisms that relate the mutation to the phenotype. A high number of genes involved in the transformation is not an issue, since the precise action of each of them will not be examined. The participation of many genes, each having a small effect on the phenotype, gives more flexibility to the model!

Mayr was right in the distinction that he made between the two types of explanations, but he was wrong in prioritizing one type of explanation instead of trying to dovetail them. Let us position these two types of explanations at the two extremities of a straight line. In an ideal world (for evolutionary biologists), there is an infinite number of possibilities to increase fitness in a given environment that are not biased by the underlying mechanisms. In a perfect world (for molecular biologists), the mechanisms to increase fitness are so rare that their description is absolutely required to understand the nature of the change and the increase in fitness. The change in the environment reveals evolutionary possibilities more than it creates them. To understand evolution is to understand the mechanisms—type of changes in the genetic material, effects of these changes on complex intracellular networks—that explain the evolutionary transformation of the organism.

The problem, but also the solution, is that in a real world the phenomena that have to be explained are located between these two extremes. The number of possibilities for evolution is neither one, nor infinite, but between the two. This intermediary situation is often interpreted as the result of constraints, historical and structural, and is revealed by the existence, common but not universal, of convergent processes. It means that both types of explanations are always required, and that their relative weight and place will have to be appreciated in each situation. I will give examples later.

I have decided to start this presentation by positioning the two types of explanations in a general framework: it is the only way to explain the recurrent attitudes of both molecular biologists and evolutionary biologists, from the 1960s to the present time. It explains the constant attempts of molecular biologists to find in one particular molecular mechanism "the" explanation of evolution. As often in science, the mechanisms that were favored by molecular biologists were the most recently discovered, and frequently the most unexpected. On the other side, evolutionary biologists tried to show that the evolutionary theory in its present form was sufficient to explain evolutionary facts, and that molecular descriptions were useless.

Here, I will describe the observations and models that laid the ground for the development of Evo-Devo, but also the numerous puzzling observations that were immediately seized upon to elaborate more or less heterodox mechanisms of evolution.

The root of Evo-Devo, of the description of the molecular developmental mechanisms and of their involvement in evolutionary transformations, is found in the distinction introduced by Jacob and Monod in 1959 between structural genes and regulatory genes (Jacob and Monod 1959). The former permit the synthesis (expression) of the molecular components required for the structures and functions of cells and organisms. The latter control the expression of the former.

In the operon model, one regulatory gene controls an ensemble of structural genes. A mutation in a regulatory gene has an effect on different structural genes. This differential effect of mutations on the two categories of genes did not escape the notice of Jacob and Monod, who briefly mentioned it in one of their publications (Jacob and Monod 1962). Therefore, it is not a surprise that Jacob considered Richard Goldschmidt, who distinguished micro- and macromutations, and micro- and macroevolution, as one of the most important geneticists (Goldschmidt 1940; Jacob, personal communication). These ideas were not developed any more by Monod and Jacob.

Similar ideas were expressed by the molecular biologists Crick and Sydney Brenner, who were convinced that heterochronic mutations, occurring in regulatory genes controlling the tempo of development, had a major effect. The geneticist Allan Wilson was also convinced that mutations in regulatory genes had a major role in evolution. In 1975, in collaboration with Mary-Claire King, he published a now famous paper demonstrating the short genetic distance between humans and chimpanzees (King and Wilson 1975). They concluded their article by stating that the huge morphological difference between humans and chimpanzees could only be explained by the mutation of a small set of regulatory genes having a major role in the control of morphogenesis, which had not yet been characterized.

Unfortunately, the conclusion of the article was neglected, and the only message that was remembered was the small genetic distance between humans and chimpanzees. In addition, Wilson became convinced that the regulatory mutations involved in morphogenetic changes were associated with chromosome rearrangements, a hypothesis that was not supported by most biologists. The importance of regulatory genes in development and evolutionary transformations guided the work of a small group of Drosophila geneticists, Antonio Garcia-Bellido, Gines Morata, David Hogness, Peter Lawrence, and Walter Gehring (Morange 2011). The functions of selector genes involved in the formation of different compartments during Drosophila development were described as those of regulatory genes. In a few years, the "genetic framework" for Drosophila development was established (Baker 1978). In parallel, Drosophila geneticists contributed to the development of molecular tools adapted to this organism. Regulatory genes controlling the development of Drosophila were isolated at the beginning of the 1980s and rapidly sequenced.

Another important contribution was that of Eric Davidson. He was one of the founders of molecular embryology and, with Roy Britten, the author in 1969 of the famous Britten–Davidson model of gene regulation (Britten and Davidson 1969). Davidson made two major contributions. The first was his decisive role in convincing embryologists that explaining development is explaining the mechanisms that control gene expression during development (Davidson 1968). And the second was to convince molecular biologists that mutations in these regulatory mechanisms drive evolution (Britten and Davidson 1971). These two assumptions were the pillars of Evo-Devo.

The years 1960–1980 were, therefore, crucial in establishing the foundations of Evo-Devo, which developed only at the beginning of the 1980s, when the genetic engineering tools entered laboratories and permitted the isolation and characterization of developmental genes (see also MacCord and Maienschein 2017).

The absence during all these years of tools adapted to the study of higher organisms and the overinterpretation of the ambiguous results produced by techniques such as molecular hybridization (McCarthy and Church 1970; Suarez 2001; Suarez-Diaz 2013; Morange 2014) explain the proliferation of more or less heterodox mechanisms of evolution that disappeared as rapidly as they had peaked. After his demonstration that retroviruses integrate their RNA genome into their host genome by copying RNA into DNA, Howard Temin proposed that similar mechanisms operate in uninfected cells, permitting selective amplification of parts of the genome (Temin 1971). Although the mechanisms permitting the transmission of these modifications to the progeny were never described in eukaryotes, this favored the emergence of Lamarckian evolutionary models, which were discussed up to the 1980s (Morange 2015). It has been shown that retrotranscription of RNAs into DNA, and their insertion into the genome, has occurred many times during evolution, but there is no argument to give this phenomenon a general evolutionary significance. The abundance of repetitive sequences in the genome of eukaryotes also gave rise to numerous speculations. Davidson gave them an important evolutionary role (Britten and Davidson 1971). The hypothesis proposed in 1980 that these repetitive sequences were selfish DNA elements (Doolittle and Sapienza 1980; Orgel and Crick 1980) led to a partial, but never total, disavowal of an evolutionary role for these sequences. It was observed that some of these sequences had been recruited during evolution as regulatory elements or even functional coding sequences. This does not mean that the abundance of repetitive sequences in eukaryotes, and in particular in mammals, is at the core of the mechanisms that were involved in their evolution. Quite similarly, the movement of genetic elements within the genome, or between the genomes of different species, first described by Barbara McClintock (Comfort 2001) and fully acknowledged in the 1970s (Cohen 1976), received a lot of attention. Was it not proof that the genome is able to engineer itself in order to adapt and evolve? The same previous message remains valid: these phenomena exist, but they do not represent the long-awaited mechanism of evolution, the one that would explain all so far unexplained evolutionary phenomena, the equivalent of dark matter for physicists!

Epigenetics is somehow the present legacy of these strange overinterpreted phenomena that permitted a certain return of lamarckian scenarios of evolution. Despite the fact that its mechanisms are much more precisely described, it has superseded previous efforts to discover "the" mechanism of evolution. Epigenetic marks can be transmitted through generations, and they can be modified by the environment. The importance of this new mechanism of inheritance would have been masked by the dominant position of genetics. The fate of the previous mechanisms urges caution. The roles and characteristics of epigenetic marks differ from one part of the evolutionary tree to another. In mammals, epigenetic marks are rarely transmitted from one generation to another, and when transmitted they remain unstable. Most of the epigenetic modifications due to the environment are not adaptive. The number of instances of epigenetic inheritance seems more limited by the rarity of their occurrence than by any bias on the part of experimenters.

During the period 1960–1980, evolutionary biologists not only had to negotiate with molecular biologists, they also had to answer the criticisms leveled by some of their own kind against the excessive power attributed to natural selection. Stephen Jay Gould was the most visible among them and emphasized the irregular rate of evolution and the role of historical contingency (Eldredge and Gould 1972; Gould and Lewontin 1979; see also Shanahan 2017; Turner 2017). The debate came too early, at a time when molecular biologists were not yet able to provide valid experimental data. Despite the high reputation of Gould, the impact on evolutionary biology was limited.

5 The Rise of Evo-Devo

The rise of Evo-Devo was the direct consequence of the first molecular characterization of developmental genes in 1984, and the unexpected discovery that these genes had been conserved during evolution between Drosophila, mice, and humans.

This conservation was a complete surprise, as was acknowledged by Jacob (1994). The best proof that this discovery was not anticipated is in the writings of researchers working on developmental genes in the years that preceded their characterization. In 1978, Ed Lewis published in *Nature* an important article on the homeotic genes of the *bithorax* complex of Drosophila. It is impossible to find in this article any reference to the potential role of similar genes in organisms other than insects (Lewis 1978). The development of Drosophila and insects was considered as totally different from that of mammals. It was not possible to imagine that similar mechanisms operated in the two!

In the same years, the *MyoD* gene involved in the formation of muscle cells was isolated by a totally different experimental approach, the ex vivo study of the mechanisms of differentiation. This discovery was important because the dominant role of this gene in the differentiation of muscle cells could be directly established (Davis et al. 1987), whereas many years were needed to demonstrate the role of developmental genes isolated by their homology to Drosophila genes: it required the development of knockout technology (inactivation of a gene through the replacement of a normal copy by an inactive one by homologous recombination) and was often obscured by the redundancy of these genes. Some years later, the genes involved in the development of plants were characterized and shown to have also been conserved during plant evolution (Meyerowitz et al. 1991). It was this conservation that convinced biologists that these genes were major players in the processes of differentiation and development and for this reason ought to be preferred targets for evolutionary mutations. In higher organisms, evolution is obviously an evolution of form and, therefore, of the developmental mechanisms that create this form.

What were the consequences for evolutionary biology? The first was to create a hierarchy between genes: between those whose study is important to understand evolution and those that play a minor role in it. Minds had been prepared by the previous distinction between structural and regulatory genes. Paradoxically, the list of developmental gene products included transcription factors, products of regulatory genes such as the homeobox-containing genes, but also signaling molecules and receptors that are the products of structural genes. Such a hierarchy did not exist in the Modern Synthesis. Even more problematic was the fact that it led to the distinction between two forms of evolution. The first was the evolution of complex organisms—an evolution of form due to the mutation of developmental genes. The second was the evolution of simpler organisms, their adaptation to changing environments that could be explained by the traditional models of population genetics, and could result from the mutation of any gene in the genome. This separation was made even more significant by the different types of mutations involved in the two types of evolution. In the evolution of complex organisms, most of the mutations are supposed to be regulatory, affecting the level of expression or the localization of the expression in the organism, whereas in simpler organisms most mutations are supposed to modify the structure of the gene products (Carroll 2008). The last issue generated by the rise of Evo-Devo was the question discussed at the beginning: What is the true explanation of evolution? Where is its motor? In the genes or in the adaptation to the environment?

As we have seen, these questions predated the rise of Evo-Devo. The attempts to explain evolution by mutation in one or a small group of genes never ceased. Human evolution has always been a target of these recurrent efforts. One example will easily illustrate these attempts, as well as their repeated failures: the isolation of the *FoxP2* gene, the so-called gene of language. A mutation was initially described in Pakistani families that causes speech alterations through the difficulty patients had in using grammatical rules. The gene that was isolated some years later had all the characteristics that could be expected (Enard et al. 2002). It encoded a transcription factor regulating the expression of dozens of genes expressed during the development of the brain. This gene has been conserved in evolution, but two mutations affecting its regulatory function occurred during the formation of modern humans (they were present in Neanderthals). How to resist the hypothesis that these mutations permitted early hominids to speak and human evolution to proceed?

But this attractive scenario did not resist additional observations and closer examination of what has to be explained to account for human evolution. Patients with a mutated form of the gene have a global impairment of their cognitive abilities and not a simple and precise alteration of language. The *FoxP2* gene is highly expressed in brain areas that control the production of songs in birds. If the gene is involved in animal communication in general, it becomes difficult to explain by its modifications the specific characteristics of human language. Beyond these biological observations, human evolution cannot be reduced to the acquisition of language—even if language and reason have often been considered as unique to humans. Plenty of modifications have occurred during the formation of modern humans (Carroll 2003; see also Van Arsdale 2017): modifications in the rate of development, increase in the size of the brain, different anatomical modifications, and also changes in metabolism and in the immune system. We should not forget also the case of feral children who do not speak, demonstrating that language is also a social construction!

It is probable that proposals of similar naive evolutionary scenarios will decrease in the future, and that the idea that multiple mutations were involved in human evolution as in any evolutionary process will progressively dominate. This does not mean that the explanation of evolution will no longer be looked for in genes and in the nature of their mutations. Two examples, one past and one present, will illustrate the ongoing debate on the causes of evolution.

After the wave of enthusiasm generated by the characterization of developmental genes, some researchers proposed to explain directly with them one major event in the evolutionary history of life, the Cambrian explosion, which occurred over a period of 30 million years, that is between 540 and 510 million years ago. For the first time in the history of life, during this period fossils are observed that exhibit all the body plans still present in extant organisms. It was hypothesized that the appearance of developmental genes was the cause of the Cambrian explosion. Molecular phylogenies did not support this simplistic scenario. Developmental genes appeared far before the Cambrian explosion (de Rosa et al. 1999). The major cause of this explosion should probably be sought in dramatic transformations of the environment. This does not mean that this toolkit of developmental genes did not participate in this explosion by permitting the development of new body plans and for this reason was one of the causes of the Cambrian explosion.

A part of the debate is focused today on the ambitious project initiated by Davidson and his colleagues to characterize the gene regulatory networks (GRNs) of organisms, i.e., the way in which regulatory genes involved in development are organized in a system (Davidson 2006): with this knowledge, they hope to account for their evolutionary history. The core of these GRNs, the "kernel" subcircuits, is highly constrained: mutations are rare, but have a major effect. It is the nature of the mutations—the modifications of the GRNs that they provoke—that determines the evolutionary consequences, whether they will give rise to a new species or to a new phylum (Davidson and Erwin 2006; Erwin and Davidson 2009). Davidson and his colleagues supposed that these mutations occurred in very small populations, in which they affected all the individuals of the population, directly or by genetic drift, and that there was no competition with the "normal" phenotype. Natural selection did not play any role. Davidson never dissimulated his distaste for the Modern Synthesis, which eventually explained the adaptation of bacteria to their environment—a process to which Davidson refused to attribute the name of evolution—but certainly not the evolution of vertebrates. For him, the description of GRNs was the first serious hope of explaining the evolutionary process. His model is utterly heterodox: it favors leaps in evolution and denies any role to natural selection.

Have we to choose between the traditional vision of evolution and the GRN model? Probably not. The right explanation is a combination of the two. In particular, it is possible to hypothesize that a dramatic change in the structure of a kernel subcircuit does not provoke a dramatic change in phenotype. It could be buffered by the regulatory mechanisms operating in organisms. However, this change in the structure of the kernel subcircuit will open up new evolutionary possibilities. It is necessary in evolutionary scenarios to distinguish the initial event, which opens up a multiplicity of possibilities, from the future evolutionary events that give these possibilities a precise face. The cause of the final result is neither uniquely in the initial event nor in the ones that followed, but in both, in the whole historical process.

6 Beyond Evo-Devo: The Functional Synthesis

There is an active debate among specialists of Evo-Devo about whether, and eventually how, the Modern Synthesis has to be extended. My personal conviction is that this extension should not be limited to developmental biology, but must be a "functional synthesis," as rightly described by Antony Dean and Joseph Thornton (2007). It must be a systematic synthesis of evolutionary history and knowledge of the structures and functions revealed by molecular studies.

The famous assertion of Theodosius Dobzhansky that "Nothing in biology makes sense, except in the light of evolution" (Dobzhansky 1973) is universally praised, but it has to be acknowledged that in many biological disciplines—biochemistry, molecular and cellular biology—most research work has nothing to do with evolutionary questioning. The only tribute to evolutionary biology is to acknowledge that the processes under study have been shaped (and optimized) by natural selection. They are perfect machines that can be studied independently of any knowledge of this optimization process. But are these machines perfect? And is it possible to understand their behaviors independently of the way they were progressively built by natural selection and, more generally, by their evolutionary history? The answer is obviously "no."

One example (among many) will help to illustrate this point. Chaperonins are large macromolecular complexes facilitating the folding of proteins. Only 5% of *E. coli* proteins require the assistance of chaperonins. Many physical–chemical studies were designed to define the structural characteristics of proteins that require the assistance of chaperonins. Contrary to expectations, the answer was not simple, and no clear correlation emerged between protein structural types and their dependence on chaperonins to fold correctly. The conclusion of this work was at odds with its initial ambitions (Kerner et al. 2005). Present observations are only a snapshot in a long evolutionary history of interactions between proteins and chaperonins. The answer to the question "Why does a protein require a chaperonin?" depends on different types of phenomena: the overall structure of the protein, the more or less complex folding pathway of the polypeptide chain, and the recent mutations of the protein. These may have generated new functions in the protein and may also have destabilized its structure and made the folding process more difficult. Other mutations may have occurred since, facilitating the folding of the protein. The answer (and the explanation) will be more historical than physical–chemical.

Even if the systems under study are perfect (or nearly perfect), understanding the path that led to this perfection—the roles of chance and necessity in the process—is not without interest. Two different, but somehow complementary, lines of research have explored these pathways. The first consists in reconstructing the evolutionary history of relatively simple systems, such as a receptor protein. One of the most accomplished studies is that of the glucocorticoid receptor. This receptor and the mineralocorticoid receptor have a common ancestor, and one of the initial objectives of the work was to characterize the properties of this ancestral receptor. The experimental strategy that was followed was, through the study of phylogenies, to deduce the sequence of the ancestral gene, to synthesize the corresponding protein, and to characterize its properties. In a second step, it became possible to determine the different mutations in the gene that have generated the extant form of receptor, their order of appearance, and by the same approach as previously to determine their effects. The first important result was the demonstration that the ancestral receptor was promiscuous, able to bind both mineralocorticoids and glucocorticoids. Further studies explained how the specificity for glucocorticoids was produced and outlined the role of neutral mutations that acquire an essential role, only when subsequent mutations have occurred (Dean and Thornton 2007).

A second line of research is that of in vitro evolution, pursued for more than 30 years by Richard Lenski. It consists in observing, during thousands of generations, how bacteria adapt to simple growth conditions. What is remarkable in this project is that when initiated, only rough observations could be made on the mechanisms of evolution. Recently, with the development of high-throughput sequencing technologies, it has become possible to reanalyze stored samples and to obtain a complete picture of the nature of the mutations that occurred and of their phenotypic effects (Tenaillon et al. 2016).

These studies require a huge amount of work, and their results may appear disappointing since they often confirm the scenarios that had been imagined by evolutionary biologists. But they add flesh to what hitherto were abstract scenarios. They demonstrate the permanent interactions between random variations and the action of natural selection and give a basis to historical contingency (Harms and Thornton 2014), a posthumous tribute to Stephen Jay Gould. They have also yielded puzzling new observations that are difficult to interpret with current models. These studies may appear to be luxuries, giving their authors the feeling of being close to a total description of minuscule episodes of the history of life. They also have immediate practical effects when applied to the evolution of pathogenic viruses and help describe the mechanisms by which, for instance, the avian influenza virus is able to adapt to humans. They also lend important support to a scientific approach to evolutionary phenomena.

7 Conclusion

One of the reasons for the (relative) success of the idea of an Intelligent Design was the vacuum that had been created between the abstract models of evolutionary biologists and the enthusiastic descriptions of "wonderful" molecular devices. The sophisticated functioning of these macromolecular machines was explained, but their evolutionary history was fully ignored: the temptation was great for some to fill the gap with "external" causes. The detailed studies in functional synthesis that I have described, which combine evolutionary models and molecular reconstructions, are the only effective way to replace ideologies by scientific explanations. From a philosophical point of view, the conception of life that emerges from the progressive growth of the functional synthesis is that what characterizes life is the historical dimension of its objects. Biology is not the only science to study objects with a history: chemical elements were progressively produced in stars, and geology is the description of the historical transformations of Earth (and other planets). What remains unique to biology is the diversity of the "histories of life." The description of the evolutionary mechanisms that have permitted these historical transformations—natural selection, genetic drift—is not sufficient to account for their richness, which can only be reached through molecular studies (and descriptive studies at other levels of organization). The molecularization of evolutionary biology is still in its infancy!

References

Baker WK (1978) A genetic framework for *Drosophila* development. Annu Rev Genet 12:451–470

Britten RJ, Davidson EH (1969) Gene regulation for higher cells: a theory. Science 165:349–357

Britten RJ, Davidson EH (1971) Repetitive and non-repetitive DNA sequences and a speculation on the origins of evolutionary novelty. Q Rev Biol 46:111–133

Carroll SB (2003) Genetics and the making of *Homo sapiens*. Nature 422:849–857

Carroll SB (2008) Evo-Devo and an expanding evolutionary synthesis: a genetic theory of morphological evolution. Cell 134:25–36

Cohen SR (1976) Transposable genetic elements and plasmid evolution. Nature 263:731–738

Comfort NC (2001) The tangled field: Barbara McClintock's search for the patterns of genetic control. Harvard University Press, Cambridge

Davidson EH (1968) Gene activity in early development. Academic, New York

Davidson EH (2006) The regulatory genome: gene regulatory networks in development and evolution. Academic, Burlington

Davidson EH, Erwin DH (2006) Gene regulatory networks and the evolution of animal body plans. Science 311:796–800

Davis RL, Weintraub H, Lassar AB (1987) Expression of a single transfected cDNA converts fibroblasts to myoblasts. Cell 51:987–1000

de Rosa R, Grenier JK, Andreeva T, Cook CE, Adoutte A, Adam M, Carroll SB, Balavoine G (1999) Hox genes in brachiopods, and priapulids and protostome evolution. Nature 399:772–776

Dean AM, Thornton JW (2007) Mechanistic approaches to the study of evolution: the functional synthesis. Nat Rev Genet 8:675–688

Dietrich MR (1994) The origins of the neutral theory of molecular evolution. J Hist Biol 27:21–59

Dietrich MR (1998) Paradox and persuasion: negotiating the place of molecular evolution within evolutionary biology. J Hist Biol 31:85–111

Dietrich MR (2007) Representing the object of controversy: the case of the molecular clock. Hist Philos Life Sci 29:161–176

Dobzhansky T (1973) Nothing in biology makes sense, except in the light of evolution. Am Biol Teach 35:125–129

Doolittle WF, Sapienza C (1980) Selfish genes, the phenotype paradigm and genome evolution. Nature 284:601–603

Eldredge N, Gould SJ (1972) Punctuated equilibria: an alternative to phyletic gradualism. In: Schopf TJM (ed) Models in paleobiology. Freeman, San Francisco, pp 82–115

Enard W, Przeworski M, Fisher SE, Lai SCL, Wiebe V, Kitano T et al (2002) Molecular evolution of *FOXP2*, a gene involved in speech and language. Nature 418:869–872

Erwin DH, Davidson EH (2009) The evolution of hierarchical gene regulatory networks. Nat Rev Genet 10:141–148

Florkin M (1949) Biochemical evolution. Academic, New York

Goldschmidt R (1940, 1982) The material basis of evolution. Yale University Press, New Haven

Gould SJ, Lewontin RC (1979) The spandrels of San Marco and the Panglossian paradigm: a critique of the adaptationist programme. Proc R Soc Lond B 205:581–598

Harms MJ, Thornton JW (2014) Historical contingency and its biophysical basis in glucocorticoid receptor evolution. Nature 512:203–207

Jacob F (1977) Evolution and tinkering. Science 196:1161–1166

Jacob F (1994) L'irrésistible ascension des gènes Hox. Médecine/Sciences 10:145–148

Jacob F, Monod J (1959) Gènes de structure et gènes de régulation dans la biosynthèse des protéines. C R Acad Sci Paris 249:1282–1284

Jacob F, Monod J (1961) Genetic regulatory mechanisms in the synthesis of proteins. J Mol Biol 3:318–356

Jacob F, Monod J (1962) Sur le mode d'action des gènes et leur régulation. Pontificia Acad Sci Scripta Varia 22:89–95

Kamp AF, La Rivière JWM, Verhoeven W (1959) Albert Jan Kluyver: his life and work. Interscience, New York, p 20

Kerner MJ, Naylor DJ, Ishihama Y, Maier T, Chang H-C, Stines AP, Georgopoulos C et al (2005) Proteome-wide analysis of chaperonin-dependent protein folding in Escherichia coli. Cell 122:209–220

Kimura M (1968) Evolutionary rate at the molecular level. Nature 217:624–626

Kimura M (1983) The neutral theory of molecular evolution. Cambridge University Press, Cambridge

King JL, Jukes TH (1969) Non-Darwinian evolution. Science 164:788–798

King M-C, Wilson AC (1975) Evolution at two levels in bacteria and chimpanzees. Science 188:107–116

Lewis EB (1978) A gene complex controlling segmentation in *Drosophila*. Nature 276:565–570

Luria SE, Delbrück M (1943) Mutations of bacteria from virus sensitivity to virus resistance. Genetics 28:491–511

Lwoff A (1944) L'évolution Physiologique. Etudes des pertes de fonction chez les micro-organismes. Hermann, Paris

MacCord K, Maienschein J (2017) Cells, development, and evolution: teeth studies at the intersection of fields. In: Delisle RG (ed) The Darwinian tradition in context: research programs in evolutionary biology. Springer, Cham, pp 289–308

Mayr E (1961) Cause and effect in biology. Science 134:1501–1506

McCarthy BJ, Church RB (1970) The specificity of molecular hybridization reactions. Annu Rev Biochem 39:131–150

Meyerowitz EM, Bowman JL, Brockman LL, Drews GN, Jack T, Sieburth LE, Weigel D (1991) A genetic and molecular model for flower development in *Arabidopsis thaliana*. Dev Suppl 1:157–167

Morange M (2009) When functional biologists propose mechanisms of evolution. J Biosci 34:373–376

Morange M (2011) Evolutionary developmental biology: its roots and characteristics. Dev Biol 357:13–16

Morange M (2014) Molecular hybridization: a problematic tool for the study of differentiation and development (1960–1980). J Biosci 39:29–32

Morange M (2015) Reverse transcriptase and Lamarckian scenarios of evolution. J Biosci 40:3–6

Morgan G (1998) Emile Zuckerkandl, Linus Pauling, and the molecular evolutionary clock, 1959–1965. J Hist Biol 31:155–178

Ohno S (1970) Evolution by gene duplication. Springer, Berlin

Orgel LE, Crick FHC (1980) Selfish DNA: the ultimate parasite. Nature 284:604–607

Peluffo AE (2015) The "genetic program": behind the genesis of an influential metaphor. Genetics 200:685–696

Shanahan T (2017) Selfish genes and lucky breaks: Richard Dawkins' and Stephen Jay Gould's: divergent Darwinian agendas. In: Delisle RG (ed) The Darwinian tradition in context: research programs in evolutionary biology. Springer, Cham, pp 11–36

Smocovitis VB (1992) Unifying biology: the evolutionary synthesis and evolutionary biology. J Hist Biol 25:1–65

Suarez E (2001) Satellite-DNA: a case-study for the evolution of experimental techniques. Stud Hist Philos Biol Biomed Sci 32:31–57

Suarez E, Barahona A (1996) The experimental roots of the neutral theory of molecular evolution. Hist Philos Life Sci 18:55–81

Suarez-Diaz E (2013) Variation, differential reproduction and oscillation: the evolution of nucleic acid hybridization. Hist Philos Life Sci 35:39–44

Temin HM (1971) The protovirus hypothesis: speculations on the significance of RNA-directed DNA synthesis for normal development and for carcinogenesis. J Natl Cancer Institute 46:3–7

Tenaillon O, Barrick JE, Ribeck N, Deatherage DE, Blanchard JL, Dasgupta A, GC W et al (2016) Tempo and mode of genome evolution in a 50,000-generation experiment. Nature 536:165–170

Turner DD (2017) Paleobiology's uneasy relationship with the Darwinian tradition: stasis as data. In: Delisle RG (ed) The Darwinian tradition in context: research programs in evolutionary biology. Springer, Cham, pp 333–352

Van Arsdale A (2017) Human evolution as a theoretical model for an extended evolutionary synthesis. In: Delisle RG (ed) The Darwinian tradition in context: research programs in evolutionary biology. Springer, Cham, pp 333–352

Zuckerkandl E, Pauling L (1965) Evolutionary divergence and convergence in proteins. In: Bryson V, Vogel H (eds) Evolving genes and proteins. Academic, New York, pp 97–166

Cells, Development, and Evolution: Teeth Studies at the Intersection of Fields

Kate MacCord and Jane Maienschein

Abstract Early in the twentieth century, biology was seen as grounded in the dual foundations of cells and evolution. Cells provided the most basic living unit, and evolution provided a way for cells to become established in different organisms. However, as the twentieth century progressed, cells and cellular level phenomena became embedded in different research traditions within developmental biology with varying connections to an evolutionary framework. While researchers focusing on differentiation could continue to link their research to evolution through heredity, those focused on morphogenesis largely gave up any evolutionary perspective. Morphogenetic research programs continued, without evolution, until late into the twentieth century, when fruitful new insights brought development back into the process of evolution. This chapter takes teeth as an exemplary case study for these changes with special focus on the enamel knot, now thought of as the morphogenetic control center of the developing tooth. Once development, and especially cellular level phenomena, was seen in the light of evolution, the enamel knot became the central component of a new paradigm in evolutionary developmental biology—one that, to this day, continues to provide a means of understanding the development and evolution of teeth. The intersection of cells and "the Darwinian tradition" is a complex relationship. This chapter offers an alternative history of the ways in which development, evolution, and cells were brought together throughout the twentieth century and challenges the common conception that genes are the sole locus of explanation for research at the intersection of development and evolution.

Keywords Development • Evolution • Enamel knot • Teeth • Morphogenesis

K. MacCord (✉)
Marine Biological Laboratory, 7 MBL Street, Woods Hole, MA 02543, USA
e-mail: kmaccord@mbl.edu

J. Maienschein
Marine Biological Laboratory, 7 MBL Street, Woods Hole, MA 02543, USA

School of Life Sciences, Arizona State University, Tempe, AZ 85287-4501, USA
e-mail: maienschein@asu.edu

R.G. Delisle (ed.), *The Darwinian Tradition in Context*,
https://doi.org/10.1007/978-3-319-69123-7_13

1 Nineteenth Century Evolution and Development

Charles Darwin dominates discussion of evolution in the nineteenth century, of course, because of his theory by which species evolve through natural selection. Yet Darwin was not a cell biologist, said only a little about embryology—mostly that it provided important evidence in favor of evolution—and had a very limited understanding of heredity. Ernst Haeckel said much more about the ways that development at the individual and species levels intersect, but he also remained focused on theory and did less with mechanistic details of processes in cells, development, and heredity. Neither Darwin nor Haeckel looked closely at causal explanations of morphogenesis, and though Haeckel's contemporary Wilhelm His did, His did not draw on evolution in his accounts (see Richards 2008; Hopwood 2002, 2015).

August Weismann is surely the figure who most energetically sought to bring together these different and mostly divergent approaches to understanding life. Frederick Churchill's masterful study of *August Weismann. Development, Heredity, and Evolution* (2015) provides an excellent starting point for reflecting on the context in which twentieth century researchers began. As Churchill shows persuasively, Weismann had one foot in the natural historical richness of the nineteenth century and another in the experimental search for causal explanations of the twentieth century. Weismann's interpretation of neo-Darwinism provided an excellent foundation for a unified study of life.

Though Weismann lacked detailed knowledge about how heredity works, biochemical and mechanical details of development, and understanding of the role of cells, he nonetheless understood the importance of all those contributions to living organisms. He had a vision of how the pieces could intersect and fit together to explain both organization of an individual and change over time. Germ plasm and chromosomes provided a basis for heredity, development, response to environment, and therefore also evolution through selection. In many ways, it is only now in the twenty-first century as researchers bring together evolution and development into developmental evolutionary biology (devo-evo or evo-devo) that we are able to realize the goals of developmental evolutionary accounts of life that Weismann set out.

As Churchill shows, Weismann was surely one of the most important biologists of any time. Weismann wanted to understand life and address questions such as how an individual comes into existence, grows, acquires the right kind of form, and then gives rise to new generations. Development and heredity both matter. But each individual is part of a species, and therefore evolution also matters. How do species evolve? What counts as evidence? What theoretical interpretations fit the facts and also lead us forward to new observations and interpretations? What do we learn from observing butterflies and so many other organisms that Weismann studied? For Weismann, these were not isolated questions. Biology must address them all, and all at once.

Weismann studied a number of organisms, but he was especially intrigued with butterflies. What causes the differences in structure, behavior, and other details, he asked. What causes morphogenesis of the different forms? He was convinced that chromosomes are involved and that the environment provides selective pressure so that some of the "determinants" that make up chromosomes win in the struggle for existence, as he put it. He did not know how they do that; he did not have a theory of morphogenesis. Yet he knew that such a theory, which could provide explanations for the differences in organic forms, was central to biology. He did his best, but knew he had a missing developmental piece.

As Churchill put it, "Weismann's contribution was to articulate the controversies, to sketch them in a lineal sequence as an artist might do, breaking new and clearer boundaries for discussion." Weismann, above all, "appeared always to have kept the bigger picture in mind" (Churchill 2015: 572, xii). Evolution, development, cells, heredity, and all the rest of phenomena of living organisms make up biology, and any account of life must embrace them all for Weismann. It took a century for biologists to articulate the questions, develop the methods, and clarify the theoretical framework grounded in developmental genetics and evolution to succeed with Weismann's goals.

In this chapter, we first look briefly at two major movements that sought to explain evolution in the twentieth century: what has come to be known as the Modern Synthesis and at what is called evolutionary developmental (or developmental evolutionary) biology ("evo-devo"). We note how both of these research programs attempted to bring together fields within the life sciences in order to explain evolution, and how these programs understood and utilized genetics in different ways.

After discussion of these two movements, we focus on a case study that does not fit the standard narrative of evolution in the twentieth century: that of tooth development. Teeth are extremely important to evolutionary biology, paleontology, forensics, anthropology, and any time we look for concrete evidence of past structural variation (morphological differences within a species) and diversity (morphological differences between species). Our story about teeth tells a history of shifting views of how to investigate and explain morphogenesis during the twentieth century. Morphogenesis has been largely neglected by historical narratives about evolution throughout the twentieth century. Yet it is the process that gives shapes to developing organs and organisms and as such is extremely important at the intersection of development and evolution. Therefore, understanding the changing approaches to studying tooth morphogenesis gives us insight into the central biological questions that Weismann raised and sheds light on how the biological synthesis that Weismann called for can be achieved in modern research.

2 Purported Modern Syntheses

The most well-known effort at suggesting a synthesis is surely the one proposed by Julian Huxley in *Evolution: The Modern Synthesis* (1943). Dedicated to Thomas Hunt Morgan, "many-sided leader in biology's advance," this volume opened with Huxley's reflection that:

> Evolution may lay claim to be considered the most central and most important of the problems of biology. For an attack upon it we need facts and methods from every branch of the science—ecology, genetics, paleontology, geographical distribution, embryology, systematics, comparative anatomy—not to mention reinforcements from other disciplines such as geology, geography, and mathematics. Biology at the present time is embarking upon a phase of synthesis after a period in which new disciplines were taken up in turn and worked out in comparative isolation. Nowhere is this movement towards unification more likely to be valuable than in this many-sided topic of evolution; and already we are seeing the first-fruits in the re-animation of Darwinism" (Huxley 1943: 13).

Many biologists and historians have commented on Huxley's synthesis, noting the extent to which it actually synthesized or failed to do so (see Smocovitis 1996; Cain and Ruse 2009; Delisle 2009, 2011, 2017). Joe Cain argued in 2009 that historians should give up the concept "evolutionary synthesis" in favor of a more robust understanding of themes in evolutionary biology (Cain 2009). That same year, Richard Delisle argued that epistemic and metaphysical pluralism within the modern synthesis was so rampant that at least three epistemic frameworks could be identified (Delisle 2009). Cain and Delisle are surely right that there is more to evolutionary biology than the synthesis. Yet the fact that Huxley invoked the idea of a synthesis and so many others took up that idea is worth noting.

Huxley's thinking is clear in his 1927 popular volume of lectures, *The Stream of Life*. In a series of lectures, he explained that all of life is essentially a stream. Evolution connects all the different kinds of organisms back to a beginning point, and development connects one generation to the previous and subsequent generations. Reproduction of individuals involves a "stream" of hereditary material interacting with the environment, one after another through time. Heredity, development, cells, and physiology all operate within the context of evolution. Scientists ask what, how, and also why the world is the way it is. What do we see? Many different instances of life. Why is there so much diversity of forms and functions? Because of evolution. How does each individual arise? Through heredity and development. Understanding evolution also allows us to improve, Huxley urged. And, thus: "Let us not forget that we men are the trustees of evolution, and that to refuse to face this problem is to betray the trust put into our hands by the powers of the universe" (Huxley 1927: 63).

In *The Modern Synthesis* in 1943, Huxley presented a much more detailed discussion of the different approaches to understanding life. Betty Smocovitis has discussed Huxley's motivations and the impact of his rallying call for synthesis (Smocovitis 1996: 138–153) Clearly, he sought both to summon and motivate the energy and attention of biologists to work on evolutionary studies and also to attract external attention for biology generally and evolution in particular. Huxley was always both a scientist and a publicist. As Smocovitis notes, Huxley was accused of having left out important areas of biology, development among them. The impact of Huxley's call to arms was to give evolutionary biologists a manifesto to which they could point. Over time, it also served as a lightning rod for dissent. Critics within genetics and developmental biology especially saw Huxley's view as limited and as leaving out their favorite fields. They saw him as providing less a real synthesis than a summary of the field of evolutionary biology alone.

Yet some leading evolutionary biologists, including Ernst Mayr, saw Huxley as less important than Theodosius Dobzhansky, whose population genetics grounded evolutionary explanation. For Mayr, the evolutionary synthesis had two central features: evolution as a gradual process based on small genetic changes, and the effects of population genetics and reproductive isolation in changing environments (Mayr 1982: 567). Focusing in this way on genetics as causing variation had the effect of largely ignoring development. Mayr's influence also had the effect of directing generations of researchers and resources towards the study of population genetics as a means to explain evolution. As Depew (2017) points out, "tracking gene frequencies is useful, even indispensable, in bringing evolutionary biology's *explanada* into view, but it cannot identify evolutionary causes. . .." Thus, while the population genetics movement pushed so forcefully by Mayr could track the progress of traits (e.g., genes and alleles) throughout populations (i.e. track evolution), it lacked the ability to explain how those traits got there in the first place. Mayr's account failed to incorporate the importance of embryological development and the processes of morphogenesis. In effect, development seems in Mayr's type of evolutionary account almost just to happen when the genes are in place.

3 Evo-Devo and the Return of Development?

By the late twentieth century, it had become clear that development does a lot more work than just transcribing and expressing genes. As seen in *From Embryology to Evo-Devo. A History of Developmental Evolution*, embryologists had always had much to say about both development and evolution throughout the twentieth century (Laubichler and Maienschein 2007).

Yet study of development had followed its own course, separate from the specialization of evolutionary biology. Thus came the call by the 1980s for integration of evolution and development, known as evo-devo. Some called for evolutionary development and others for developmental evolution, with somewhat different emphases. A symposium in 2000 took place at the Society for Integrative and Comparative Biology (formerly American Society of Zoologists). The session formally introduced the new Division of Evolutionary Developmental Biology, chaired by Rudolf Raff. In his introduction to the session, philosopher of science Richard Burian pointed to research a century earlier, when study of cell division, embryology, evolution, heredity, and so on were connected. "These problems were generally held to be intimately interconnected, so much so that many biologists thought of them as inseparable, forming a single nexus" (Burian discussed in Maienschein and Laubichler 2014: 157). After a century of specialization and divergence, it was time to reconnect and reunify biological thinking—but how was this unification to be accomplished?

Pigliucci (2017) and Depew (2017) illuminate the conceptual framework of evo-devo. As an addendum to these eloquent contributions, we would like to add a further point: that evo-devo is a theoretically and epistemically diverse field. Evo-devo researchers utilize many methods, embrace many ways of explaining developmental phenomena, and often enter the field after being trained in disciplines such as molecular biology, genomics, paleontology, and developmental biology. While the goal of evo-devo has always been to unite development and evolution, and specifically to use development to explain evolution, it has struggled since its inception to achieve this goal—a point that we will return to in the conclusion.

If the hallmark of the Modern Synthesis was tracking gene frequency changes within and between populations, then the hallmark of evo-devo has been tracing the roles of genes during development. In effect, evo-devo traded in population genetics for developmental genetics. At first, this meant identifying genes and recognizing their temporal–spatial distributions throughout embryos. Early on, however, researchers recognized that genes are not just expressed; they are also controlled. From this recognition came the search for regulation, and uncovering gene regulatory networks that underlie developmental phenomena has become a mainstay of the field.

4 Genes, Cells, and Unifying Development and Evolution

So far we have seen how August Weismann laid out a unified vision of biology that brought development, heredity, and evolution together. Weismann, as we have noted, was unable to bring this vision to fruition because he had no theory of morphogenesis. That is, Weismann lacked the ability to move from the genetic determinants in the germ plasm to the level of cells in order to explain how the germ plasm can account for development. We have also seen how the Modern Synthesis of the mid twentieth century, as dictated by Ernst Mayr, broke Weismann's unified vision of biology by discounting development as relevant to its goals. Investigators working within the Modern Synthesis framework turned to population genetics and tracking changes in gene and allele frequencies through time. The field of evo-devo, which emerged in the last quarter of the twentieth century, brought development back into questions about evolution, and did so by shifting from population genetics to developmental genetics, and more recently, by searching for gene regulatory networks that underpin development.

In this history leading from Weismann to the Modern Synthesis to evo-devo, genes have remained central. They have been depicted as the locus of explanatory value for evolutionary biologists from the mid-twentieth century to the present.

Within the history of science, this depiction is common. Historians who have dealt with evolutionary biology, developmental biology, and evo-devo have tended to privilege the molecular aspects of these fields (MacCord and Maienschein 2017). While genes have had an undeniable impact on these movements, a great deal of research has taken an alternative path: following the cellular processes that build the embryo, e.g., morphogenesis.

We mentioned in the previous section that evo-devo has struggled since its inception to find a way to achieve the goal of using development to explain evolution. This, we argue, has in part been due to an overemphasis on the role of genes as the locus of explanation for both development and evolution. Cells and cellular processes build morphology and traits within the developing embryo. Cells are also not under the sole control of genes. Thus, a gene-centric perspective of development, morphology, and evolution cannot completely account for the development and evolution of morphological traits. In order to do so, the cellular processes of development (morphogenesis) need to be taken into account.

In the following section, we turn to our particular case study that allows us to follow twentieth century research programs that traced morphogenesis, leading to a modern example of merging explanations that include genes and cells, and development and evolution. By following a series of research programs from 1913 to the year 2000 that revolved around individual tooth development, we show how researchers interested in tooth development initially made a conscious decision to ignore evolution and the theoretical musings inherent to Darwinian evolution at the turn of the twentieth century. Tooth development research continued throughout the century, largely untouched by the Darwinian paradigm or by molecularization. However, in the 1990s, when researchers invested in understanding the morphogenetic development of teeth and how they achieve their morphological diversity embraced a new developmental and evolutionary biology, they reintegrated the Darwinian paradigm and genetics back into their research. This case study ends with a modern research program in evo-devo that has succeeded in using development to explain evolution by building a theory of development that takes both genetic and morphogenetic processes into account. Thus, this case both traces a history of morphogenetic research as it changed throughout the twentieth century and also indicates how morphogenesis now intersects with the Darwinian tradition.

5 A Case Study: The Enamel Knot

5.1 Why Teeth?

Teeth play an oversized role in our understanding of mammalian evolution, in part because they are the bits of the body that fossilize the best and in part because their morphology varies so widely and distinctively across species. The morphological diversity witnessed in teeth has long been used to identify species and construct phylogenies of both extinct and extant organisms. But why are they so distinctive?

For answers, we need to look closely at development and specifically at morphogenesis. How does each tooth emerge and gain its distinctive morphological features? It took the full twentieth century to work out fundamentals of tooth morphogenesis, which required accumulation of different kinds of evidence and different approaches to interpreting that evidence. We tell that story below, in brief, drawing on Kate MacCord's extensive research on this topic. A small cluster of cells within the developing tooth, called the enamel knot, plays a central role, and tracing its history shows us how morphogenetic research changed throughout the twentieth century (MacCord 2017).

5.2 Background

The part of teeth that we see is called the crown. The surfaces of crowns that meet (or occlude) are covered in bumps called cusps. Mammals have an enormous range of sizes, shapes, and numbers of cusp configurations on their teeth—far more than any other clade. In the nineteenth century, paleontologists and embryologists came up with a number of theories to account for this phenotypic diversity by relying on discussions of both development and evolution.

Paleontologists Edward Drinker Cope and Henry Fairfield Osborn developed the tritubercular theory, which held that the tritubercular molar (a three-cusped structure where the cusps are arranged in a triangle) is the common type from which all subsequent mammalian molar diversity arose. As an alternative, the embryologist Carl Röse elaborated the concrescence theory—a late nineteenth century theory that held that teeth are formed by the fusion during development of initially independent cusps. Both the tritubercular theory of Cope and Osborn and the concrescence theory of Röse were evolutionary at their roots, building on interpretations of how evolution and development work to shape the morphological diversity of mammalian teeth (MacCord 2017).

In the twentieth century, however, embryologists interested in teeth shifted their gaze from dental diversity to individual tooth development, and in the process evolution faded into the background. This shift stemmed in part from a growing distaste for appeals to evolution for the explanatory framework of development (Ahrens 1913). To provide what they considered an adequate account of tooth development, embryologists sought to explain the processes by which teeth go from being a small clump of cells to a fully grown organ with characteristic form. This is the problem of morphogenesis.

Morphogenesis within the research programs of these embryologists changed over time in terms of how it was investigated and explained. From the early twentieth century, where morphogenesis was understood as the dynamic shifts in the relationships between tissues and structures during development, to the mid-twentieth century focus on the movement and proliferation of cells, researchers understood morphogenesis in different ways. This shifting history of morphogenesis becomes particularly interesting in light of the history of interpretation of what came to be known as the enamel knot. The enamel knot is a cluster of cells within early stages of developing teeth. The cells of the enamel knot do not divide, and yet they signal for cells in the surrounding tissues to proliferate. This combination of an inert cluster of cells surrounded by rapidly proliferating cells shapes the developing tooth. Today the enamel knot is at the core of explanations of how teeth develop their characteristic forms. And yet, over the course of the 100+ years since its discovery in 1913, the enamel knot has moved into and out of the explanatory framework of tooth development.

Shifts in the enamel knot's explanatory value are tied to shifting concepts of morphogenesis. This story is, therefore, one of changing scientific choices about methods and concepts, and it shows the ways evolution variously faded into the background or was seen as playing a central causal role. A focus on four research programs throughout the twentieth century shows how the enamel knot emerged and was understood in different ways depending upon how researchers investigated and explained morphogenesis; together they offer a case study of twentieth century research in a Darwinian world, as well as insights into evo-devo's problem of how to use development to explain evolution.

5.3 Discovery of the Enamel Knot: Hans Ahrens

Hans Ahrens's research on tooth development culminated in an article in 1913 (Ahrens 1913). Here, Ahrens does not work with a single question in mind so much as with a suite of questions derived from gaps he saw in the literature. Ahrens wanted a more detailed understanding of the morphogenesis of the developing cells and tissues, which he was convinced required closely observing developing teeth rather than relying on theory (Ahrens 1913: 172). He sought to challenge established theories of dental development—especially those that had relied heavily on appeals to evolution such as Röse's concrescence theory.

Through working with local clinics and hospitals around Munich, Ahrens amassed an astonishing sample of human fetal and postnatal remains. He fixed his materials in formalin, applied a number of contrast dyes, and made sections for every stage of development (Ahrens 1913). Ahrens was primarily concerned with characterizing the fine morphological changes through which the tooth forms. Research into how teeth develop had not yet taken the experimental turn that had characterized many historical depictions of embryology at this time, as Garland Allen (1975, 1979) describes. Nor did evolution play an explanatory role for Ahrens.

Ahrens firmly believed in the necessity of reconstructing structures through serial sections and wax models (Ahrens 1913: 170), and he modified the wax modeling technique developed and made famous by Gustav Born (1883). He was exquisitely careful with his preparations, pressing each section between writing paper saturated with pure formalin and rubbing it with his thumb before running it through an alcohol and toluene series. Using this method, Ahrens was able to create impeccable serial sections and sequences that afforded him a view into changing relationships between different parts of the oral tissues. He then used serial sections of younger specimens following techniques of earlier researchers to get at development from the earliest stages (Born 1883; discussed by Hopwood 1999 and Radlanski 1995).

As part of his broad study of tooth development, Ahrens discussed development of what he called the enamel organ (Ahrens 1913: 184), which forms the enamel of the tooth crown and gives the tooth its characteristic form. Because the folding of tissues during the early stages of the enamel organ development is complex, Ahrens used his serial sections and wax models to help him visualize the shifting forms of tooth morphogenesis. Through these reconstructions of early stage teeth, he first noticed a cluster of cells that he called the "enamel knot" (Ahrens 1913: 188, 192). Ahrens did more than simply establish that there is such a structure. He also noticed that the enamel knot bulges out of the enamel organ into the surrounding tissue, causing the formation of two grooves. These "enamel grooves," Ahrens believed, ultimately become the cusps. Ahrens thus placed the enamel knot as a main factor in explaining individual tooth development.

Ahrens's techniques and research helped organize previously diffuse studies of teeth. He meticulously traced development from the earliest appearance of tooth germs through to their final forms and placed all these observations within a single publication. For Ahrens, an adequate explanation was a description of the fine morphological changes through which teeth form. He relied on his sections and at no point did he feel it essential to observe living tissues directly. He also understood that development occurs through cellular-level processes, but he made no appeal to those processes such as mitosis, cell death, or cell migration.

Ahrens's histological work on tooth development built on the research of previous authors to give a more accurate depiction of the processes of tooth development. His work, unlike that of many of his predecessors like Cope, Osborn, and Röse, created an understanding of tooth development without evolutionary explanations. Evolutionary explanations did continue after Ahrens's 1913 publication, but researchers concerned with dental development did not often appeal to evolution as explanatory nor use dental development to test evolutionary hypotheses. Explanation came from the local details of morphological change, not through some distant evolutionary past. The era was post-Darwinian, but the biology ignored evolution because it did not seem to add anything to explanation.

5.4 Erwin Reichenbach, 1926/1928

Despite Ahrens's advances, there was still much to learn about how teeth develop from an initial clump of cells. As Erwin Reichenbach noted, "While in the field of tooth development, the research has mainly, through the work of Ahrens, come to a certain conclusion, the researchers have chiefly worked on dental histology, but cannot sufficiently clarify all the problems of this difficult issue" (Reichenbach 1926: 524).

Other researchers added histological studies, but these accumulating reports became a point of contention because even though they helped characterize fine morphological changes through which the tooth forms, they remained scattered, schematized, and often had contradictory results. In the mid 1920s, Reichenbach, an assistant at the Dental Institute of the University of Munich, attempted to give an account of tooth development built on and extending beyond Ahrens's studies. Like Ahrens, Reichenbach set aside evolutionary considerations as unimportant in his focus on details of dental development.

In line with the research program outlined by D'Arcy Thompson in his famous *On Growth and Form* (1917), Reichenbach believed that tooth development had to be explained by describing transformations of cells and especially the pressures that shape and move them throughout ontogeny. Reichenbach called for mechanical accounts in particular: "Apart from the purely biological factors whose analysis today is hardly accessible... mechanical forces can also have a formative influence on shaping the tooth crown. The change in liquids inside of the enamel organ along with the unequal differentiation of the enamel pulp result in specific points of localized proliferation within the enamel epithelium, which in turn stretch out other sections" (Reichenbach 1928: 53). Reichenbach thus shifted discussion of tooth development from characterization of fine morphological detail of tissues to characterization of cells and forces that shape them. In doing so, he also shifted the way in which morphogenesis was to be investigated and explained within developing teeth.

In his two-part Habilitationsschrift for the University of Munich, published in 1926 and 1928, Reichenbach investigated morphogenesis during development of pig teeth, seeking to give a biomechanical account for how the tooth goes from a small cluster of cells (known as a germ) into a fully formed organ (Reichenbach 1928: 494). Reichenbach amassed and processed his own collection of pig teeth, gathering specimens, creating sections, and applying several types of contrast dyes. Within his search for a biomechanical explanation of development, Reichenbach took special interest in active elements of the developing tooth—movements and mitoses of cells and fluids and the pressures that shape them. Reichenbach's publications read as a direct response to many of Ahrens's claims, particularly about the formation and role of the enamel knot.

Reichenbach was interested in questions like where do the cells of the enamel knot come from and how do they coalesce into a cluster? What is the relationship of the enamel knot to surrounding tissues? And, what happens to the enamel grooves that Ahrens deemed the precursors to cusps? From observations of his serial sections, Reichenbach concluded that the enamel knot was not so distinct a structure as Ahrens had thought. He had trouble clearly distinguishing it, especially in later stages, from the underlying tissue.

Reichenbach also had difficulty determining how the enamel knot formed. If it was through passive properties rather than increased mitosis or cell movement, then he reasoned that there should be evidence of a localized increase in individual cell bodies (Reichenbach 1928: 494). He found little evidence for this. Nor did he find evidence of active properties such as mitosis or cell movement. Thus, Reichenbach concluded that formation of the enamel knot was "due most likely to passive aggregation of existing cell material" (Reichenbach 1928: 495, 535). His observations led him to conclude that the enamel knot plays no active role in shaping the developing tooth, but might serve as a temporary reservoir of cells (Butler 1956).

To Reichenbach, an adequate explanation of development had to account for the biomechanical activities that shape the tooth—i.e., morphogenesis was conceived of as differential mitosis, and cell and intercellular fluid movements. He found value in work like Ahrens's that traced fine morphological changes in structures, but he saw such accounts as inadequate to explain tooth development and morphogenesis. Reichenbach used techniques very similar to those of Ahrens but saw them differently. Whereas Ahrens had used his serial sections and wax models to infer dynamic relationships between tissues and structures, Reichenbach used his materials to look for mitosis and movement. Like Ahrens, he did not have direct access to the living processes but drew inferences from observing appearances of cells and changes that he believed provided evidence for mitosis and movement.

Reichenbach's biomechanical understanding of tooth development put the emphasis on active properties of development, and physically passive properties of tissues were deemed impediments to growth that only indirectly affected overall form. Given Reichenbach's understanding of how to investigate and explain morphogenesis, it is unsurprising that the enamel knot played no explanatory role for him—his observations of the structure rendered it devoid of the active properties necessary to explain development.

5.5 *Nozue and Colleagues*

Following Reichenbach, tracing cell proliferation within early stages of tooth development became an important problem for dental embryologists because they considered differential mitosis a main factor for shaping tooth development. Thus, knowing which cells were dividing and where was important. In light of conflicts in the literature, Tetuo Nozue, a member of the Faculty of Medicine in the Anatomy Department of the University of Tokyo, decided to investigate the enamel knot more closely and discern "whether or not mitoses are found in this structure" (Nozue 1971a: 1).

Nozue gathered human fetal remains, and fixed, sectioned, and stained his sample. Using these materials, he found that while cells within the enamel knot did not divide and proliferate, the cells immediately adjacent to the enamel knot experienced increased mitosis. (Nozue 1971a: 4).

In his next study, Nozue gathered both fetal human and mouse specimens (Nozue 1971b). Both the human and mouse specimens were fixed, sectioned, and stained, but for mice, Nozue used a wider array of stains that would allow him to examine different properties of the cells and tissues. Nozue concluded that cell death accounted for some of the observations, which was an important finding because cell death, along with cell proliferation and migration, were considered the main processes of morphogenesis that work together to shape development.

Nozue continued his investigations by teaming up with two other dental researchers in Tokyo—Tadao Kirino and Motohiko Inoue. The team used experimental methods in ways that previous researchers had not, allowing them to intervene in the course of normal development to determine the influence of the enamel knot on shaping surrounding tissues. By the early 1970s, using experimental techniques to look at tooth development was a well-established practice. Yet no researcher had looked at the enamel knot experimentally and little was known about the role that the enamel knot played in the morphogenetic processes that shape the developing tooth (Kirino et al. 1973).

The group devised an experiment in which they injected a chemical, called Mitomycin C, into pregnant mice. Mitomycin C was known from previous work to interrupt the communication between tissue layers that are adjacent during development, called epithelium and mesenchyme (Tanimura 1968). Teeth, like many other organs, develop through epithelial–mesenchymal interactions, and the enamel knot (an epithelial structure) was likely to be affected by this chemical if it had a role in these epithelial–mesenchymal interactions. The results of this experiment indicated in two ways the crucial role that the enamel knot plays in tooth morphogenesis. First, the group noticed that in cases where the chemical had prevented the enamel knot from forming, the subsequent development of the tooth was interrupted. That is, without an enamel knot, tooth formation stalled. Second, mitosis was extremely low and the cells were irregularly arranged in these specimens without enamel knots, indicating that the enamel knot played a role in cell proliferation and cell arrangement.

Nozue and colleagues represent an important change in the history of research on the enamel knot. While they recognized the importance of tracking morphological processes at the cellular level, like mitosis and cell death, and incorporated these observations into their explanatory framework, they also were the first to utilize experimental methods to test the role of the enamel knot in tooth development. This testing grew out of the increased interest and activity in dental research surrounding the roles of epithelium and mesenchyme in directing morphogenesis. In turning to experimentation, the group sought to define the enamel knot in terms of its signaling capacity, i.e., whether or not it could direct morphogenesis in surrounding tissues. Thus, Nozue and colleagues still considered morphogenesis in terms of moving and dividing cells, but they also understood that tissue interactions, i.e., signaling between tissues, could direct tissue growth and cell proliferation.

Despite their advances, their experimental methods granted them only indirect access to evidence about the enamel knot's role in tooth development. They could not determine what caused the inductive phenomenon between the enamel knot and adjacent tissue—that is, they could not identify what signals were producing the effects they witnessed or how these signals were operating. Yet their work nonetheless implied the possibility that it had this property. Thus, the enamel knot gained new value for explaining individual tooth development.

5.6 The Enamel Knot's Finnish Renaissance: Jukka Jernvall

Investigators of tooth development utilized experimental techniques but did not look extensively at the enamel knot or explore its explanatory value. While the research of Nozue and colleagues represents an important shift in the way investigators understood the enamel knot and used it to explain development and tooth morphogenesis, their work went almost completely unnoticed, possibly because of their publication within obscure Japanese journals. Even at the time, the enamel knot was largely relegated to typological obscurity—existing almost entirely within the confines of oral histology texts.

The enamel knot's fate began to change only in the early 1990s when Jukka Jernvall, a doctoral candidate at the University of Helsinki, took an interest in understanding tooth development. Jernvall began his investigations at a time when developmental biology was undergoing massive changes. The first fluorescent in situ hybridization was conducted in 1980, and by the end of the decade its application had become widespread within the developmental community. Developmental biologists using this technique sought spatial information regarding gene activity in the developing embryo in order to get clues about the functions of newly cloned genes (see Koopman 2001). The possibility of locating genes in situ had profound implications for developmental biology—after a century of searching for the formative signals of development, the presence of differentiating signals (e.g., gene expression) could be localized and recorded in temporal-spatial parameters according to the development of the organism.

Jernvall's work on tooth development grew out of this period of in situ hybridizations and the search for gene expression patterns. Importantly, though, his investigations were also influenced by his training in paleontology. His graduate fieldwork at a Miocene site in Peshawar, Pakistan, gave Jernvall insights into teeth as biological and species indicators. This work, Jernvall acknowledged in personal communication, gave him an appreciation of form and pushed him to explore in his dissertation experiments the morphogenetic potentials of cell populations within the developing tooth in order to understand better how teeth gain their characteristic forms.

Jernvall's move towards utilizing the enamel knot to explain tooth development began with an accidental finding. He began his research program with no idea of what an enamel knot was, which is not surprising given that it had been marginalized for decades. Jernvall was interested in the problem of how teeth develop their characteristic forms. To him, this was a question of morphogenesis, a phenomenon composed of the processes of cell death, cell proliferation, and cell migration, all of which had genetic underpinnings. Jernvall began his research on tooth development and morphogenesis by asking, as Reichenbach and Nozue had, where is mitosis happening within the developing tooth?

In order to track mitosis, Jernvall devised an experiment to label cells in developing mouse embryos that were actively undergoing mitosis. When the labeled specimens were harvested and sectioned, Jernvall was able to observe where cells were proliferating at the different stages of development. Using this technique, Jernvall found areas of enhanced cell proliferation surrounding a ball of cells that showed no mitotic activity—a finding reminiscent of Nozue (1971a, b). This finding indicated to Jernvall a flawed methodology—he was not yet familiar with the structure called the enamel knot and thought that the presence of a static area within a rapidly proliferating tissue was a possible artifact of his labeling technique.

While puzzling over what he had found, Jernvall came across an article by Lee Niswander and Gail R. Martin that looked broadly at the expression of the gene FGF-4 throughout the developing mouse embryo (Niswander and Martin 1992). They had found FGF-4 expression in the location where Jernvall had discovered the inert cluster of cells—a structure that they labeled the enamel knot.

Jernvall's understanding of how to investigate and explain morphogenesis appealed to more than just the physical forces that Reichenbach had sought; he also understood that development could be characterized by revealing the genes that could cause the movements and mitosis that Reichenbach had understood development to be. To Jernvall, an understanding of development required both of these perspectives. Because of his commitment to approaching morphogenesis from both of these perspectives, Jernvall decided to replicate Niswander and Martin's experiment within the teeth and looked for gene expression at different stages of tooth development. He found that FGF-4, which is a gene that greatly enhances cell proliferation, was expressed by the non-proliferating cells of the enamel knot. This led Jernvall to consider the possibility that the enamel knot, by both not dividing itself and by expressing genes like FGF-4 that cause heightened cell proliferation in the surrounding tissue, could be shaping the developing tooth.

In order to track the possible connection between FGF-4, cell proliferation, and the enamel knot more closely, Jernvall made computer-assisted 3-D reconstructions of his serial sections that incorporated his data on cell proliferation and gene expression into the models. By combining this data within a single 3-D model of each of the stages of mouse molar development in which the enamel knot was present, Jernvall was able to recognize the tight spatial and temporal relationship between the enamel knot, FGF-4 gene expression, cell proliferation in surrounding tissues, and the emergence of the tooth cusps. Thus, the enamel knot gained a central role in explaining tooth development, and it did so because Jernvall brought together morphogenesis, cellular phenomena, and genetics.

Jernvall's work demonstrates the emergence of a way of understanding tooth development and morphogenesis wherein both cellular processes and gene expression are necessary. In his 1994 paper, Jernvall referred to the enamel knot as a potential control center rather than a signaling center. In doing so, Jernvall sought to clarify that the enamel knot did not merely act in terms of a signaling capacity.

Rather, the enamel knot both directs surrounding cells to proliferate and through its own static properties shapes the outgrowth and transformation of the tissues surrounding it. Thus, tooth morphogenesis to Jernvall was the result of the physical forces that come from cells rapidly proliferating around a static object, like the enamel knot, as well as the genes that cause the rapid proliferation. This research became the basis of Jernvall's enamel knot theory, which holds that the development and cusp patterning of mammalian molars are driven by morphogenetic control centers called enamel knots.

Over the next six years, Jernvall and his lab worked to expand knowledge of the enamel knot in terms of its role in shaping teeth (Jernvall 1995, 2000; Jernvall et al. 1998, 2000; Keränen et al. 1998, 1999; Pispa et al. 1999; Vaahtokari et al. 1996). They sought information about what signals the enamel knot expresses throughout its life cycle as well as what roles these signals have on the cellular processes shaping the surrounding tissues.

In directing this line of research, Jernvall's goal was not simply to understand how the enamel knot shapes a mouse tooth; rather his goal was to understand how the enamel knot can underlie the enormous diversity of molar forms that had provoked Cope, Osborn, and Röse in the nineteenth century to devise their theories of tooth development and evolution. Thus, Jernvall's group built their theory of the enamel knot's role initially by looking at mice (the traditional model organism of tooth development), but also took a comparative approach by checking whether the same processes were at play in the development of vole teeth (a close relative of mice) and asking how the morphological differences between the two species could be achieved by altering the temporal and spatial arrangement of enamel knots (Jernvall 1995; Keränen et al. 1998).

This comparative work came to fruition in 2000, when Jernvall's lab tied their detailed analyses of the processes that produce dental morphology to evolution (Jernvall et al. 2000). By comparing the relationship between enamel knot gene expression patterns and emerging morphology in developing teeth across mice and voles, the group turned the enamel knot theory of tooth development into a theory of both development and evolution. Through research that took into account cellular processes and gene regulation, Jernvall was able to develop a theory of tooth development that could be used to explain tooth evolution, thus achieving the longstanding goal of evo-devo.

6 A New Synthesis of Development and Evolution: Bringing Cells Back

Weismann's vision was to unite development, heredity, and evolution. He was committed to a comprehensive biological research program and believed that these three areas necessarily intersect and could and should be addressed together. The frustration for Weismann's program was that he did not have the ability to move from his germ plasm theory of inheritance up to the level of cells or morphogenetic processes in order to provide an adequate account of how the germ plasm could give rise to form. He did not understand enough about development or genetics and did not have ways to tie those processes to evolution sufficiently.

The history of biology in the twentieth century, as we have noted, has often been told from the perspective of the gene—of the molecular determinants of inheritance and development. Looking at genes allows us to ask crucial questions about our bodies and our place in nature but does not give complete answers.

Other chapters in this volume have referred to the reintegration of development into evolutionary theory in the last third of the twentieth century, in the movement that came to be called evolutionary developmental biology ("evo-devo") (Depew, Pigliucci). Pigliucci points out that at the core of evo-devo is an emphasis on linking genes to morphology through development and that, "... one of its major contributions so far has been a marked shift of emphasis in the study of morphology and development, from the sort of classical population genetic studies focused on structural genes to an emphasis on regulatory genes and their potential to help us build a credible theory of the origin of evolutionary novelties." Pigluicci points us towards gene regulation as a way of surmounting the divide between genes, development, morphology, and evolution.

During development, genes interact. Through these interactions, genes guide and help build the cells into tissues and organs that will become an organism. As this process unfolds, cells multiply and divide, changing shapes and identities depending on their location within the emerging body, their neighbors, the physical forces that they encounter, and the genetic signals they receive. As Lewis Wolpert put it, "genes control development by controlling cell behavior" (Wolpert 1994). But, are these complex and multivariate shapes that we see throughout organisms, as well as the processes that give rise to them, solely the output of gene regulation?

Stuart Newman and colleagues in 2006 pointed out that while cellular activities may be largely governed by gene regulation, the results (e.g., compressive forces, cellular asymmetry, etc.) may produce mechanical responses within the cells and tissues that are not governed by genes, but affect organismal form (Newman et al. 2006). Cells, then, and the forces that surround and shape them into tissues and traits are likely not completely subservient to gene regulation. One excellent recent example of this is research revealing that mechanical forces (e.g., stretching) control cell division in epithelia (Gudipaty et al. 2017). Research such as this shows us that more than networks of gene regulation are necessary in order to connect genes, development, morphology, and evolution.

The process of development is one of shifting cells, changing forms, and genetic regulation, all of which interact to produce an organism. This outcome, or the phenotype of the organism, is what comes into contact with the world and is subjected to evolutionary pressures. Phenotypes, guided by developmental processes, vary within species, and this variation gives natural selection something upon which to act. Development, then, is both the source of variation and the source of evolutionary diversity. Development is also a process that requires information about how genes direct and regulate morphogenesis, as well as information about how morphogenetic processes produce emerging forms, in order to be explained. This later component—the tracking of morphogenetic processes—has been much neglected by historians of science, but not by scientists.

In this chapter, we have seen how investigators tracked the morphogenetic processes that give rise to teeth throughout the twentieth century. They did so, at first, by breaking Weismann's unified vision of biology. In 1913, Ahrens made a conscious decision to depart from the nineteenth century trend of explaining individual tooth development by appeals to evolution as part of his explanatory framework. Even as researchers such as Reichenbach in the late 1920s focused on morphogenesis and the mechanics of cellular forces as the key to explaining how teeth form, evolution lay by the wayside. This trend continued within studies of tooth development—whether from the perspective of morphogenesis or otherwise.

Jernvall, working in the tradition of closely studying the behavior of cells and their morphogenetic properties, provided the bridge that Weismann lacked between the mechanisms of inheritance (i.e., the genes) and the phenotype. The research framework that Jernvall built for tooth development—focused on form and understanding the mechanisms (both cellular and genetic) that produce it—extended easily to incorporate evolutionary thinking. As Jernvall wrote in his dissertation, "Mammalian teeth are morphologically diverse structures whose shapes reflect developmental and ecological processes. By using a comparative approach combining new morphological, embryological and molecular evidence, this paper addresses molar tooth shape diversity, and how changes in molecular mechanisms can produce observed diversity patterns" (Jernvall 1995: 1). In other words, understanding the ways in which form arises throughout development, and the mechanisms (both cellular and genetic) that produce this form, can be extended through comparisons across species to give us insights into how we get so many different patterns of cusps throughout mammalian teeth.

Study of the enamel knot and Jernvall's developmental evolutionary approach gives us a new synthesis that brings Weismann's vision to reality. It also shows different ways that a Darwinian paradigm has affected research—from the nineteenth century desire to elaborate explanations that were both developmental and evolutionary in nature, through the twentieth century setting aside of evolutionary concerns, up to Jernvall's reinvestment in developmental evolution. Jernvall's approach also shows us that when it comes to bridging the divide between genes, morphology, development, and evolution, one should look both to genes and to cells. Cells, after all, are what build morphological characters. Only with Jernvall were the pieces put back together again for teeth. Weismann would have been pleased.

References

Ahrens H (1913) Die Entwickelung der menschlichen Zähne. Anatomische Hefte 48:167–266

Allen G (1975) Life science in the twentieth century. Wiley, New York

Allen G (1979) Thomas Hunt Morgan. The man and his science. Princeton University Press, Princeton

Born G (1883) Die Plattenmodellimethode. Arch Mikrosk Anat 22:584–599

Butler P (1956) The ontogeny of molar pattern. Biol Rev Camb Philos Soc 31:30–69

Cain J (2009) Rethinking the synthesis period in evolutionary studies. J Hist Biol 42:621–648

Cain J, Ruse M (2009) Descended from Darwin. Insights into the history of evolutionary studies, 199–1970. American Philosophical Society, Philadelphia

Churchill F (2015) August Weismann. Development, heredity, evolution. Harvard University Press, Cambridge

Delisle R (2009) The uncertain foundation of neo-Darwinism: metaphysical and epistemological pluralism in the evolutionary synthesis. Stud Hist Philos Biol Biomed Sci 40:119–132

Delisle R (2011) What was really synthesized during the evolutionary synthesis? A historiographic proposal. Stud Hist Philos Biol Biomed Sci 42:50–59

Delisle RG (2017) From Charles Darwin to the evolutionary synthesis: weak and diffused connections only. In: Delisle RG (ed) The Darwinian tradition in context: research programs in evolutionary biology. Springer, Cham, pp 133–168

Depew DJ (2017) Darwinism in the 20th century: productive encounters with saltation, acquired characteristics, and development. In: Delisle RG (ed) The Darwinian tradition in context: research programs in evolutionary biology. Springer, Cham, pp 61–88

Gudipaty S, Lindblom J, Loftus P, Redd M, Edes K, Davey C, Krishnegowda V, Rosenblatt J (2017) Mechanical stretch triggers rapid epithelial cell division through Piezo1. Nature 543:118–121

Hopwood N (1999) Giving body' to embryos: modeling, mechanism, and the microtome in late 19th-century anatomy. Isis 90:462–496

Hopwood N (2002) Embryos in wax: models from the Ziegler Studio. Whipple Museum of the History of Science, Cambridge

Hopwood N (2015) Haeckel's embryos. Images, evolution, and fraud. University of Chicago Press, Chicago

Huxley J (1927) The stream of life. Harper and Brothers, New York

Huxley J (1943) Evolution. The modern synthesis. Harper and Brothers, New York

Jernvall J (1995) Mammalian molar cusp patterns: developmental mechanisms of diversity. Acta Zool Fenn 198:1–61

Jernvall J (2000) Linking development with generation of novelty in mammalian teeth. Proc Natl Acad Sci 97:2641–2645

Jernvall J, Kettunen P, Karavanova I, Lawrence M, Thesleff I (1994) Evidence for the role of the enamel knot as a control center in mammalian tooth cusp formation: non-dividing cells express growth stimulating *Fgf-4* gene. Int J Dev Biol 38:463–469

Jernvall J, Åberg Y, Kettunen P, Keränen S, Thesleff I (1998) The life history of an embryonic signaling center: BMP-4 induces *p21* and is associated with apoptosis in the mouse tooth enamel knot. Development 125:161–169

Jernvall J, Keränen S, Thesleff I (2000) Evolutionary modification of development in mammalian teeth: quantifying gene expression patterns and topography. Proc Natl Acad Sci 97:14444–14448

Keränen S, Åberg T, Kettunen P, Thesleff I, Jernvall J (1998) Association of developmental regulatory genes with the development of different molar tooth shapes in two species of rodents. Dev Genes Evol 208:477–486

Keränen S, Kettunen P, Åberg T, Thesleff I, Jernvall J (1999) Gene expression patterns associated with suppression of odontogenesis in mouse and vole diastema regions. Dev Genes Evol 209:495–506

Kirino T, Nozue T, Inoue M (1973) Deficiency of enamel knot in experimental morphology. Okajimas Folia Anat Jpn 50:117–131

Koopman P (2001) In situ hybridization to mRNA: from black art to guiding light. Int J Dev Biol 45:619–622

Laubichler M, Maienschein J (2007) From embryology to evo-devo: a history of developmental evolution. MIT Press, Boston

MacCord K (2017) Development, evolution, and teeth: how we came to explain the morphological evolution of the mammalian dentition. PhD dissertation at Arizona State University
MacCord K, Maienschein J (2017) The historiography of embryology and developmental biology. In: Dietrich M, Borrello M, Harman O (eds) Historiography of biology. Springer, New York
Maienschein J, Laubichler M (2014) Exploring development and evolution on the tangled bank. In: Thompson P, Walsh D (eds) Evolutionary biology: conceptual, ethical, and religious issues. Cambridge University Press, Cambridge, pp 151–171
Mayr E (1982) The growth of biological thought. Diversity, evolution, and inheritance. Harvard University Press, Cambridge
Newman S, Forgacs G, Müller G (2006) Before programs: the physical origination of multicellular forms. Int J Dev Biol 50:289–299
Niswander L, Martin G (1992) Fgf-4 expression during gastrulation, myogenesis, limb and tooth development in the mouse. Development 114:755–768
Nozue T (1971a) Chronological study of enamel knot with special reference to mitoses in enamel knot. Okajimas Folia Anat Jpn 48:1–13
Nozue T (1971b) Specific spindle cells and globular substances in enamel knot. Okajimas Folia Anat Jpn 48:139–151
Pigliucci M (2017) Darwinism after the modern synthesis. In: Delisle RG (ed) The Darwinian tradition in context: research programs in evolutionary biology. Springer, Cham, pp 89–104
Pispa J, Jung H, Jernvall J, Kettunen P, Mustonen T, Tabata M, Kere J, Thesleff I (1999) Cusp patterning defect in *Tabby* mouse teeth and its partial rescue by FGF. Dev Biol 216:521–534
Radlanski R (1995) Morphogenesis of human tooth primordia: the importance of 3D computer-assisted reconstruction. Int J Dev Biol 39:249–256
Reichenbach E (1926) Die Umwandlung der Schmelzpulpa und der Schmelzepithelien während der Entwicklung des Zahnes. I Untersuchungsmethoden und eigene Befunde. Z ges Anat I Z Anat EntwGesch 80:524–546
Reichenbach E (1928) Die Umwandlungen der Schmelzpulpa und der Schmelzepithelien während der Entwicklung des Zahnes. II und III. Z Anat Entwicklungsgesch 85:490–540
Richards R (2008) The tragic sense of life. Ernst Haeckel and the struggle over evolutionary thought. University of Chicago Press, Chicago
Smocovitis B (1996) Unifying biology: the evolutionary synthesis and evolutionary biology. Princeton University Press, Princeton
Tanimura T (1968) Effects of mitomycin C administered at various stages of pregnancy upon mouse fetuses. Okajimas Folia Anat Jpn 44:337–355
Thompson D (1917) On growth and form. Cambridge University Press, Cambridge
Vaahtokari A, Åberg T, Jernvall J, Keränen S, Thesleff I (1996) The enamel knot as a signaling center in the developing mouse tooth. Mech Dev 54:39–43
Wolpert L (1994) Do we understand development? Science 266:571–572

Symbiogenesis and Cell Evolution: An Anti-Darwinian Research Agenda?

Ulrich Kutschera

Abstract In 1905, Constantin S. Mereschkowsky (1855–1921) proposed that the green organelles (chloroplasts) of algae and land plants evolved from ancient, once free-living cyanobacteria. This endosymbiotic hypothesis was based on numerous lines of evidence. In a 1910 paper, Mereschkowsky argued that the time has come to introduce a new theory on the origin of living beings; since Darwin's era, so many new findings have accumulated that now an alternative, anti-selectionist theory of evolution has to be established. Based on the principle of symbiosis (i.e., the union of two different organisms whereby both partners mutually benefit), Mereschkowsky coined the term "symbiogenesis theory," which is based on an analogy between the feeding process of amoebae and cellular events that may have occurred in the ancient oceans. Mereschkowsky's symbiogenesis hypothesis explains the origin of chloroplasts from archaic cyanobacteria, with respect to plant evolution. In 1927, the Russian cytologist Ivan E. Wallin (1883–1969) proposed that the mitochondria of eukaryotic cells are descendants of ancient, once free-living bacteria. Here, I outline the origin and current status of the Mereschkowsky–Wallin concept of symbiogenesis (primary and secondary endosymbiosis) and explain why it is compatible with the Darwin–Wallace principle of natural selection, which is described in detail. Nevertheless, largely due to the work of Lynn Margulis (1938–2011), symbiogenesis is still considered today as an Anti-Darwinian research program. I will summarize evidence indicating that symbiogenesis, natural selection, and the dynamic Earth (plate tectonics) represent key processes that caused major macroevolutionary transitions during the 3500-million-year-long history of life on Earth.

Keywords Charles Darwin • Dynamic Earth • Macroevolution • Natural selection • Symbiogenesis • Synade model

---

U. Kutschera (✉)
Institute of Biology, University of Kassel, Kassel, Germany
e-mail: kut@uni-kassel.de

R.G. Delisle (ed.), *The Darwinian Tradition in Context*,
https://doi.org/10.1007/978-3-319-69123-7_14

1 Introduction

The German botanist Andreas F. W. Schimper (1856–1901) is well known for his elegant studies on the microstructure of the "chlorophyll bodies" in plant tissues, notably in green leaves of crop species. In a landmark paper of 1883, published in the *Botanische Zeitung*, Schimper introduced the idea of a symbiotic origination of plastids (chloroplasts) as follows:

> Should it turn out definitely that the plastids are not formed de novo in the egg cells, then their relation to the organism, which contains them, would reasonably remind one of a symbiosis. It is likely that the green plants have their origin in the unification of a colourless organism with another living being, which was evenly coloured green by chlorophyll (Schimper 1883: 105).

This brief statement, consisting of only two sentences (a footnote to the main text), marks the origin of the concept of symbiogenesis, i.e., the theory that green organisms (algae, bryophytes, ferns, etc.) originated via the fusion of microbial cells, which developed a symbiotic relationship that later gave rise to a new class of organisms, i.e., the land plants (embryophytes) (Hagemann 2007).

Interestingly, this idea was already in the mind of another German botanist, before Schimper (1883) published his remark. In his famous *Vorlesungen über Pflanzen-Physiologie* (*Lectures on the Physiology of Plants*, 1882), Julius Sachs (1832–1897) referred to earlier publications of Schimper. Based on his own observations, Sachs (1882) wrote that the "chlorophyll bodies" (chloroplasts) in the moss *Funaria hygrometrica* multiply via divisions, as if they were independent microbes living within the protoplasmic space of foreign cells (Fig. 1). However, this conclusion was largely ignored over the following two decades. Due to the work of the Russian biologist Constantin S. Mereschkowsky (1855–1921), who published in 1905 a general hypothesis on the endosymbiotic origin of chloroplasts from once free-living cyanobacteria (Fig. 2), this idea became popular among biologists.

In 1890, the German physician and histologist Richard Altmann (1852–1900) proposed that subcellular particles we today call mitochondria ("bioplasts") may be descendants of once free-living microbes (Altmann 1890). This idea was elaborated and extended by the Russian biologist Ivan Wallin (1883–1969), who published a major monograph on this topic (Wallin 1927).

The novel concept of the emergence of new body plans via the unification of archaic cell types (symbiogenesis, a term coined by Mereschkowsky in 1910) was rejected by cell biologists of the 1920s but four decades later revived and corroborated by independent evidence. Moreover, it was viewed early on as being at odds with the dominant "evolutionary force" of natural selection. Hence, the idea that emerged with Schimper, Sachs, Altmann, Mereschkowsky, and Wallin must be interpreted as an "Anti-Darwinian" concept of evolution (Mereschkowsky 1920). Accordingly, in the next section, I will outline the idea of natural selection, as envisioned by Darwin and Wallace, in order to make clear why symbiogenetic thinking became a major research agenda that claimed to be an alternative view,

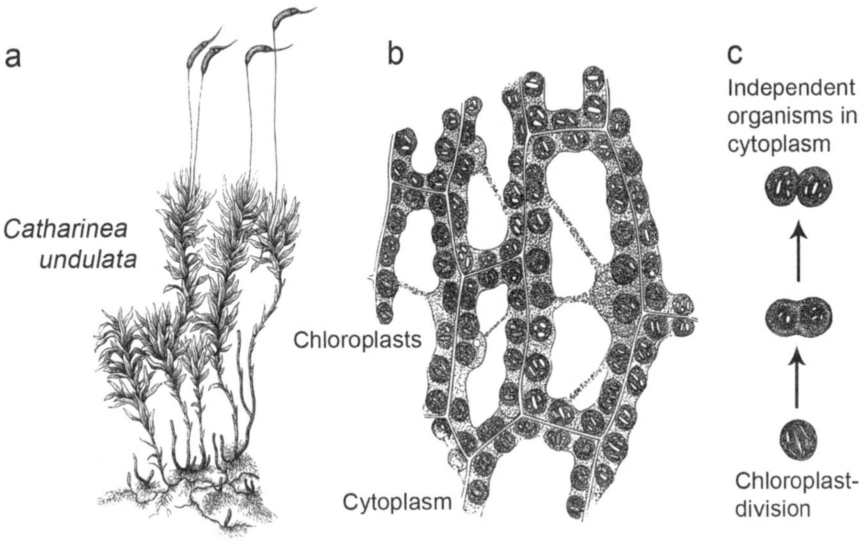

Fig. 1 Illustration of chloroplast division in the green leaves of the gametophyte of the moss *Catharinea undulata* by Julius Sachs. Adult plantlet (gametophyte) with sporophytes (**a**), cytoplasm (leaf) with numerous large chloroplasts (**b**), and photosynthetic organelle in the process of division (organellokinesis) (adapted from Sachs 1882)

regarded to be incompatible with the "Darwinian (selectionist)" mode of evolution (see Margulis 2010).

2 Darwin and Wallace: Natural Selection and the Elimination of the Unfit

It has been shown repeatedly that natural selection (or the survival of the fittest) was not a "one-man idea" of Charles Darwin (1809–1882) but rather originated independently in the brains of two naturalists: Darwin, summarized in his *Origin of Species* (1859), and the much lesser-known Alfred Russel Wallace (1823–1913) (see Depew 2017). Hence, it is fair to describe this idea as the "Darwin–Wallace principle of natural selection" (Kutschera 2003, 2008a, b, 2009a, b) (Fig. 3). Since Wallace has written extensively on this topic, notably in books and articles published after Darwin's death (Kutschera and Hossfeld 2013), we will summarize some of Wallace's descriptions of this key process that brings about evolutionary change, with reference to sexual selection and related issues in humans. The quotes in the next section are adapted from Smith (2012), where the original references are listed.

In a 1866 letter to Darwin, Wallace wrote that "Natural selection . . . does not so much select special variations as exterminate the most unfavourable ones." In 1877,

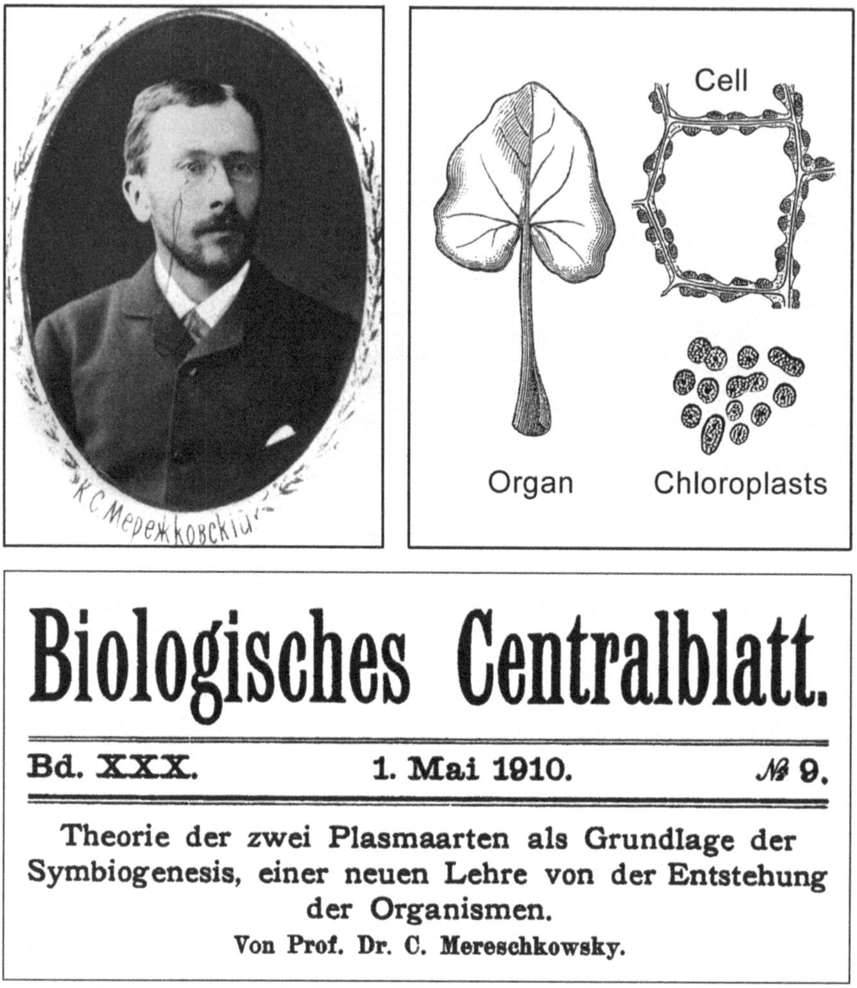

Fig. 2 Portrait of the Russian biologist Constantin Mereschkowsky (1855–1921). His green model organisms (land plants), with leaf cell and chloroplasts, are depicted. In addition, the title of his key publication of 1910 is shown

Wallace argued, with reference to insects, as follows: "In the case of butterflies the argument becomes even stronger, because the fertility is so much greater, and the weeding out of the unfit takes place, to a great extent, in the egg and the larvae state." In his famous monograph entitled *Darwinism*, Wallace (1889) argued that "Natural selection... acts perpetually and on an enormous scale in weeding out the 'unfit' at every stage of existence, and preserving only those which are in all respects the very best." One year later (1890), Wallace wrote in an article that "The survival of the fittest is really the extinction of the unfit."

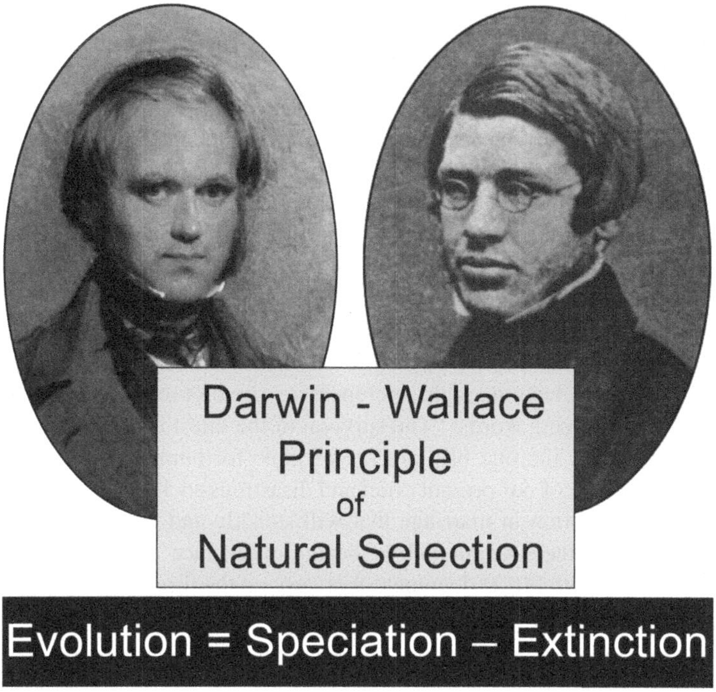

Fig. 3 Portraits of the British naturalists Charles Darwin (1809–1882) and Alfred Russel Wallace (1823–1913) and their major discovery (natural selection). In addition, the formula, "evolution equals speciation minus extinction," is added to the scheme

It is well known that Wallace was a champion of women's rights (Kutschera 2015b). Accordingly, he discussed this gender issue with reference to the principle of natural selection. In 1893, he argued that "The survival of the fittest is really the extinction of the unfit. … In order to cleanse society of the unfit we must give to woman the power of selection in marriage, and the means by which this most important and desirable end can be attained will be brought about by giving her such training and education as shall render her economically independent." One year later (1894), the British naturalist expressed this idea as follows: "I believe that the unfit will be gradually eliminated from the race, and human progress secured, by giving to the pure instincts of women the selective power in marriage."

In 1896, Wallace summarized natural selection as follows: "Accepting, then, these facts of variation, and always keeping in mind the severity of the struggle for existence, nine tenths at least of the progeny of the higher animals perishing annually before reaching maturity, thus leading to a systematic and continual weeding out of the less fit …"

In the year 1900, Wallace addressed the "women's issue" with reference to natural selection again, in the following words: "It would operate, not as among the lower animals and plants by the actual destruction of the unfit, but by their less rapid increase, since, under equal conditions of education and mode of life, it is certain that marriage would be delayed till some industrial success had been reached by both parties."

In an article of 1909, Wallace referred to the philosopher Herbert Spencer (1820–1903), who coined the phrase "survival of the fittest," adapted by Darwin and used in later editions of his *Origin of Species* (Darwin 1872). Accordingly, Wallace wrote that "Spencer suggested the term 'survival of the fittest', as more closely representing what actually occurs; and it is undoubtedly this survival, by extermination of the unfit, combined with universally present variation, which brings about that marvelous adaptation to the ever-varying environment, which is an essential feature of every living creature which survives to produce offspring."

In the year of his death, the 90-year-old biologist summarized the principles of natural and sexual selection, with reference to women's choices in selecting a husband, in the following words: "The survival of the fittest is really the extinction of the unfit; and it is the one brilliant ray of hope for humanity that, just as we advance in the reform of our present cruel and disastrous social system, we shall set free a power of selection in marriage that will steadily and certainly improve the character, as well as the strength and the beauty of our race" (Wallace 1913).

These quotes (Smith 2012) document that, around the time when the concept of symbiogenesis emerged, natural selection was interpreted to be largely a "destructive" process, i.e., a form of species "extinction" (see the equation evolution = speciation – extinction in Fig. 3). However, due to the work of Wallace (1889), and notably that of August Weismann (1834–1914), it is shown that purifying selection removes phenotypes that are not well adapted to a stable environment, whereas dynamic natural selection under gradually changing conditions "creates" new forms of life, via the emergence and propagation of those variants that are adapted to novel environments for survival and reproduction (Weismann–Schmalhausen principle of dynamic selection; see Kutschera 2009a, b). In the next section, we will explore in detail how the anti-(neo)Darwinian concept of symbiogenesis emerged and spread among scientists and philosophers in Europe.

3 Historical Roots and Elaboration of Symbiogenesis

In excellent review articles on the origin of symbiogenetic theorizing in biology and philosophy, Carrapiço (2010, 2015) summarized the achievements of several eminent thinkers not mentioned above. Beginning with Anton de Bary (1831–1888), who introduced, in 1878, the term "symbiosis" at the 51st *Congress of German Naturalists and Physicians* in Kassel, Germany (Kutschera 2011a, b), the following key figures should be recognized.

In 1902, the Russian author, anarchist, and politician Peter Kropotkin (1842–1921) published a book entitled *Mutual Aid. A Factor of Evolution*. In this monograph, he argued that, contrary to the Darwin–Wallace concept of natural selection, cooperation between organisms rather than competition should be viewed as the key factor that has driven biological evolution. However, since Kropotkin (1902) was no biologist and argued from a purely philosophical perspective, his work has been largely ignored.

Three years later (1905), Constantin Mereschkowsky published his landmark paper on the "Nature and Origin of Chromatophores in the Plant Kingdom" that will be discussed in detail below.

In 1915, the British biologist Hermann Reinheimer (1872–1950s?) published a monograph entitled *Symbiogenesis: The Universal Law of Progressive Evolution*. In this book, the term symbiogenesis was used, but without reference to Mereschkowsky's papers on this topic (notably his article of 1910) (Fig. 2). It is likely that Reinheimer was not aware of Mereschkowsky's works, which were published in German. On the other hand, Reinheimer understood the German language—unfortunately, we cannot reconstruct anymore whether or not he had read Mereschkowsky's article. In his first book (*Symbiogenesis*), Reinheimer (1915) defined "symbiosis" as a physiological partnership between individuals of different species, exactly as Anton de Bary had introduced this key term into the biological sciences some decades earlier. Reinheimer's definition of "symbiogenesis" reads as follows:

> By symbiogenesis I mean the production and increase of values throughout organic life by means of a symbiotic principle of co-operation or reciprocity between different organs of the individual, by evolved and complex body, as well as between different organisms in a species or different species, genera, orders, etc., even in the last and most fundamental way between plant and animal in the web of life (Reinheimer 1915: 156).

It is obvious that this very broad and inclusive definition of "symbiogenesis" did not impress the biologists of Reinheimer's time, because, in the natural sciences, only concepts and ideas that have an unequivocal meaning are taken seriously and are discussed openly in the peer-reviewed literature. Nevertheless, it is important to acknowledge that a first book with this key term in its title was published before Mereschkowsky had written the last of his three major contributions on this topic (Mereschkowsky 1905, 1910, 1920).

In the year when Mereschkowsky's last symbiogenesis article appeared in print, Reinheimer (1920) published a monograph entitled *Symbiosis. A Socio-Physiological Study of Evolution*. In this work, the interaction of organisms during development and evolution is described in detail. The author regarded all organisms in combination as a kind of "world society," composed of many species and families of plants and animals that represent individuals of this collection of living beings. Again, this work was more of a philosophical than a scientific nature, so that Reinheimer (1920) was largely ignored by evolutionary researchers of his time. A short note on Reinheimer is not out of place here: he was born in 1872 in Germany (Hesse) and became a British citizen in 1901. Reinheimer lived in London as a self-employed stock broker and died during the 1950s (the exact date of his death is unknown). Reinheimer published his books via Editors and Companies that were associated with alternative-esoteric views of life, such as vegetarianism, theosophy, anarchism, or metaphysics. Based on our limited knowledge, it is likely that he had never been affiliated with any academic institution in England or Germany and may have been (like Alfred R. Wallace) a self-educated private person working in the area of organismic biology.

In 1923, the Russian biologist Ivan Wallin (1883–1969) published an article on the origin of mitochondria, and 4 years later, his important monograph *Symbionticism and the Origin of Species* appeared in print (1927), wherein the work of Altmann (1890) was acknowledged. Like Mereschkowsky, Wallin (who had emigrated to the United States) was a "hands-on biologist" who published original work on several biological topics (see below).

Finally, it is worth mentioning the book of the Russian biologist Boris N. Kozo-Polyansky (1890–1957), who published, around the time when Wallin released his most important work, a monograph entitled *Symbiogenesis*: *A New Principle of Evolution* (in Russian). In this work, Kozo-Polyansky (1924) argued that symbiogenesis, defined *sensu* Mereschkowsky (1910), must be regarded as an important driving force during evolution; in contrast to most of his contemporaries, Kozo-Polyansky accepted the Darwinian principle of natural selection (Fig. 3). In addition to these basic insights, the Russian biologist introduced the ecological concept of the organism as a "consortium" (Kozo-Polyansky 1924). Recently, Margulis (2010) argued that the work of Kozo-Polyansky was more important than previously assumed and that this biologist should be credited with being one of the founding fathers of this anti-Darwinian research agenda.

In summary, this historic review documents that the basic ideas of symbiosis and symbiogenesis (Figs. 1 and 2), respectively, were very popular at a time when the Darwinian principle of natural selection (Fig. 3) was eclipsed by the erroneous theory of "Mutationism," i.e., the hypothesis that new species emerge as a result of macro-mutations in populations of parental organisms, without any role of natural selection (Kutschera and Niklas 2004). It should be stressed again that most of the authors cited above were "anti-selectionists"; they did not accept the Darwin–Wallace principle as a positive force in the "creation" of new species and body plans.

In the next section, we will summarize the contributions of Mereschkowsky and other scientists that shaped our current view of the evolutionary process. Finally, I will address the work of Lynn Margulis (1938–2011) and Margaret Dayhoff (1925–1983) and provide an integrative general scheme of macroevolution (see Figs. 6 and 8).

4 The Mereschkowsky–Wallin Principle of Symbiogenesis

As noted, the work of the lesser-known symbiogenesis theorists referred to above was of limited significance. In this section, the key insights published by the leading theorists Mereschkowsky and Wallin are summarized. In his first "symbiogenesis paper" published in 1905, entitled *Über Natur und Ursprung der Chromatophoren im Pflanzenreiche* (On the nature and origin of chromatophores in the plant kingdom), C. S. Mereschkowsky concluded that the chloroplasts (plastids) of green algae and land plants (embryophytes) were once free-living cyanobacteria (Cyanophyceae) (Figs. 2 and 4; the structure of a major chloroplast is shown in Fig. 5). This endosymbiotic concept explaining the origin of plant organelles was based on several lines of evidence—data from the scientific literature, and novel microscopic observations by the Russian biologist (Mereschkowsky 1905). In his

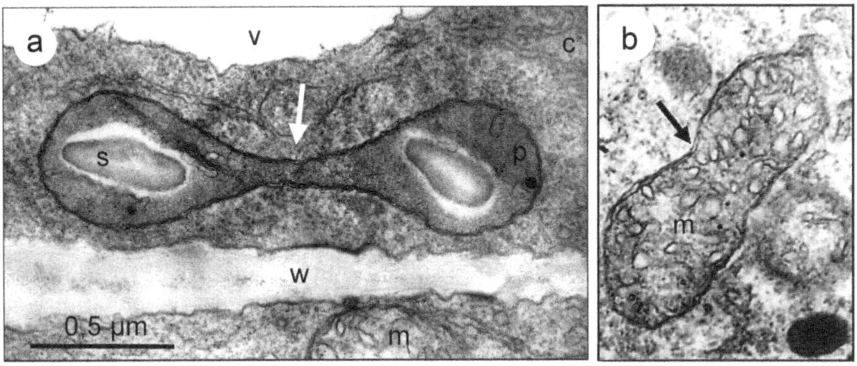

Fig. 4 Transmission electron micrographs of dividing plastids (etioplasts) (**a**) and mitochondria (**b**) in a young coleoptile of dark-grown rye seedlings (*Secale cereale*). *m* mitochondrion, *s* starch grain, *v* vacuole, *w* cell wall. The *arrow* indicates the area of organellokinesis (original micrograph)

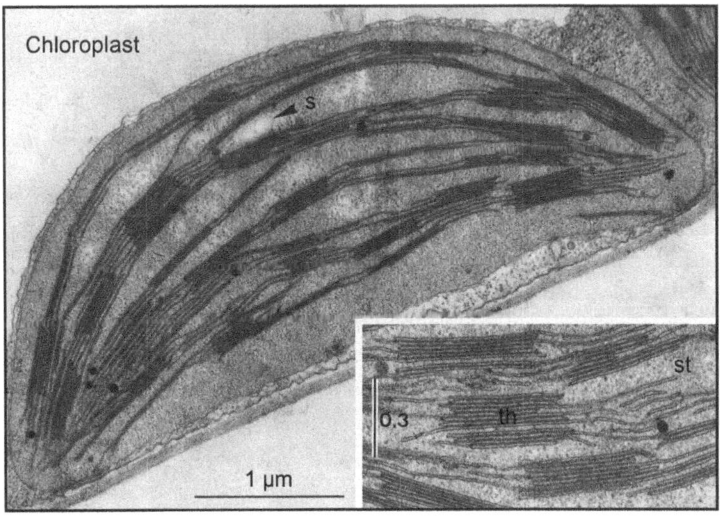

Fig. 5 Transmission electron micrograph of a mature, fully developed chloroplast in the mesophyll of a green leaf (bean, *Phaseolus vulgaris*) s starch, st stroma (adapted from Kleinig and Sitte 1986)

second article entitled *Theorie der zwei Plasmaarten als Grundlage der Symbiogenesis, einer neuen Lehre von der Entstehung der Organismen* (Theory of two species of plasmas as basis of symbiogenesis, a new concept of origin of organisms), Mereschkowsky (1910) wrote that his intention was to publish a new theory on the evolutionary development of living beings on Earth. In this key publication (Fig. 2), Mereschkowsky stated that the attempts of Charles Darwin, Ernst Haeckel (1834–1919), and others to solve this problem have been without success, because not all pertinent facts were available when these naturalists published their most influential books (Darwin 1859, 1872; Haeckel 1866, 1877). The work of Alfred Russel Wallace on natural selection, notably his popular monograph *Darwinism* (Wallace 1889), was ignored by Mereschkowsky, which may be due to the fact that the British naturalist usually played down his true achievements, compared to those of his mentor Charles Darwin (Kutschera 2008a, b). In this paper, Mereschkowsky (1910) argued that many novel findings in the areas of biochemistry, cytology, and physiology have accumulated since the time of Darwin and Haeckel, notably with respect to unicellular organisms that occur in aquatic as well as terrestrial habitats. Accordingly, Mereschkowsky (1910) boldly claimed that it is now necessary to propose a new theory on the origin of species, with a focus on plants.

With reference to the de Baryan concept of symbiosis (i.e., the union of two different organisms whereby both partners mutually benefit), Mereschkowsky (1910) introduced the term "symbiogenesis theory." The basic idea of symbiogenesis, as envisioned by him, can be interpreted as an analogy between the uptake of small particles or bacteria (i.e., phagocytosis) of amoebae, which are eukaryotic unicellular microbes, and hypothetical processes that may have occurred millions of years ago in the oceans of the young Earth. Mereschkowsky's symbiogenesis hypothesis attempted to account for the origin of the chloroplasts from ancient cyanobacteria and hence provided insight into the first steps in the evolution of the Kingdom Planta, notably that of the land plants (embryophytes) (Kutschera and Niklas 2005, 2008). In his third and less influential symbiogenesis paper, Mereschkowsky (1920) published a tentative scheme illustrating his idea as to how land plants may have evolved from green algae (Sapp et al. 2002; Geus and Höxtermann 2007).

Six years after Mereschkowsky's death, the Russian cytologist Ivan E. Wallin proposed that the mitochondria of eukaryotic cells may be descendants of ancient, once free-living bacteria (Wallin 1927). In addition, the author suggested that the primary source of genetic novelty for speciation events may have been a periodic, repeated fusion of bacterial endosymbionts with eukaryotic host cells. However, this second hypothesis of Wallin, which was, decades later, elaborated by Margulis and Sagan (2002) is not supported by convincing data (Kutschera and Niklas 2005, 2008).

5 Evolutionary Origin of Multicellular Organisms

As mentioned above, two scientists, Lynn Margulis and Margaret Dayhoff, have greatly contributed to our understanding of symbiogenesis and cell evolution. Whereas the work of Margulis has been acknowledged in many details (see Carrapiço 2010, 2015; Cavalier-Smith 2013), the key insights of Dayhoff remained less popular. In a recent article, Martin and Cerff (2017) summarized the elegant molecular work of Dayhoff (DNA-sequence analyses, reconstruction of phylogenetic trees, etc.) (Figs. 4, 5), which led to the definitive proof that chloroplasts and mitochondria descended, with modification, from once free-living cyanobacteria and alpha-proteobacteria, respectively. In the following section, we summarize the pertinent cellular events that led to the emergence of eukaryotic cells (eukaryogenesis).

Ancient endosymbiotic processes (i.e., symbiogenesis) that occurred ca. 2100–1600 million years ago (mya) in the oceans (i.e., after the Great Oxygenation Event, ca. 2300 mya) gave rise to the first eukaryotic cells. Today, these key events in the history of life are explained within the framework of the "serial primary endosymbiosis theory" for cell evolution, which is supported by a solid body of empirical data (see Kleinig and Sitte 1986; Margulis 1993 for a classic review, and Kutschera and Niklas 2004, 2005, 2008; Zimorski et al. 2014; Archibald 2014; Speijer et al. 2015; Martin et al. 2015; Martin and Cerff 2017 for more recent accounts).

The capture of an ancient alpha-proteobacterium by a host cell that resembled an extant (a-mitochondriate) Archaeon occurred probably only once during evolution (Fig. 6). Evidence for this major conclusion is largely based on the finding that the protein import machineries (TIM/TOM in mitochondria, TIC/TOC for plastids) of these organelles are uniform in all Kingdoms of life. After subsequent intracellular domestication events, the once free-living alpha-proteobacterium was reduced to an organelle, which produces and exports energy-rich adenosine triphosphate (ATP, intra-cytoplasmatic concentration ca. 5 mM). This "energy currency of the cell" has not only the well-known function to permit biochemical processes to occur but also to stabilize proteins in the "crowded" protoplasm (Rice and Rosen 2017).

In a subsequent primary endosymbiotic event, an ancient cyanobacterium was engulfed, domesticated-incorporated, and finally reduced to a photosynthetic, green organelle (chloroplast). After the domestication of these ancient microbes, horizontal gene transfer to the nucleus occurred in both mitochondria and plastids, so that today these "enslaved" organelles contain a "miniaturized" genome (Zimorski et al. 2014). These alpha-proteobacterial and a cyanobacterial endosymbionts (i.e., the ancestral mitochondrion and chloroplast, respectively) multiply in the cytoplasm by binary fission, like their free-living ancestors (Fig. 4). In most organisms, they are inherited, during sexual reproduction, via the egg cell.

All multicellular organisms consist of eukaryotic cells, which are much more complex than prokaryotes (archaea, bacteria, cyanobacteria). From an energetic point of view, the ATP level per gene is at least 1000-fold higher in eukaryotic cells, due to mitochondrial activity, compared to prokaryotes. As detailed by Martin and Cerff (2017), under aerobic conditions, heterotrophic eukaryotic cells can produce (due to the presence of mitochondria) theoretically 30–40 ATP per glucose. Under real-world conditions, the number is ca. 32 ATP per metabolized glucose molecule. Prokaryotic microbes, on the other hand, create only about 4 ATP per glucose metabolized, using fermentation under oxygen-limiting conditions.

Without the emergence of mitochondria- and chloroplast-containing (photosynthetic) complex cells via serial primary endosymbiosis (i.e., symbiogenesis), the extant biosphere would exclusively be inhabited by prokaryotes, but no heterotrophic protists, chlorophytes (green algae), and their multicellular descendants would be present. As a result, animals (including humans), fungi, and plants evolved as a consequence of ancient invasions of prokaryotes into an Arachaean host (Figs. 6 and 8), a concept also known as the "two primary domains of life model" (Martin et al. 2015; Kutschera 2015a, 2016; McInnerney and O'Connell 2017). This merger of two cell types to create novel unicellular organisms (the Mereschkowsky–Wallin principle of symbiogenesis) was a key macro-evolutionary process leading to the development of complex organisms on Earth (Archibald 2014; Niklas 2016; Kutschera and Niklas 2005, 2008; Kutschera 2015a, 2016, 2017).

In addition, at least three independent ancient secondary endosymbiotic events, i.e., the incorporation and enslavement of unicellular algae by heterotrophic eukaryotic host cells, resulted in chimeric "monster organisms" (such as euglenids and dinoflagellates). Today, these photosynthetic protists represent the majority of extant phytoplankton in marine and freshwater ecosystems of the Earth (Figs. 6 and 7). They are the dominant photosynthetic primary producers in the oceans and account for ca. 40–50% of primary photosynthetic activity in the biosphere (Cavalier-Smith 2000, 2013; Knoll 2003; Kutschera and Niklas 2008; Martin and Quigg 2012).

According to Mereschkowsky (1905, 1910, 1920), Margulis (1993, 2010), Margulis and Sagan (2002), and other symbiogenesis researchers (see Carrapiço 2010, 2015), primary and secondary endosymbiosis is an evolutionary process incompatible with natural selection. For a discussion of this argument, we briefly recapitulate some achievements of Alfred Russel Wallace.

In his popular book *Darwinism*, Wallace (1889) discussed the "problem of the Origin of Species" and coined the phrase "the Darwinian theory of natural selection." Moreover, he wrote that "I am the advocate of pure Darwinism" (p. 12). His own significant contributions to the development of the "Darwin–Wallace principle of natural selection" (Fig. 3) are only briefly described in his book. However, Wallace (1889) rejected the Lamarckian–Darwinian concept of an inheritance of acquired characteristics and incorporated the discoveries and theoretical principles of the German zoologist August Weismann into his theoretical concepts. Hence, Wallace became one of the founding fathers of the Neo-Darwinian theory of

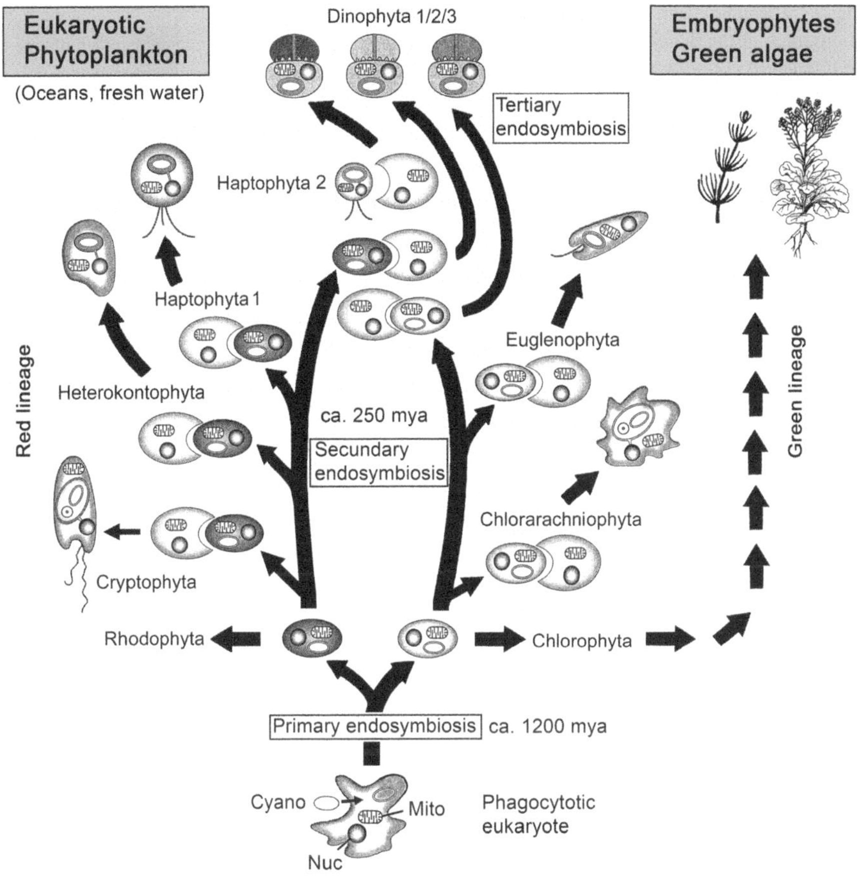

Fig. 6 Symbiogenesis and the origin of green algae/land plants (embryophytes) and the phytoplankton of freshwater/marine ecosystems, respectively. Primary endosymbiosis gave rise to the unicellular chlorophytes, which evolved into land plants. Secondary endosymbiotic events led to the origination of planktonic organisms (red and green lineages, respectively) that are important primary producers in the oceans. *mya* million of years ago (adapted from Kutschera 2015a)

biological evolution, which later gave rise to the Synthetic Theory (Mayr 1984, 2001; Kutschera and Niklas 2004, 2008; Kutschera and Hossfeld 2013; Kutschera 2015a; 2017).

The key concept of this theory of the 1940s, the "Darwin–Wallace principle of natural selection," which may be interpreted as a process resulting from biological differences among individuals in expanding populations, has been confirmed in numerous field and laboratory studies, ranging from bacteria to humans and plants (Endler 1986; Bell 1997; Mayr 1984, 2001; Kutschera and Niklas 2004; Carroll 2006; Gregory 2008; Niklas 2016). Natural selection also operates in variable populations of unicellular eukaryotic microbes that originated from primary and secondary endosymbiotic events (i.e., via symbiogenesis), such as diatoms (Fig. 7).

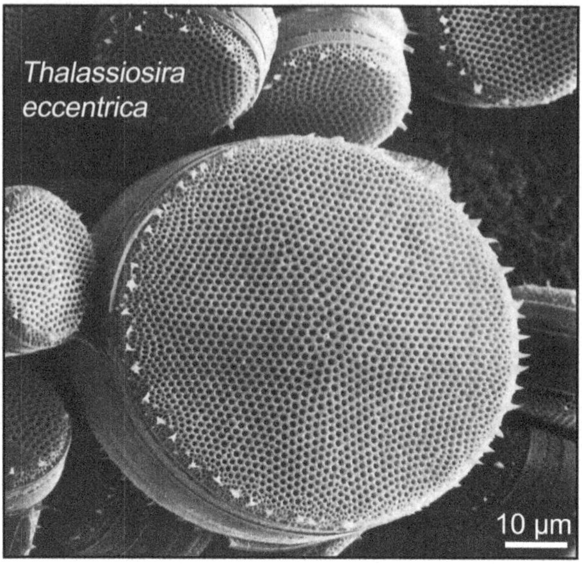

Fig. 7 Scanning electron micrograph of the diatom *Thalassiosira eccentrica*, a photosynthetically active member of the marine phytoplankton (the two solid shells of the microorganism are shown). This eukaryotic microbe originated via secondary endosymbiosis (adapted from Kleinig and Sitte 1986)

Hence, the argument that symbiogenesis and natural selection are contradictory, mutually exclusive processes is invalid. In the next section, we briefly discuss geologic events that were responsible for the long-term creation of new environments and hence major selection pressures over millions of years of organismic evolution.

6 The Snider–Wegener Concept of Shifting Continents

In 1669, Nicolaus Steno (1638–1686) established some of the fundamental principles of paleontology and stratigraphy by identifying fossils as remnants of once-living organisms, and the proposal that rock strata are analogous to the pages in a history book. Accordingly, Steno concluded that the surface of the Earth is not static, but dynamic, and that the fossil record represents a chronology of living beings that inhabited our planet in different eras of Earth's history (Cutler 2003).

Despite Steno's early insights, which indicated that the surface of our planet may be in slow motion, the idea of a static Earth prevailed. In 1858, when Darwin and Wallace published their papers on natural selection (Fig. 3), Antonio Snider-Pellegrini (1802–1885) proposed that identical plant fossils found in European and North American coal deposits may be explained by the idea that the two continents were once connected together during the Pennsylvanian period. In his book *The Creation and its Mysteries Unveiled*, Snider-Pellegrini (1858) published two maps of the Earth, depicting the continents before and after separation. Although the author referred to fossils with reference to continental drift, Snider-Pellegrini's primary source of inspiration was the book of Genesis in the Bible (LeGrand 1988).

As expected, Snider-Pellegrini's (1858) fanciful maps did not convince the geologists of his time. Accordingly, the idea of a static Earth prevailed again. Due to the work of the German scientist Alfred Wegener (1880–1930), convincing evidence for a purely naturalistic theory of continental drift was proposed that later revolutionized geology. Like Snider-Pellegrini (1858), Wegener examined the maps of the globe and suggested that most of the extant continents may fit together like a puzzle. For instance, the West African coastline fits into the east coast of South America and the Caribbean Sea; a similar fit is apparent across the Pacific. Even more striking results were obtained when the submerged continental shelves were compared rather than the continents. In 1915, the first edition of his book was published. Subsequently, Wegener revised and considerably extended the text, so that the 4th edition of 1929 represents the definitive version of this important monograph.

In his book, *The Origin of the Continents and Oceans* (1929), Wegener proposed that the current, isolated continents were once compressed into a single proto- or supercontinent: Pangaea ("all lands"), which covered about half of the Earth's surface, was surrounded by one giant ocean called Panthalassa ("universal sea"). Wegener's drift theory provided a novel explanation for the formation of mountains via the compression and upward folding of the edges of moving continents. Moreover, he postulated that earthquakes and volcanism are definitively caused by continental drift (Wegener 1929).

Wegener's inability to provide an adequate explanation for the physical forces responsible for the possible drift of the continental land masses, and the prevailing assumption that the Earth was immovable (static), resulted in the ignorance and dismissal of his theory. In the late 1960s, Snider's and Wegener's forgotten ideas were rediscovered, supported by independent lines of evidence from geology and paleobiology, and expanded into the concept of the dynamic Earth, also known as the theory of plate tectonics (LeGrand 1988; Irving 2005; Nield 2007; Kutschera 2009a; Mallard et al. 2016) (Fig. 8).

Independent lines of evidence document that the dynamic Earth has not only created and modified most terrestrial and aquatic habitats through the eons of geological time but also destroyed entire groups of organisms via volcanic eruptions and the associated release of poisonous gases (SO_2, SO_3, CO_2, etc.) leading to mass extinctions. Hence, plate tectonics has been responsible for the creation of new ecological niches, as well as the destruction of populations, and therefore naturally selected those individuals in variable populations that propagated "their kind" under new environmental conditions (survival-reproduction of the most suitable individuals) (Kutschera 2017). In 1915, when these ideas were proposed for the first time by Wegener, the Russian biologist Mereschkowsky was working on his last symbiogenesis paper that was published 5 years later. In this major contribution, Mereschkowsky (1920) further expanded the organismic view of evolutionary theory by outlining a multi-kingdom perspective that is summarized below.

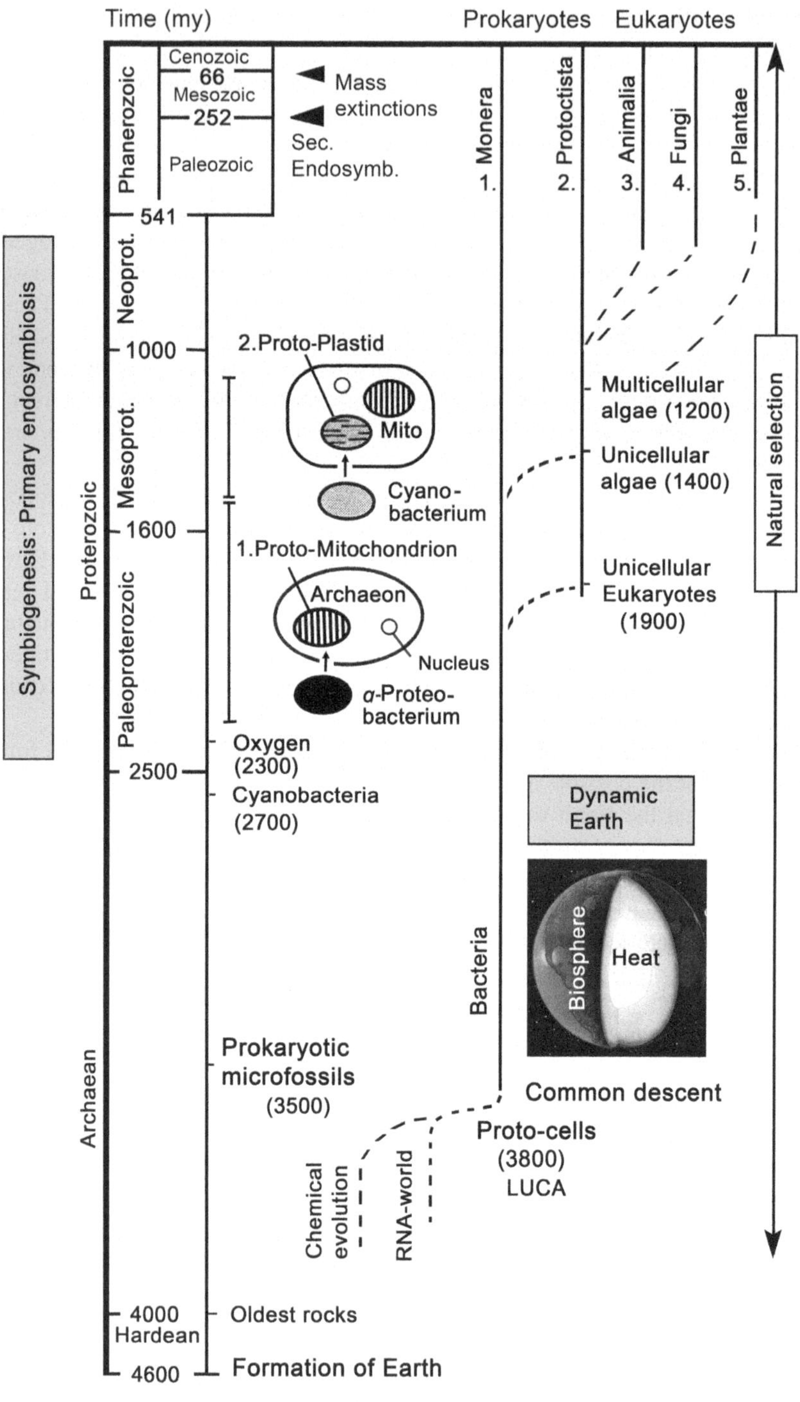

Fig. 8 General scheme of organismic macroevolution during the history of life on Earth, with special reference to symbiogenesis (primary endosymbiosis). Chemical evolution resulted in the Last Universal Common Ancestor (LUCA). Via directional natural selection, under changing environmental conditions (dynamic Earth), members of all five Kingdoms of life evolved (Monera, Protoctista, Animalia, Fungi, Plantae) (adapted from Kutschera 2015a)

7 Constantin Mereschkowsky and the Five Kingdoms of Life

In the nineteenth century, when Darwin (1859, 1872) published his *Origin of Species*, and Wallace (1889) summarized and extended these revolutionary ideas, systematic biology (taxonomy) was not very well developed. Throughout these great works, which provided the organizing principle of modern biology—descent with modification (organismic evolution)—Darwin and Wallace referred to animals and plants, i.e., multicellular macroorganisms. Only at the end of the last chapter, Darwin (1859, 1872) briefly mentioned "lower algae." From such animal-plant-like intermediate forms of life (freshwater flagellates of the genus *Euglena*), all organic beings may have descended (Darwin 1872). It has been shown that Darwin's classical "Proto-Euglena hypothesis" is no longer acceptable (Kutschera and Niklas 2008). Hence, in Darwin's and Wallace's time, animals, plants, and very few "infusoria" were the model organisms of choice to explain the principle of evolution via natural selection. Bacteria, amoebae, and many other microorganisms described by nineteenth-century naturalists are not mentioned by Darwin (1859, 1872) and only briefly addressed by Wallace (1889, 1913). It was Ernst Haeckel who introduced the "Protista" and "Bacteria," microbes he studied from a taxonomic point of view (Haeckel 1866, 1877). However, his concepts concerning their mode of evolution remained speculative and unconvincing (Hossfeld 2010; Kutschera 2011a, b, 2016).

The Russian botanist and cytologist Mereschkowsky (1905, 1910, 1920) was one of the first to integrate, in addition to animals and plants, bacteria, cyanophytes (i.e., cyanobacteria), green algae, amoebae (Protists), fungi, and other "lower organisms" into an evolutionary scenario that he called symbiogenesis—the origination of new forms of life by the combination of two or several unicellular living beings which enter into symbiosis. Hence, Mereschkowsky—notably in his paper of 1920—was one of the founding fathers of a "numerous Kingdoms principle" that incorporated all known forms of life into an evolutionary framework. Today, the organisms on Earth are classified according to the "Five-Kingdom System" (Barnes 1998; Margulis and Schwartz 1998):

1. Monera (Bacteria or Prokaryotae)
2. Protoctista (protists, like diatoms [Fig. 7], algae, and amoebae)
3. Animalia (animals, including humans)
4. Fungi (molds, yeasts, and mushrooms)
5. Plantae (bryophytes, ferns, and seed plants)

According to this classification scheme of the living world, we distinguish between prokaryotic microbes, unicellular microorganisms that consist of small bacterial cells (Kingdom 1), and the eukaryotes (Kingdoms 2–5). These micro- and macroorganisms are composed of much larger eukaryotic cells, which are defined by the presence of a nucleus and mitochondria (see Fig. 8).

8 Evolution of Life in a Bacterial World: Animals and Plants as Superorganisms

During the 1950s, it became obvious that bacteria dominate the biosphere. At that time, it was estimated that about 50% of protoplasmic biomass on Earth may be composed of prokaryotic microbes (essentially aquatic cyanobacteria, plus archaea and eubacteria). Decades later, this concept solidified so that it is now generally accepted that we live on a "planet of the microbes" (Whitman et al. 1998; Kutschera 2009a, b, 2011a, b, 2015a). The "unseen majority" of bacteria inhabit, for instance, the gut of animals/humans, where they are important symbionts for digestion and health of the eukaryotic host organism (Charbonneau 2016).

However, it is not widely recognized that the growth of land plants (embryophytes), Mereschkowsky's model organisms, is regulated and modified to some extent by microbes: plant growth-promoting rhizobacteria (PGPRs) and pink-pigmented facultative methylotrophic bacteria (PPFMs, also called, "methylobacteria"). Among the PGPRs, we distinguish between symbiotic microbes that live inside the plant body and free-living bacteria that inhabit the rhizosphere (region around the roots) of their host organism. The most prominent PGPRs are bacteria of the genus *Rhizobium* that induce symbiotic root nodules in leguminous crop plants, such as pea, lupines, etc., and fix atmospheric nitrogen (N_2). Of similar importance are free-living bacteria of the genera *Azobacter*, *Azotobacter*, *Bacillus*, *Phyllobacterium*, *Pseudomonas*, and *Streptomyces*. These root-associated rhizobacteria promote the growth of crop plants (cucumber, wheat, rice, sunflower, maize, strawberries, potato, Indian lilac, etc.) by the production/ secretion of phytohormones (auxins, cytokinins), the solubilization of mineral nutrients (potassium, phosphate, etc.), or the production of antibiotics (prevention of plant diseases). Since, for instance, sugarcane plants harbor in their intercellular spaces large populations of endophytic bacteria (*Beijerinckia*, *Herbaspirillum*, etc.), and, in addition to the PGPRs, the PPFMs or methylobacteria (genus *Methylobacterium*) likewise live attached to these green organisms (from the flowers via the leaves/stem down to the root tips), it is fair to interpret land plants as superorganisms. The well-known soil-borne mycorrhizas (fungi associated with the root system) should also be mentioned in this context, since Mereschkowsky (1905, 1910, 1920) discussed these organisms in some detail and published a scheme illustrating their possible evolutionary development (Kutschera 2007; Kutschera and Khanna 2016).

9 Conclusions: Symbiogenesis as the "Big Bang" in Organismic Evolution

The most ancient traces of microbial life on Earth are about 3800 mya old (Allwood 2016).

After the emergence of living units via the occurrence of hypothetical proto-cells about 4000 mya (LUCA, i.e., the last universal common ancestor), bacteria, and later cyanobacteria, dominated the aquatic habitats on the young Earth (Knoll 2003; Schopf 2006; Kutschera 2017). As a result of the evolutionary "invention" of oxygenic photosynthesis, ancient cyanobacteria created the O_2-containing atmosphere that emerged about 2200 mya (Knoll 2003; Zimorski et al. 2014). Symbiogenesis (i.e., primary endosymbiosis) was the key macro-evolutionary event (or the "big bang") that gave rise to the first eukaryotic microorganisms via the fusion of an archaeon (host) and a bacterium (guest). This "two primary domains of life" model is in accordance with Ernst Haeckel's (1866) idea that all living beings on planet Earth originated from bacteria (Kutschera 2016; McInnerney and O'Connell 2017). The first cell chimeras were heterotrophic, mitochondria-containing units without photosynthetic organelles; later, cells with photoautotrophic microbes, i.e., domesticated/enslaved cyanobacteria (which later became chloroplasts), evolved (Figs. 6 and 8).

Natural selection of those individuals best adapted to the corresponding environment in growing populations of pro- and eukaryotic micro- and macroorganisms not only "shaped" the evolving phenotypes but was also responsible for the diversification of life (Dobzhansky 1955; Mayr 1984, 2001; Bell 1997; Klingsolver and Pfennig 2007; see also Pigliucci 2017). The dynamic Earth (i.e., plate tectonics) resulted in the formation of mountains and deep oceans and caused volcanism (Mallard et al. 2016). Hence, via these geological processes, new habitats and niches for evolving populations of organisms in all five Kingdoms of life were created. In addition, the climate of the planet has been modified via changes in oceanic and atmospheric chemistry, as well as global topography. Mass extinctions were to a large extent caused by plate tectonics/volcanism, although extraterrestrial causes, such as meteorite impacts, may also have elicited these global catastrophes.

Figure 8 illustrates that symbiogenesis, natural selection, and the dynamic Earth were the key processes or dominant "evolutionary factors" that caused the origination as well as extinction of organisms on this ever-changing planet. This integrative "synade model" of macroevolution, which is a general theory of organismic evolution that consists of a set of fundamental biogeological principles, does not make specific predictions as to the phylogeny of any group of organisms. Neither Darwin (1859, 1872) and Wallace (1889, 1913) nor the architects of the Synthetic Theory of the 1950s had incorporated symbiogenesis (and plate tectonics) into their corresponding explanatory framework of evolutionary change (Dobzhansky 1955; Mayr 1984, 2001, 2004; Gould 2002; Haffer 2007; see also Depew 2017; Pigliucci 2017). These "driving forces" of biological evolution were rediscovered and refined during the post-synthesis era of evolutionary thought (the modern theory of biological evolution as an expanded synthesis; see Kutschera and Niklas 2004; Kutschera 2008a, b, 2009a, b, 2011a, b, 2017). The implications of this extended view of the evolving geo-biosphere can be summarized as follows.

Without the internal heat in the center of the Earth (Fig. 8), which is driven primarily by radioactive decay of heavy, naturally occurring elements such as Uranium, no continental land masses would have been created via plate tectonic events. It follows that without the dynamic Earth, life would probably still be unicellular and restricted to the oceans—no land plants and terrestrial animals would ever have had a chance to evolve. Other processes, notably natural selection under changing environmental conditions, were likewise of major importance during the about 3500 million years of history of life on Earth. All organisms produce more progeny than the environment can support. Nevertheless, symbiogenesis (primary endosymbiosis) was the "big bang" in cell evolution that gave rise to all eukaryotic organisms on Earth, from amoeba to animals and land plants. Later in the history of life, secondary endosymbiotic events led to the origin of the eukaryotic phytoplankton that represents the dominant organismic component of the oceans (Figs. 6 and 8).

To sum up, symbiogenesis and the corresponding focus on cell evolution considerably broadened our perspective of the modes and mechanisms of organismic evolution. As a result, an integrative view is emerging that goes far beyond what Darwin and Wallace, as well as the architects of the synthetic theory, ever have imagined when they published their groundbreaking monographs on the origin and phylogenetic development of life on Earth (Darwin 1859, 1872; Wallace 1889, 1913; Dobzhansky 1955; Gould 2002; Mayr 1984, 2001, 2004).

Acknowledgements I thank two anonymous reviewers for helpful comments on an earlier version of the manuscript and the Alexander von Humboldt-Stiftung (Bonn, Germany) for financial support (AvH-fellowship 2013, Stanford, CA, USA).

References

Allwood AC (2016) Geology: evidence of life in Earth's oldest rocks. Nature 537:500–501

Altmann R (1890) Die Elementarorganismen und ihre Beziehungen zu den Zellen. Verlag von Veit, Leipzig

Archibald JA (2014) One plus one equals one: symbiosis and the evolution of complex life. Oxford University Press, Oxford

Barnes RSK (ed) (1998) The diversity of living organisms. Blackwell, Oxford

Bell G (1997) Selection: the mechanism of evolution. Chapman and Hall, New York

Carrapiço F (2010) How symbiogenic is evolution? Theory Biosci 129:135–139

Carrapiço F (2015) Can we understand evolution without symbiogenesis? In: Gontier N (ed) Reticulate evolution. Interdisciplinary evolution research, vol 3. Springer, Cham, pp 81–105

Carroll SB (2006) The making of the fittest. DNA and the ultimate forensic record of evolution. WW Norton, New York

Cavalier-Smith T (2000) Membrane heredity and early chloroplast evolution. Trends Plant Sci 5: 174–182

Cavalier-Smith T (2013) Symbiogenesis: mechanisms, evolutionary consequences and systematic implications. Annu Rev Ecol Evol Syst 44:145–172

Charbonneau MR (2016) A microbial perspective of human developmental biology. Nature 535: 48–55

Cutler A (2003) The seashell on the mountaintop. Dutton, New York

Darwin C (1859) On the origin of species by means of natural selection, or the preservation of favoured races in the struggle for life. John Murray, London

Darwin C (1872) The origin of species by means of natural selection, or the preservation of favoured races in the struggle for life, 6th edn. John Murray, London
Depew DJ (2017) Darwinism in the 20th century: productive encounters with saltation, acquired characteristics, and development. In: Delisle RG (ed) The Darwinian tradition in context: research programs in evolutionary biology. Springer, Cham, pp 61–88
Dobzhansky T (1955) Evolution, genetics, and man. Wiley, New York
Endler JA (1986) Natural selection in the wild. Princeton University Press, Princeton, NJ
Geus A, Höxtermann E (Hg) (2007) Evolution durch Kooperation und Integration. Zur Entstehung der Endosymbiosetheorie in der Zellbiologie. Basilisken-Presse, Marburg
Gould SJ (2002) The structure of evolutionary theory. Harvard University Press, Cambridge
Gregory TR (2008) Evolution as fact, theory and path. Evol Educ Outreach 1:46–52
Haeckel E (1866) Generelle Morphologie der Organismen, vol I/II. Georg Reimer, Berlin
Haeckel E (1877) Anthropogenie oder Entwicklungsgeschichte des Menschen. W. Engelmann, Leipzig
Haffer J (2007) Ornithologie, evolution, and philosophy. The life and science of Ernst Mayr 1904–2005. Springer, Berlin
Hagemann R (2007) The reception of the Schimper-Mereschkowsky endosymbiont hypothesis on the origin of plastids—between 1883 and 1960—many negative, but a few relevant positive reactions. Ann Hist Philos Biol 12:41–59
Hossfeld U (2010) Ernst Haeckel. Biographienreihe absolute. Orange, Freiburg i. Br
Irving E (2005) The role of latitude in mobilism debates. Proc Natl Acad Sci USA 102:1821–1828
Kleinig H, Sitte P (1986) Zellbiologie. Ein Lehrbuch. 2. Auflage. Verlag Gustav Fischer, Stuttgart
Klingsolver JG, Pfennig D (2007) Patterns and power of phenotypic selection in nature. Bioscience 57:561–572
Knoll AH (2003) Life on a young planet. The first three billion years of evolution on earth. Princeton University Press, Princeton, NJ
Kozo-Polyansky BM (1924) Symbiogenesis: a new principle of evolution. Marshall University, Huntington, WV
Kropotkin P (1902) Mutual aid. A factor of evolution. Free Press, London
Kutschera U (2003) A comparative analysis of the Darwin-Wallace papers and the development of the concept of natural selection. Theory Biosci 122:343–359
Kutschera U (2007) Plant-associated methylobacteria as co-evolved phytosymbionts: a hypothesis. Plant Signal Behav 2:74–78
Kutschera U (2008a) Darwin-Wallace principle of natural selection. Nature 453:27
Kutschera U (2008b) From Darwinism to evolutionary biology. Science 321:1157–1158
Kutschera U (2009a) Symbiogenesis, natural selection, and the dynamic earth. Theory Biosci 128: 191–203
Kutschera U (2009b) Charles Darwin's *Origin of Species*, directional selection, and the evolutionary sciences today. Naturwissenschaften 96:1247–1263
Kutschera U (2011a) Darwiniana Nova. Verborgene Kunstformen der Natur. LIT, Berlin
Kutschera U (2011b) From the scala naturae to the symbiogenetic and dynamic tree of life. Biol Direct 6(33):1–20
Kutschera U (2015a) Evolutionsbiologie. Ursprung und Stammesentwicklung der Organismen. 4. Auflage. Verlag Eugen Ulmer, Stuttgart
Kutschera U (2015b) A prescient view of women in evolution. Nature 523:35
Kutschera U (2016) Haeckel's 1866 tree of life and the origin of eukaryotes. Nat Microbiol 1: 16114
Kutschera U (2017) Evolution. Reference module in life sciences. Article 06399. Elsevier, pp 1–5
Kutschera U, Hossfeld U (2013) Alfred Russel Wallace (1823–1913): the forgotten co-founder of the Neo-Darwinian theory of biological evolution. Theory Biosci 132:207–214
Kutschera U, Khanna R (2016) Plant gnotobiology: epiphytic microbes and sustainable agriculture. Plant Signal Behav 11(12):e1256529, 1–4

Kutschera U, Niklas KJ (2004) The modern theory of biological evolution: an expanded synthesis. Naturwissenschaften 91:255–276
Kutschera U, Niklas KJ (2005) Endosymbiosis, cell evolution, and speciation. Theory Biosci 124: 1–24
Kutschera U, Niklas KJ (2008) Macroevolution via secondary endosymbiosis: a Neo-Goldschmidtian view of unicellular hopeful monsters and Darwin's primordial intermediate form. Theory Biosci 127:277–289
LeGrand HE (1988) Drifting continents and shifting theories. Cambridge University Press, Cambridge
Mallard C, Coltice N, Seton M, Muller RD, Tackley PJ (2016) Subduction controls the distribution and fragmentation on Earth's tectonic plates. Nature 353:140–143
Margulis L (1993) Symbiosis in cell evolution. Microbial communities in the Archean and Proterozoic eons, 2nd edn. WH Freeman, New York
Margulis L (2010) Symbiogenesis. A new principle of evolution rediscovery of Boris Mikhaylovich Kozo-Polyansky (1890–1957). Paleontol J 44:1525–1539
Margulis L, Sagan D (2002) Acquiring genomes: a theory of the origin of species. Basic Books, New York
Margulis L, Schwartz KV (1998) Five kingdoms. An illustrated guide to the phyla of life on earth, 3rd edn. WH Freeman, New York
Martin WF, Cerff R (2017) Physiology, phylogeny, early evolution and GAPDH. Protoplasma 254:1823–1834
Martin RE, Quigg A (2012) Evolving phytoplankton stoichiometry fuelled diversification of the marine biosphere. Geosciences 2:130–146
Martin WF, Garg S, Zimorski V (2015) Endosymbiotic theories for eukaryote origin. Philos Trans R Soc B 370:20140330
Mayr E (1984) The growth of biological thought. Diversity, evolution and inheritance. Harvard University Press, Cambridge
Mayr E (2001) What evolution is. Basic, New York
Mayr E (2004) What makes biology unique? Considerations on the autonomy of a scientific discipline. Cambridge University Press, Cambridge
McInnerney JO, O'Connell MJ (2017) Microbiology: mind the gaps in cellular evolution. Nature 541:297–299
Mereschkowsky C (1905) Über Natur und Ursprung der Chromatophoren im Pflanzenreiche. Biol Centralbl 25:593–604, 689–691
Mereschkowsky C (1910) Theorie der zwei Plasmaarten als Grundlage der Symbiogenese, einer neuen Lehre von der Entstehung der Organismen. Biol Centralbl 30:278–303, 321–347, 353–367
Mereschkowsky C (1920) La plante considerée comme un complexe symbiotique. Bull Soc Sci Nat Fr 6:17–98
Nield T (2007) Supercontinent. Ten billion years in the life of our planet. Harvard University Press, Cambridge
Niklas KJ (2016) Plant evolution. An introduction to the history of live. University of Chicago Press, Chicago, IL
Pigliucci M (2017) Darwinism after the modern synthesis. In: Delisle RG (ed) The Darwinian tradition in context: research programs in evolutionary biology. Springer, Cham, pp 89–104
Reinheimer H (1915) Symbiogenesis: the universal law of progressive evolution. Knapp, Drewett, London
Reinheimer H (1920) Symbiosis: a socio-physiological study of evolution. Headley, London
Rice AM, Rosen MK (2017) Perspective: ATP controls the crowd. Science 356:701–702
Sachs J (1882) Vorlesungen über Pflanzen-Physiologie. Wilhelm Engelmann, Leipzig
Sapp J, Carrapico F, Zolotonosov M (2002) Symbiogenesis: the hidden face of Constantin Merezhkowsky. Hist Philos Life Sci 24:413–440

Schimper ATW (1883) Über die Entwicklung der Chlorophyllkörner und der Farbkörper. Botanische Zeitung 41:105–114, 121–131, 137–146, 153–162
Schopf WJ (2006) Fossil evidence of Archaean life. Philos Trans R Soc B 361:869–885
Smith CH (2012) Alfred Russel Wallace and the elimination of the unfit. J Biosci 37:203–205
Snider-Pellegrini A (1858) La Création et ses mystères dévoilés. Franck et Dentu, Paris
Speijer D, Lukes J, Elias M (2015) Sex is a ubiquitous, ancient, and inherent attribute of eukaryotic life. Proc Natl Acad Sci USA 112:8827–8834
Wallace AR (1889) Darwinism. An exposition of the theory of natural selection with some of its applications. MacMillan, London
Wallace AR (1913) Social environment and moral progress. Cassell, London
Wallin IE (1927) Symbionticism and the origin of species. Bailliere, Tindall & Cox, London
Wegener A (1929) Die Entstehung der Kontinente und Ozeane. 4. Auflage. F. Vieweg & Sohn, Braunschweig
Whitman WB, Coleman DC, Wiebe WJ (1998) Prokaryotes: the unseen majority. Proc Natl Acad Sci USA 95:6578–6583
Zimorski V, Ku C, Martin WF, Gould SB (2014) Endosymbiotic theory for organelle origins. Curr Opin Microbiol 22:38–48

Paleobiology's Uneasy Relationship with the Darwinian Tradition: Stasis as Data

Derek D. Turner

> *[S]tasis is one of the most neglected theoretical problems in evolutionary biology.*
>
> T.F. Hansen and D. Houle (2004: 130)

Abstract During the late twentieth century and up to the present, paleobiologists' thinking about evolution has had an uneasy relationship with the Darwinian tradition. In this chapter, I use the concept of stasis as a guiding thread for exploring these tensions. Beginning in the 1970s, paleobiologists put stasis on the agenda of evolutionary biology, and in doing so, they challenged Darwinian tradition in at least three ways: (1) famously, the theory of punctuated equilibria implied that stasis, rather than gradual, selection-driven change, is the hallmark of evolutionary history; (2) subsequent attempts to explain evolutionary stasis have shown the limits of neo-Darwinian explanations that emphasize the power of stabilizing selection, pointing to the need for a more hierarchical approach; and (3) the issue whether stasis *vs.* change should be the default expectation for evolutionary systems remains unsettled.

Keywords Darwinism • Macroevolution • Paleobiology • Punctuated equilibria • Stasis

1 Introduction

During the 1970s and 1980s, paleontology went through an exciting period that some have called "revolutionary" (Ruse 2009; Sepkoski 2012). In this interval, a number of scientists began to entertain ideas that placed them in an uneasy relationship to the neo-Darwinian modern synthesis. As Pigliucci (2017) notes,

D.D. Turner (✉)
Department of Philosophy, Connecticut College, New London, CT, USA
e-mail: derek.turner@conncoll.edu

R.G. Delisle (ed.), *The Darwinian Tradition in Context*,
https://doi.org/10.1007/978-3-319-69123-7_15

these developments in paleontology helped provide some impetus to later calls for an extended evolutionary synthesis (Laland et al. 2014). To give just one example: at a time when many mainstream evolutionary biologists were rushing to embrace the gene selectionism of George Williams and Richard Dawkins, paleontologists such as Stephen Jay Gould, Elizabeth Vrba, and Steven Stanley were developing and defending models of species selection.[1] Shanahan (2017) explores this contrast between Dawkins and Gould in greater detail. In this chapter, I suggest that we can improve our understanding of paleontology's relationship with the modern synthesis by focusing on the concept of stasis.[2] In the early 1970s, paleontologists placed stasis back on the agenda of evolutionary biology. But as Jonathan Kaplan has written, "the solution to the problem of evolutionary stasis still awaits" (2009: 808). The renewal of interest in stasis did much to destabilize the modern synthesis.

Probably the first person to point to stasis as a problem was Hugh Falconer.[3] In 1863, just a few years after the publication of the *Origin*, Falconer and Darwin were cordial correspondents. And Falconer's sympathy for Darwin's new way of thinking was growing. However, in that year, he published a work on fossil elephants that included the following theoretical challenge:

> If we cast a glance back on the long vista of physical changes which our planet has undergone since the Neozoic Epoch, we can nowhere detect signs of a revolution more sudden and pronounced, and more important in its results, than the intercalation and subsequent disappearance of the Glacial period. Yet the mammoth lived before it, and passed through the ordeal of all the hard extremities which it involved, bearing his organs of locomotion and digestion all but unchanged (Falconer 1863: 252–3).

Falconer saw that according to Darwin's picture, environmental change should drive evolutionary change. It's hard to imagine a more significant environmental change than the advance and retreat of ice age glaciers. But the fossil record for woolly mammoths told a story of stability across a time interval that saw extreme climatic fluctuation. At the very least, there was a puzzle here for Darwin.[4]

Falconer's point about mammoths presaged the theory of punctuated equilibria (hereafter, PE), which was first proposed by Niles Eldredge (1971), and then presented collaboratively by Eldredge and Gould (1972) in the paper that became the canonical statement of the theory. According to PE, most lineages in the fossil record are characterized by morphological stasis throughout most of their history. Eldredge and Gould argued that most evolutionary change happens in conjunction with speciation events and that speciation is often geologically instantaneous. New forms show up suddenly in the fossil record, persist for a while with little or no

[1]For an overview of the species selection debate, see Turner (2011, Chapters 3 and 4). Jablonski (2008) surveys the scientific work on species selection over the last couple of decades. See Stanley (1975) for an especially clear articulation of the idea of species selection.

[2]Lidgard and Hopkins (2015) provide an invaluable annotated bibliography on the topic of evolutionary stasis.

[3]Gould (2002: 746–8) credits Falconer with first noticing that stasis is a problem.

[4]And the puzzle remains today. Prothero et al. (2012), who quote Falconer, see a pattern of stasis in mammals and birds from the La Brea tar pits during the same time interval.

change, and then disappear. Eldredge and Gould's famous slogan was that "Stasis is data" (Gould and Eldredge 1977). By that they simply meant that we should take stasis in the fossil record more or less at face value, as evidence that nothing much is happening during long stretches of evolutionary history. Section 2 develops these ideas in greater detail. The literature on PE is quite vast, but I shall tell the story in a somewhat unusual way here by focusing on the issue of stasis, rather than on the punctuations, which often get more attention.[5]

The very notion of "evolutionary stasis" can seem like an oxymoron. In ordinary language, when people talk about "evolution," they often just mean "gradual change." Not only that but the standard definition of "evolution" as any change in gene (or trait) frequencies in a population—the definition associated with the modern synthesis—seems to equate stasis with a lack of evolution. Nevertheless, one early criticism of Eldredge and Gould's PE model was that they had offered no good explanation for stasis (e.g., Gingerich 1984). If stasis is the dominant pattern of evolutionary history, then one might reasonably wonder what sorts of evolutionary mechanisms could work to maintain morphological stability for long stretches of time. In the event, a number of scientists went on to propose different models for explaining stasis. Section 3 surveys some of these proposals. We'll see that some of the proposals—like stabilizing selection—offer explanations of stasis that fall squarely within the modern synthesis, thus apparently diminishing the challenge that PE might pose to Darwinian tradition. But some other proposals invoke mechanisms that extend the modern synthesis in various ways.[6] So this survey will reveal one of many respects in which paleobiology's relationship with the Darwinian tradition has been an uneasy one. We should bear in mind, though, that just what "the Darwinian tradition" amounts to is very much up for grabs. For example, Delisle (2017) raises some critical questions about the relationship between Darwin and the biologists who orchestrated the modern synthesis.

Explaining stasis is one challenge, but there are deeper issues in play, which I take up in Sect. 4. When they first introduced PE, Eldredge and Gould made much of the issue of expectations: What should we expect to see when we look at the fossil record? The traditional view, which they called "phyletic gradualism," tells us that we should see evidence of gradual, directional, cumulative morphological change. PE tells us that we should see stasis. This way of framing the issue raises the question whether stasis or gradual change should be the default expectation or zero-force condition for our evolutionary models. This question is very much on the table in some current discussions of evolutionary theory. I'll argue that

[5]Gould (1977, 1980) at times flirted with saltationism. As Peter Bowler (this volume) observes, saltationism was one of several non-Darwinian ideas that found some support during the early twentieth century. Depew (this volume) also discusses Darwinism's relationship with saltationism. Gould's embrace of saltationism helps explain why the punctuational part of PE was so controversial.

[6]Thus, recent work in paleontology suggests the need for a "rethink" of the modern synthesis, along the lines suggested by Laland, Sterelny, Müller, and others in the recent debate in Nature (Laland et al. 2014).

paleontology's greatest challenge to Darwinian tradition has been to raise such fundamental questions about the structure of evolutionary models.

2 Stasis in the Early Punctuated Equilibria Debate

When Eldredge and Gould first introduced PE, it's not entirely clear how much they wanted to challenge the modern synthesis. They presented their new model as a straightforward consequence of Mayr's story about allopatric speciation—a story which was a central part of the modern synthesis. "Allopatric" here means "different territory," and Mayr had previously argued that speciation often occurs when a subpopulation gets geographically isolated from the rest of the species. The isolated subpopulation will then find itself with somewhat different environmental conditions. Mayr's view represented a bit of a departure from Darwin, who (though he didn't use this terminology) tended to think of speciation as sympatric. Sympatric speciation occurs when different variants in a population gradually adapt to different conditions (say, by exploiting different food sources) in the same geographic territory.

In the opening pages of the 1972 paper, Eldredge and Gould make it clear that their target is phyletic gradualism in paleontology. Suppose that speciation is usually allopatric. What would that mean for the fossil record? To start with, the allopatric speciation story seems to imply that a lot of evolutionary change occurs rapidly during speciation events. When some segment of a population becomes geographically isolated from the rest, the new subpopulation is likely to be small, so random drift will be a powerful factor. There is much discussion among philosophers of biology about how to conceptualize drift, but as a first pass, we can think of it as statistical sampling error. For example, if you flip a fair coin ten times, you would expect to get a ratio of 5:5 heads to tails. Similarly, in a biological population with two variants, you might expect—if natural selection were the only factor—to get a certain proportion of variants in the next generation. But suppose you get seven heads after tossing the coin ten times. One way to explain the deviation from the expected result is to point out that the sample size is quite small. If you tossed the coin 100 or 1000 times, the frequency of heads would approach 0.5. A small biological population likewise means that sampling error—i.e., drift—will be a bigger factor.

An isolated subpopulation is also likely to find itself in different environmental conditions. So you might expect rapid evolutionary change in that population. Also important is the fact that the new population will be in a different place. Paleontologists many millions of years in the future, looking back at an allopatric speciation episode, would very likely see a new species with distinctive morphology show up all of a sudden in the fossil record.

A committed phyletic gradualist might chalk the sudden appearance of a new type of fossil up to the incompleteness of the fossil record. Maybe the new morphology resulted from a long and gradual evolutionary process that simply never got recorded in the geology of that location. A local event such as a river changing course and delivering sediment to a different place can mean that the fossil record for a given location contains a gap of millions of years. In other words, the phyletic gradualist might treat the geologically sudden appearance of the new form as an illusion due to the gappiness of the fossil record. Eldredge and Gould argued that we should take seriously the possibility that the appearance of the new form really was geologically sudden, as suggested by Mayr's allopatric speciation model. They also argued that once a new species arises, its morphology typically does not change much, at least not in any kind of uniform, cumulative way. Nearly all the evolutionary change happens rapidly (at least, from a geological perspective) during speciation episodes. When we see morphological stasis in the fossil record, that is not something to be explained away—say, by supposing that there is gradual change happening offstage, during episodes that did not get recorded. Rather, the fossil evidence should be taken at face value, indicating that for long stretches of evolutionary history, there is not much going on.

In advancing PE, Eldredge and Gould were also taking stands on two further issues: the completeness of the fossil record as well as paleontology's relevance to evolutionary theory. First, there was an epistemic side to the debate about PE. In the *Origin*, Darwin had famously emphasized the incompleteness of the fossil record, largely in order to fend off potential critics who were concerned about the absence of transitional fossils. But if you go too far in lamenting the incompleteness of the fossil record, you start to raise doubts about whether studying fossils can tell us much about how evolution works. In the early 1970s, those Darwinian doubts about the usefulness of fossils were still very much alive, and Eldredge and Gould hoped to answer them by making a decisive move. PE straightforwardly implies that the fossil record is more complete than anyone had realized. Second, during the late nineteenth and early twentieth centuries, paleontologists had done a lot of theorizing about macroevolution, or about the big picture of the history of life on Earth. Most scientists use the term "macroevolution" to refer to patterns and processes at large scales, above the level of species, whereas "microevolution" refers to changes (usually changes in gene or trait frequencies) that occur within species. Much of that early theorizing about macroevolution was non-Darwinian, and much of it was marginalized by the modern synthesis. Indeed, Simpson's *Tempo and Mode in Evolution* (1944), which is often cited as an effort to bring paleontology into the modern synthesis, represented a major theoretical turn away from earlier ideas such as orthogenesis, the theory of "racial senescence" (occasionally offered as an explanation of the extinction of the dinosaurs), and the neo-Lamarckian theorizing about macroevolution that one finds in the work of Edward Drinker Cope.[7] Still one big issue that Simpson's work had not entirely resolved was whether paleontology would have much more to contribute to neo-Darwinian evolutionary theory. PE represents an effort to give a decisively affirmative answer to that question.

[7]Other contributions to this volume discuss orthogenesis and Lamarckism in greater detail. See, for example, Loison and Herring on the influence of Lamarckism in French biology.

Although Eldredge and Gould did not in the first instance see themselves as challenging the modern synthesis, but rather as drawing out the consequences of the allopatric speciation model, their ideas do have two implications that suggest less direct challenges.

First, PE raises serious questions about the overall importance of within-lineage, selection-driven evolutionary trends. If stasis is the dominant theme of evolutionary history, then natural selection seems rather less important than we thought. For most lineages, throughout most of their life spans, natural selection is not driving any significant directional evolutionary change. Selection does matter, but it works primarily in spurts, during speciation episodes, and in small populations where drift is also a major factor. This downplaying of natural selection coheres well with some of the other ideas coming out of paleontology in the 1970s, such as Gould and Lewontin's (1979) famous critique of adaptationism, as well as the much earlier decision to leave natural selection out of the early MBL model of macroevolution.[8] "MBL" refers to the Marine Biological Laboratory in Woods Hole, Massachusetts, where a group of scientists (including not only Gould but also David Raup, Dan Simberloff, Thomas Schopf, and Jack Sepkoski) convened in the early 1970s. One product of that meeting was an early computer simulation of large-scale evolutionary processes—the MBL model. The model simulated the evolutionary fates of whole lineages, including extinction and speciation. But it left out processes like natural selection that take place within lineages.

Second, if stasis is the norm *within* lineages, and if we still see larger-scale trends in evolutionary history when we focus on aggregate measures, such as mean body size, then it's not clear that we can explain those larger-scale trends in terms of natural selection. Something else must be going on, such as the sorting of whole lineages. Thus, stasis within lineages suggests the need for some other mechanism operating at larger scales, such as species selection. And as I hinted at the beginning of this chapter, species selection—or what is sometimes called the hierarchical expansion of evolutionary theory—really does pose a significant challenge to the picture handed down by the modern synthesis.[9] PE does not logically entail species selection, but it does suggest the possibility (Turner 2010). That also goes a long way toward explaining some of the more brash claims that Gould, in particular, made in the 1980s.

So, PE had two soft implications: that natural selection is less important than we thought and that species selection might be more important than we thought. But I want to focus more on two scientific challenges for PE. One of these is empirical and one theoretical.

First, the empirical challenge: Eldredge and Gould (1972) were making a bold empirical claim about patterns in the fossil record. The claim is that most of what

[8]On the MBL model, see Huss (2009) as well as Sepkoski (2012, Chap. 7).

[9]Hierarchical thinking is very much alive and well in paleontology. For further exploration of what a hierarchical approach to evolutionary theory might entail, see the papers collected in Eldredge et al. (2016).

we see is stasis, punctuated by brief spurts of speciation and rapid morphological change. Critics of PE were fairly quick to marshal counterexamples—cases of well-documented gradual morphological trends in the fossil record. And the friends of PE responded that they never meant to claim that stasis is the *only* thing we see in the fossil record, but that stasis is nevertheless a dominant theme or pattern in evolutionary history. Every pattern involves a certain amount of noise. It's easy to see how the empirical issue would quickly become a disagreement of emphasis, with the friends of PE arguing that stasis is relatively more important in evolutionary history and the foes trying to downplay its significance. This is a classic example of what Beatty (1997) has called a "relative significance debate" in biology. As in other cases, this particular relative significance debate gave impetus to new empirical work, with scientists doing large statistical studies of fossil samples in order to identify the dominant pattern. Jackson and Cheetham (1999), in one of the best known studies of this sort, did find that stasis is important. They conclude that "most cases of speciation in the sea over the past 25 my (million years) show prolonged morphological stasis punctuated by geologically sudden morphological shifts at cladogenesis" (1999: 76). Empirically speaking, PE has fared reasonably well.

Owing to the nature of the disagreement over the reality of a pattern, it may not make much sense to talk about confirmation or disconfirmation of PE in the traditional sense.[10] However, empirical studies of large fossil samples—usually of marine invertebrates, since their record is the most complete—can help to bring the larger patterns into sharper focus. Hunt (2007) affords one nice example of this type of empirical work. He studied 250 sequences of evolving traits, from 53 different lineages (mostly marine invertebrates). He compiled the results of previously published studies on those sequences and found that only 5% showed directional (probably selection-driven) evolutionary change in morphology. Of the remaining cases, about half exhibited stasis, while the rest exhibited morphological change that fit best with a random walk model. A random walk model differs both from stasis and from selection-driven change. It provides a way of thinking about how change can occur even in the absence of directional natural selection. For example, suppose (as seems plausible) that when mammals first evolved, they started out very small—near the minimum size for animals with a mammalian body plan. Suppose that over evolutionary time, size increases and decreases occur randomly. Because there is, in a sense, "nowhere to go but up," we might expect the average body size of mammals to increase, even without any help from natural selection. The picture that we get from studies like Hunt's (2007) suggests that stasis is an important evolutionary phenomenon.

Second: the theoretical challenge: If we look at the fossil record and see gradual, cumulative, directional evolutionary change within lineages, there is no mystery about how to explain the pattern. Neo-Darwinian theory gives us a perfectly good

[10]For a more detailed discussion of scientific disagreements about the reality of patterns, see Turner (2011, Chap. 6).

understanding of how natural selection can generate such patterns. Early on, though, critics of PE quite reasonably wondered what sort of evolutionary mechanism could explain stasis. Although scientists have done a lot of work on this issue in the meantime—I'll survey some of that work in Sect. 3—Eldredge and Gould were a bit slow at first to meet the explanatory challenge. In the closing pages of their 1972 paper, they introduced the rather murky claim that species are homeostatic systems. Generally speaking, a homeostatic system is one that maintains internal stability in the face of external perturbations. For example, the human body dissipates heat by sweating, thus maintaining a stable temperature on a hot day. Eldredge and Gould hinted that species might also exhibit developmental and genetic homeostasis, though it wasn't entirely clear what genetic homeostasis, for example, might mean. (And critics, such as Gingerich 1984, quickly pounced on this.) To skeptics it looked like Eldredge and Gould were just redescribing the problem: Somehow, species remain stable even in the face of major environmental change. But how? Or if genetic homeostasis was supposed to be some kind of mechanism, it sounded both new and imprecise, and it was hard to see how such a mechanism might cohere with the modern synthesis. In some ways, this early idea of genetic homeostasis foreshadowed Gould's later flirtation with more radical ideas, such as Richard Goldschmidt's notion of "hopeful monsters" or macromutations that occasionally give rise to entirely new species.

One concept that was conspicuously absent from Eldredge and Gould's 1972 paper was stabilizing selection. Stabilizing selection occurs when natural selection disfavors the extreme variants in a population. Eldredge and Gould's omission is surprising since, for many evolutionary biologists, stabilizing selection has always been the fallback explanation of stasis. Even today, people routinely invoke stabilizing selection as a model for explaining stasis. There's nothing mysterious about stabilizing selection, and it fits easily within the neo-Darwinian modern synthesis picture. To give just one example, the physiological mechanism that controls blood clotting in humans is extraordinarily complex and perched on something of an evolutionary knife edge. If your blood does not clot readily or quickly enough (i.e., if you have hemophilia), then that is a serious problem. On the other hand, if your blood clots too readily, then you are at risk of getting blood clots in your legs even when you've suffered no injury—also bad news, since blood clots are potentially fatal if they migrate to your heart or lungs. In a case like this, natural selection will work strongly against any variants that fall outside a fairly narrow range. With respect to this particular trait, we should expect stasis.

It's not clear why Eldredge and Gould did not invoke stabilizing selection in their original paper, especially since Gould (2002) later came around and allowed that it might be a factor. If they had, the subsequent debate about PE might have taken a somewhat different shape. I mentioned above that one way in which PE challenged evolutionary biology was by downplaying natural selection. But if stabilizing selection turns out to be the best explanation of stasis, then far from downplaying natural selection, PE would seem to depend on it. Invoking stabilizing selection would also have insulated Eldredge and Gould from the potential objection that they were introducing mysterious new ideas into evolutionary theory. But by the same token, it would have made PE seem less revolutionary.

3 Models for Explaining Stasis

Stabilizing selection is just one of several models that have been proposed for explaining stasis.[11] Some evolutionary biologists have argued that stabilizing selection is the only mechanism we need (Estes and Arnold 2007). It's also worth bearing in mind that the different models for explaining stasis do not necessarily compete. It could be that stabilizing selection maintains stability in most cases, but that there are exceptional ones where other forces come into play. Or it could be that multiple forces are in play in virtually every case, but with stabilizing selection playing the most significant role. Or it could be that stabilizing selection is somewhat less important than many scientists think. These empirical questions remain very much up for grabs. We also have here the makings of another Beatty-style relative significance debate. The first one (discussed in Sect. 2) concerned evolutionary *patterns*. Is stasis the dominant pattern? This second relative significance debate concerns underlying evolutionary *processes*. Is stabilizing selection the process doing all (or most) of the work? To what extent might other processes come into play. I obviously won't be able to settle this issue here, but it will be instructive to survey some of the going models, because they have different implications concerning paleobiology's relationship with neo-Darwinism. My survey will not be completely exhaustive (but see Lidgard and Hopkins 2015).[12]

Gould sometimes frames his work as an extended argument against "extrapolationism." Extrapolationism is basically a view about the relationship between microevolution and macroevolution; we might contrast it loosely with "expansionism" or hierarchical thinking. The extrapolationist is someone who thinks that whatever macroevolutionary patterns and trends we succeed in identifying, those will prove to be mere by-products of well-understood microevolutionary processes taking place within populations. The modern synthesis gives us an adequate theory of microevolution. The neo-Darwinian synthetic theory plus extrapolationism gives us (some would say) an adequate picture of macroevolution as well. Extrapolationism is thus a version of reductionism: in a word, it's the view that macroevolution is reducible to microevolution. An awful lot of the paleobiological theory that emerged in the 1970s and 1980s was strongly anti-reductionist in

[11]Throughout the chapter, I will continue to focus on morphological stasis, since that is also what draws the attention of most paleobiologists. However, there are also some fascinating cases of genomic stasis (Tamas et al. 2002; Bomfleur et al. 2014).

[12]Development, in particular, is relevant to explaining stasis, but I will not focus on development here. Wake et al. (1983) observe that developmental plasticity could give rise to stasis. Hansen and Houle (2004) argue that stasis raises questions having to do with evolvability.

spirit.[13] But PE's place in this discussion is somewhat ambiguous. If it were to turn out that stabilizing selection is the primary force that maintains stasis over long periods, then PE would look very extrapolationist. The macroevolutionary patterns of stasis and punctuational evolutionary change would simply be a by-product of well-understood microevolutionary processes: stabilizing selection and allopatric speciation. Charlesworth, Lande, and Slatkin (1982) defend this extrapolationist view, as do Estes and Arnold (2007).

As an explanatory model, stabilizing selection has some limitations. It works extremely well for traits like the blood clotting mechanism, where variation lying outside a narrow range would likely spell trouble. However, what about a trait where we know that there is quite a range of morphological possibility, and where our background knowledge gives us no reason to think that extreme variants would be selected against. Body size might be like this in some cases. At the La Brea tar pits, in California, many mammal lineages exhibit body size stasis over a period of tens of thousands of years, even during a cold spell of peak glaciation around 20,000 years ago. It's not too plausible that stabilizing selection would maintain stasis in a trait like body size under cooling climatic conditions, since Bergmann's rule predicts that larger body size would be adaptive (Prothero et al. 2012).

Another theoretical problem worth considering is one that I will call the *deactivation problem.* The deactivation problem afflicts all the models for explaining stasis and is by no means peculiar to stabilizing selection. But it is certainly an issue for theorists such as Estes and Arnold (2007), who think that stabilizing selection is the whole story about stasis. The problem, in a nutshell, is that evolutionary change does happen. This means that whatever mechanisms or processes work to maintain stasis for long periods must occasionally be deactivated. This creates something of a new explanatory challenge: If stabilizing selection (for example) is so effective at maintaining stasis, then why does it sometimes fail to do so? Ironically, the more powerful the mechanisms for explaining stasis, the bigger the puzzle about why they sometimes get put out of commission.

A further issue to consider is that stabilizing selection can only explain stasis with the help of an additional background assumption. Hansen and Houle (2004, p. 133) point out that stabilizing selection will tend to keep variation clustered within an optimal range. Any variants that appear on either tail of the distribution will get selected against. But stabilizing selection is actually compatible with a shifting optimum. For example, the optimum (i.e., fittest) body size for a species might depend on lots of environmental factors (such as average temperatures, the local predators, etc.). As long as those factors remain stable, stabilizing selection will help to keep variation in body size clustered around that optimum. But if the optimum shifts due to changed conditions, then stabilizing selection will not prevent evolutionary change. So a fuller explanation of stasis would have to include a story about the stability of the optimum, but that is precisely where the puzzle

[13]See Grantham (2007) for helpful discussion of paleontology's challenge to reductionism about macroevolution.

lies.[14] Why should the optimum have remained stable in cases where we know that significant environmental change has occurred?

Living fossil lineages help to illustrate this last problem. We should be careful here not to make too much of particular cases; statistical studies of larger samples of fossils carry more weight. Nor is it totally clear that "living fossils" represent a biologically useful category. Nevertheless, some particular cases of striking morphological stability might make us wonder about stabilizing selection. One example is the Wollemi pine, discovered in Australia's Wollemi National Park in the 1990s. Technically not a pine tree at all, but a member of the Araucariaceae family, it closely resembles fossils from the Cretaceous period, 90 million years ago. Of course, we know that during that time interval, the Wollemi pine has experienced significant environmental upheaval, including a global calamity and mass extinction 66 million years ago, a major faunal turnover with mammals replacing any dinosaurs that might have browsed Wollemi saplings back in the Cretaceous, as well as wild temperature swings, with a global hot spell at the end of the Paleocene 55 million years ago. This is obviously just one case, and as such, it doesn't tell us how evolution works in general. Nevertheless, in this case (and perhaps others like it), it is hard to see how stabilizing selection can be the whole story. Maybe it is, but it's also *prima facie* plausible that the significant environmental changes experienced by the Wollemi pine at various times in its history would make for directional natural selection and thus evolutionary change.

As it happens, stabilizing selection is not the only potential explanation for stasis. Peter Sheldon (1996) has proposed an alternative model for explaining stasis: the *plus ça change* model. One precursor of Sheldon's approach is E.D. Cope's "law of the unspecialized." No one today would go so far as to call it a "law"; it's a pattern at best. But Cope observed that the ecological specialists of more recent eras are often the descendants of the generalists of earlier eras. One could say that over time, there is a trend toward specialization. Cope was very likely thinking about extinction selectivity or the differential extinction vs. persistence of species. He thought, for example, that mammals in all of their specialized richness are descended from Mesozoic generalists that may have spent much of their time trying to stay out of the way of the dinosaurs. So the reason why the "law of the unspecialized" holds up is that during challenging times, ecological specialists are far more vulnerable to extinction. Ecological generalists are much better positioned to ride out episodes of environmental turmoil. This is the basic idea of the *plus ça change* model: ecological generalists handle environmental change by staying more or less the same. The more environmental change, the worse the news for ecological specialists.[15] So one distinct possibility is that when we look at the

[14]Hansen (2012) develops this line of reasoning even further. In a wide-ranging discussion, he argues that what we really need in order to connect microevolution with macroevolution is an account of the dynamics of adaptive landscapes. It's not enough, he argues, to study evolution on adaptive landscapes; we have to consider the evolution of those very landscapes.

[15]Conservation biologists also take this idea seriously. See, for example, Gallagher et al. (2014).

fossil record and see long-lasting lineages with little or no morphological change, what we're looking at is the record of ecological generalists. This could happen if, for example, places with lots of environmental change, such as shallow marine zones, also happen to be well represented in the fossil record.

Sheldon's model turns upside down the traditional idea that changed environmental conditions make for evolutionary change. He does this by shifting the level of analysis from changes within the population to the differential extinction and persistence of whole species. This shift coheres well with other efforts to move toward a more hierarchical conception of evolutionary processes. And it avoids the extrapolationism implicit in the stabilizing selection model. Of course, both stabilizing selection and extinction selectivity could capture important aspects of the truth, because different mechanisms could operate at different levels to produce the same result (Grantham 1999). In another respect, though, the models do seem to conflict. Stabilizing selection will work best in a stable environment, where selection will tend to keep the population in place on a peak in the fitness landscape. But Sheldon's *plus ça change* model presupposes an unstable, changing environment.

Another approach to explaining stasis focuses on gene flow and population structure. Eldredge et al. (2005) suggest that population size and structure can help maintain stasis (compare also Lieberman and Dudgeon 1996). Consider a species, such as white-tailed deer in North America, with abundant numbers and a large geographical range. Different subpopulations may find themselves living in somewhat different environmental conditions. Deer living in the southern USA will enjoy warmer conditions, for example, than those living in Canada. This creates an impediment to evolutionary change. If a new variation arises in one subpopulation living in a particular region, the variation still has to spread to the rest of the population. But if environmental conditions vary across the species' range, and if there is gene flow across subregions, a locally advantageous trait might never be able to establish itself in the species as a whole. Natural selection might favor the trait locally, but if individuals with the locally advantageous trait migrate or interbreed with individuals living in other regions, the trait might no longer be selected for.

To complicate the picture even further, paleontologists have also documented several cases of *coordinated stasis*, where a whole community of plants and animals persists through millions of years without significant evolutionary change (Brett and Baird 1995; Brett et al. 1996; DiMichele et al. 2004). At first blush, one might think that a paleoecological phenomenon calls for an ecological explanation. It's possible that coordinated stasis could result from the independent action of stabilizing selection (or constraints, or whatever), operating within each lineage. But that doesn't really explain why the community would coalesce at a time and then dissolve quickly—where "quickly" is understood in geological terms. A change that's rapid at geological scales could still take tens of thousands of years. Stabilizing selection might explain why a particular lineage remains stable, but it also cannot explain why, for example, you don't see new species migrating in. A further appeal to ecological considerations, such as competitive exclusion, might be necessary (see Sterelny 2001 for a thorough review of the issues).

Table 1 A hierarchy of models for explaining stasis

Level	Process	Proponent
Ecosystem	Ecological processes, such as ecological locking or competitive exclusion	Morris et al. (1995); see Sterelny (2001) for excellent discussion
Across species	*Plus ça change*/extinction selectivity	Sheldon (1996)
Across population, within species	Geographically structured large metapopulation + gene flow	Lieberman and Dudgeon (1996), Eldredge et al. (2005)
Within population	Stabilizing selection	Charlesworth et al. (1982), Estes and Arnold (2007), Hendry (2007), Haller and Hendry (2013)
Within organism	Development	Wake et al. (1983), Hansen and Houle (2004)

Hopefully the foregoing is sufficient to show just how complicated the picture is in contemporary evolutionary biology. One legacy of the early punctuated equilibria debate is that morphological stasis continues to pose an explanatory challenge. One need not be a partisan of PE in order to appreciate that stasis is an important pattern. Although some evolutionary biologists continue to argue that stabilizing selection is sufficient, there are good reasons for thinking that such a narrow, selectionist view is misguided. Other explanatory approaches have just as much *prima facie* plausibility. Crucially, the approaches just surveyed invoke evolutionary processes operating at different levels (Table 1). It's entirely possible that different evolutionary processes, including stabilizing selection, could work together to maintain stasis in particular cases. Perhaps a useful image here is that a stable lineage (or perhaps a clade) is like a stationary swan: above the surface, it looks like nothing much is happening. But beneath the water, the swan is paddling against the currents and the breeze, and the paddling involves a complicated churn, with different forces exerted in different directions.

This multilevel approach to explaining evolutionary patterns sits uneasily with the modern synthesis view that macroevolutionary patterns are just by-products of microevolutionary processes. It's becoming clearer that the real challenge of explaining stasis is to show how processes operating at different levels can combine to generate patterns at larger scales. The phenomenon of stasis really does seem to call for a more hierarchical picture of evolution than anything countenanced by the modern synthesis.

The contrast between this multilevel approach and the more traditional, neo-Darwinian approach of Estes and Arnold (2007) is striking. They say explicitly that they "hope to reveal the microevolutionary underpinnings of stasis" (2007: 228). They start with a data set of evolutionary divergence times supplied by

Gingerich (2001), and then they test a number of different microevolutionary models, to see which best fits the data:

- A neutral/drift model
- A gradually moving peak in the adaptive landscape (representing steady, directional natural selection)
- A fitness peak exhibiting Brownian motion, with no directional change in its mean position.
- A peak shift model, in which the population moves from one peak in the adaptive landscape to another.
- A displaced optimum model, in which the peak in the adaptive landscape suddenly moves a small distance and the population "chases" it. After the displacement of the optimum, it remains fixed, and stabilizing selection kicks in.

Estes and Arnold found that the last of these—the displaced optimum model with stabilizing selection—fit Gingerich's data the best. The data set also included a wide range of samples involving different timescales—from the microevolutionary to the paleontological. So Estes and Arnold claim that the displaced optimum model with stabilizing selection explains what we see "on all timescales"—an apparent victory for neo-Darwinian extrapolationism.

One problem with Estes and Arnold's approach is that all of the models they are working with represent various microevolutionary processes. They do not take seriously the possibility that mechanisms operating at higher levels—think again of Sheldon's invocation of extinction selectivity—might contribute to maintaining stasis. In other words, they frame the question in narrowly neo-Darwinian terms: Which sort of population genetic model does the best job accounting for Gingerich's data? It's therefore no surprise that they should get a narrow, neo-Darwinian answer. What's more, Kaplan (2009) worries, I think rightly, that Estes and Arnold are merely describing the phenomenon to be explained. The displaced optimum model describes a pattern that is compatible with a number of different underlying processes. One way to see this is just to ask why the optimum behaves as it does. Why *doesn't* it shift steadily? Unless we can answer questions like that, we don't really have a good explanation of stasis, even if we do know that the displaced optimum model fits the data quite well.

4 Stasis, Change, and Expectation

One of the many things that Eldredge and Gould were up to when they introduced PE was to argue, *à la* Thomas Kuhn, that the expectations we bring to the fossil record can influence what we see there. For example, they worried that a background commitment to phylogenetic gradualism would prevent scientists from even seeing stasis in the fossil record, and they pushed back hard:

> We wanted to expand the scope of relevant data by arguing that morphological breaks in the stratigraphic record may be real, and that *stasis is data*—that each case of stasis has as much meaning for evolutionary theory as each example of change (Gould and Eldredge 1977: 117).

Their concern, in other words, was that paleontologists were not even seeing stasis in the fossil record, because their antecedent commitment to a gradualist Darwinian picture seemed to imply that stasis, if it occurred at all, would be an evolutionary aberration. In good Kuhnian fashion, Gould and Eldredge saw that it's not possible to do science without being committed to some sort of paradigm, so the only way to get people to see stasis in the fossil record was to replace the old gradualist picture with a new one.[16] And in the new picture, punctuated equilibria, stasis is *expected.* Whatever we think of the Kuhnian influence on their work, Eldredge and Gould's early way of framing the issues raises some deep questions about stasis, change, and expectation. When we think about evolution, should stasis or gradual change be the default expectation? There are seeds here of another challenge to the Darwinian tradition. Indeed, this challenge is arguably more fundamental than the explanatory one that I explored in Sect. 3. The earlier question was just how to explain stasis. But here the issue has more to do with the way in which our thinking about stasis and change structures evolutionary explanation.

Gould revisited the issue of default expectations in *The Structure of Evolutionary Theory*:

> Second, and even more generally, the validation of predominant stasis as a norm would impel us to recast the basic problematic of evolution itself. If, following our conventional assumptions from Darwin to now, change represents the norm for a population through time, then our task, as evolutionary biologists, lies in specifying how this unexpected and universal phenomenon operates. But if, as punctuated equilibrium suggests, stasis represents the norm for most populations at most times . . . then evolutionary change itself must be reconceptualized as the infrequent breaking of a conventional and expected state, rather than as an inherent and continually operating property of biological materials, ecologies, and populations (Gould 2002: 884–5).

Gould here is recommending a shift toward thinking of stasis as the "norm" for evolutionary systems, and as the "conventional and expected state." With this shift, he argues, would come a corresponding shift of explanatory target. Yet even though this passage comes at the end of a discussion of various strategies for explaining stasis, Gould here seems to be saying that it's not stasis, per se, that cries out for explanation, but rather the occasional disruption of stasis. At any rate, these questions about stasis, change, and expectation that Eldredge and Gould raised in the 1970s, and that Gould continued to wrestle with, are right at the center of an ongoing debate in philosophy of biology.

[16]For further discussion of Kuhn's influence on Gould, see Turner (2011, Chap. 1).

To start with, consider the role that the Hardy–Weinberg principle plays in the received view of population genetics (Sober 1984). In doing so, we have to be mindful of issues of scale. When Eldredge and Gould claimed that stasis is data, they were talking about the grand sweep of evolutionary history and about patterns of stasis in the fossil record. Population geneticists focus rather on the microlevel changes from generation to generation. Nevertheless, there is an important sense in which the Hardy–Weinberg principle treats stasis as the default expectation for biological populations. The Hardy–Weinberg principle just says (putting it loosely and informally) that if a population starts out with certain allele frequencies, then assuming that no external forces act upon it, you will see such-and-such genotype frequencies in the next generation. The allele frequencies will remain stable. In this case, the external forces include drift, selection, nonrandom mating, mutation, and migration. Sober (1984) developed an analogy with Newtonian physics here: The theoretical apparatus of population genetics treats stasis (with respect to allele frequencies) as the inertial state for a population and then describes how the operation of various forces can drive the population out of its inertial state—that is, how external forces acting on the population can cause changes in gene frequencies. For present purposes, the important thing to note is that stasis plays the role of a default expectation.

Sober's approach treats population genetic theory as involving what I will call an inertial state model (or "ISM" for short). Generally speaking, an ISM partitions the possible states of a system into two groups: a set of inertial states and all the other non-inertial states. When no external forces are acting upon the system, it will be in one of its inertial states. The ISM also specifies a set of relevant external forces. When the system is in one of its non-inertial states, that deviation from the default expectation is to be explained in terms of the operation of the external forces.

Recently, though, Dan McShea and Robert Brandon have challenged the received way of understanding the role of the Hardy–Weinberg principle as specifying the zero-force condition for biological populations. McShea and Brandon (2010) argue for a kind of *Gestalt* shift: Instead of thinking of stasis as the inertial state for biological systems, we should think of *complexity increase* as the inertial state (compare also Brandon 2010). They refer to this idea as the *ZFEL*, or the "zero-force evolutionary law." Importantly, they define complexity in purely structural terms, so that for them it means something like internal variance or heterogeneity. Very roughly, what they have in mind is something like diversity of part types. For example, multicellular organisms like us have different types of cells: we have muscle tissue, bone tissue, blood cells, skin cells, nerve cells, and so on. An organism with more types of cells has greater complexity, in McShea and Brandon's sense.

The *ZFEL* harkens back not so much to Darwin as to Herbert Spencer's notion of the "instability of the homogeneous." It clashes with the received view when it comes to the role of drift. The Hardy–Weinberg principle treats drift as an external force that can act on a population to bring about changes in gene frequencies. The *ZFEL*, by contrast, treats drift as something that occurs in the zero-force condition. This issue has been much debated in recent philosophy of biology (Brandon 2006; Barrett et al. 2012; Brandon and McShea 2012; Gouvêa 2015). The basic question is whether stasis *vs.* change should count as the inertial state, or the default expectation, for biological systems. There is a strong echo here of Eldredge and Gould's question whether we should expect to see stasis or change when we look at the fossil record.

McShea and Brandon converged on the *ZFEL* from somewhat different directions. Brandon (2006) had been thinking about drift, whereas McShea (2005) was working on the issue of how to explain large-scale macroevolutionary trends, such as complexity increase. Their view also represents a significant departure from the traditional neo-Darwinian conception of the relationship between complexity and natural selection. The assumption had always been that you need cumulative natural selection in order to explain complexity. McShea and Brandon turn this upside down, arguing that complexity tends to increase with or without the help of natural selection. They even go so far as to invoke selection in order to explain cases where complexity does not increase, as well as cases where complexity seems to decline. For example, the loss of vision in cave-dwelling species might seem, at first glance, to be a problem case for the *ZFEL*. But they propose to explain this and other cases of complexity loss by suggesting that natural selection is working against the *ZFEL*.

The *ZFEL* itself is a neo-Spencerian idea whose relationship to the Darwinian tradition is uneasy. The *ZFEL* offers a non-selectionist explanation of two of the most striking large-scale trends in evolutionary history: complexity and diversity increase. For present purposes though, the non-Darwinian roots of the *ZFEL* are less important than the debate itself. Eldredge and Gould (1972) did more than merely challenge Darwinian gradualism. They started a conversation about the role that default expectations play in evolutionary theorizing. Their way of framing the contrast between phyletic gradualism and punctuated equilibria set the stage for further exploration of the role that expectations about stasis *vs.* change should play in evolutionary theory. The central issue in the debate about the *ZFEL* is whether stasis or change should be regarded as the default expectation for biological systems. This issue, which remains up for discussion, is a close relative of the one that Eldredge and Gould took themselves to be raising with punctuated equilibria.

5 Conclusion

Using stasis as a guiding thread, I've identified three different points of tension between paleobiological research from the 1970s on and the Darwinian tradition as represented by the modern synthesis. To start with, there was Eldredge and Gould's more or less straightforward challenge to Darwinian gradualism. But their model of punctuated equilibria led to two further and even more important developments. On the one hand, it naturally suggested the need for evolutionary explanations of stasis. Macroevolutionary stasis, especially coordinated stasis, is difficult to make sense of by means of stabilizing selection alone. In coordinated stasis, for example, the "coordination" seems ecological: you have a whole community of species that seems to persist without change. But stabilizing selection can only explain, at best, why we might see stasis within a particular species. It can't explain the coordination. That is bad news for the sort of "extrapolationist" picture of evolution that Gould so detested, because it points toward the need for a more hierarchical approach that gives consideration to processes operating at different levels. This, too, places paleobiology in an uneasy relationship with neo-Darwinism. Third, Eldredge and Gould were also raising deeper questions about whether stasis vs. change should be the default expectation with respect to evolutionary systems. This issue remains very much alive in contemporary philosophy of biology, with the received view treating stasis (as described by the Hardy–Weinberg principle) as the default condition for evolving populations, with the *ZFEL* view treating change as the default.

This brief survey of some of the tensions between paleobiology and the Darwinian tradition is by no means exhaustive. There are other places where one could dig in, especially species selection and the application of random walk models to macroevolution. I've focused here a bit more narrowly on punctuated equilibria and some of the questions about stasis that it generated. Ironically, paleontologists' work on stasis may have contributed to destabilizing the neo-Darwinian modern synthesis.

Acknowledgments My work on this paper was supported by a sabbatical fellowship from the KLI, in Klosterneuburg, Austria. I'm deeply grateful to Lee Altenberg, Argyris Arnellos, Dan Brooks, James DiFrisco, Chiara Ferrario, Barbara Fischer, Yogi Jaeger, and Gerd Müller for helping me to get a bit clearer in my thinking about stasis and for providing a wonderful, collegial environment in which to work. I am also grateful to Fulbright Canada for the opportunity to spend time at the University of Calgary, where I also did some work on this paper. Thanks especially to Richard Delisle for his very helpful advice and editorial suggestions.

References

Barrett M, Clatterbuck H, Goldsby M, Hegelson C, McLoone B, Pearce T, Sober E, Stern R, Weinberger N (2012) Puzzles for ZFEL: McShea and Brandon's zero force evolutionary law. Biol Philos 27:725–735

Beatty J (1997) Why do biologists argue like they do? Philos Sci S64:231–242

Bomfleur B, McLoughlin S, Vajda V (2014) Fossilized nuclei and chromosomes reveal 180 million years of genomic stasis in royal ferns. Science 343:1376–1377

Brandon RN (2006) The principle of drift: biology's first law. J Philos 103(7):319–335

Brandon RN (2010) A Neo-Newtonian model of evolution: the ZFEL view. Philos Sci 77:702–715

Brandon RN, McShea DW (2012) Four solutions for four puzzles. Biol Philos 27:737–744

Brett CE, Baird GC (1995) Coordinated stasis and evolutionary ecology of Silurian to Middle Devonian faunas in the Appalachian Basin. In: Erwin DH, Antsey RL (eds) New approaches to speciation in the fossil record. Columbia University Press, New York, pp 285–315

Brett CE, Ivany LC, Schopf KM (1996) Coordinated stasis: an overview. Palaeogeogr Palaeoclimatol Palaeoecol 127:1–20

Charlesworth B, Lande R, Slatkin M (1982) A neo-Darwinian commentary on macroevolution. Evolution 36(3):474–498
Delisle RG (2017) From Charles Darwin to the evolutionary synthesis: weak and diffused connections only. In: Delisle RG (ed) The Darwinian tradition in context: research programs in evolutionary biology. Springer, Cham, pp 133–168
DiMichele AK, Behrensmeyer AK, Olszewski TD et al (2004) Long-term stasis in ecological assemblages: evidence from the fossil record. Annu Rev Ecol Syst 35:285–322
Eldredge N (1971) The allopatric model and phylogeny in Paleozoic invertebrates. Evolution 25 (1):156–167
Eldredge N, Gould SJ (1972) Punctuated equilibria: an alternative to phyletic gradualism. In: TJM S (ed) Models in paleobiology. Cooper, Freeman, San Francisco, CA, pp 85–115
Eldredge N, Thompson JN, Brakefield PM, Gavrilets S, Jablonski D, Jackson JBC, Lenski RE, Lieberman BS, McPeek MA, Miller W (2005) The dynamics of evolutionary stasis. Paleobiology 31(2):133–145
Eldredge N, Pievani T, Serrelli E, Temkin I (eds) (2016) Evolutionary theory: a hierarchical perspective. University of Chicago Press, Chicago, IL
Estes S, Arnold SJ (2007) Resolving the paradox of stasis: models with stabilizing selection explain evolutionary divergence on all timescales. Am Nat 169(2):227–244
Falconer H (1863) On the American fossil elephant. Nat Hist Rev 3:43–114
Gallagher AJ, Hammerschlag N, Cooke SJ, Costa DP, Irschick DJ (2014) Evolutionary theory as a tool for predicting extinction risk. Trends Ecol Evol 30(2):61–65
Gingerich PE (1984) Punctuated equilibria: where is the evidence? Syst Zool 33(3):335–338
Gingerich PE (2001) Rates of evolution on the time scale of the evolutionary process. Genetica 222:159–161
Gould SJ (1977) The return of the hopeful monster. Nat Hist 86:22–30
Gould SJ (1980) Is a new and general theory of evolution emerging? Paleobiology 6(1):119–130
Gould SJ (2002) The structure of evolutionary theory. Harvard University Press, Cambridge, MA
Gould SJ, Eldredge N (1977) Punctuated equilibria: the tempo and mode of evolution reconsidered. Paleobiology 3(2):115–151
Gould SJ, Lewontin R (1979) The Spandrels of San Marco: a Critique of the Adaptationist Programme. Proc R Soc Lond B 205(1161):581–598
Gouvêa D (2015) Explanation and the evolutionary first law. Philos Sci 82(3):363–382
Grantham T (1999) Explanatory pluralism in paleobiology. Philos Sci 66(3):236
Grantham T (2007) Is macroevolution more than successive rounds of microevolution? Paleontology 50(1):75–85
Haller BC, Hendry A (2013) Solving the paradox of stasis: squashed stabilizing selection and the limits of detection. Evolution 68(2):483–500
Hansen TF (2012) Adaptive landscapes and macroevolutionary dynamics. In: Svensson EI, Calsbeek R (eds) The adaptive landscape in evolutionary biology. Oxford University Press, Oxford, pp 205–221
Hansen TF, Houle D (2004) Evolvability, stabilizing selection, and the problem of stasis. In: Pigliucci M, Preston K (eds) Phenotypic integration: studying the ecology and evolution of complex phenotypes. Oxford University Press, Oxford, pp 130–154
Hendry A (2007) The Elvis paradox. Nature 446(8):147–150
Hunt G (2007) The relative importance of directional change, random walks, and stasis in the evolution of fossil lineages. PNAS 104(47):18404–18408
Huss J (2009) The shape of evolution: the MBL model and clade shape. In: Sepkoski D, Ruse M (eds) The paleobiological revolution: essays on the growth of modern paleontology. University of Chicago Press, Chicago, IL, pp 326–345
Jablonski D (2008) Species selection: theory and data. Annu Rev Ecol Evol Syst 39:501–524
Jackson JBC, Cheetham AH (1999) Tempo and mode of speciation in the sea. Trends Ecol Evol 14 (2):72–77

Kaplan J (2009) The paradox of stasis and the nature of explanation in evolutionary biology. Philos Sci 76:797–798

Laland K, Uller T, Feldman M, Sterelny K, Müller G et al (2014) Does evolutionary theory need a rethink? Nature 514(7521):161–164

Lidgard S, Hopkins M (2015) Stasis. Oxf Bibliogr. https://doi.org/10.1093/OBO/9780199941728-0067

Lieberman BS, Dudgeon S (1996) An evaluation of stabilizing selection as a mechanism for stasis. Palaeogeogr Palaeoclimatol Palaeoecol 127:229–238

McShea DW (2005) The evolution of complexity without natural selection: a possible large-scale trend of the fourth kind. Paleobiology 31(2 Supp):146–156

McShea DW, Brandon RN (2010) Biology's first law. University of Chicago Press, Chicago, IL

Morris PJ, Ivany LC, Schopf KM, Brett CE (1995) The challenge of paleoecological stasis: reassessing sources of evolutionary stability. PNAS 92:11269–11273

Pigliucci M (2017) Darwinism after the modern synthesis. In: Delisle RG (ed) The Darwinian tradition in context: research programs in evolutionary biology. Springer, Cham, pp 89–104

Prothero DR, Syverson VJ, Raymond KR, Madan M, Molina S et al (2012) Size and shape stasis in late Pleistocene mammals and birds from Rancho La Brea during the Last Glacial-Interglacial cycle. Q Sci Rev 56:1–10

Ruse M (2009) Has paleobiology been through a paradigm shift? In: Sepkoski D, Ruse M (eds) The paleobiological revolution: essays on the growth of modern paleontology. University of Chicago Press, Chicago, IL, pp 518–528

Sepkoski D (2012) Rereading the fossil record: the growth of paleobiology as an evolutionary discipline. University of Chicago Press, Chicago, IL

Shanahan T (2017) Selfish genes and lucky breaks: Richard Dawkins' and Stephen Jay Gould's: divergent Darwinian agendas. In: Delisle RG (ed) The Darwinian tradition in context: research programs in evolutionary biology. Springer, Cham, pp 11–36

Sheldon PR (1996) Plus ça change: a model for stasis and evolution in different environments. Palaeogeogr Palaeoclimatol Palaeoecol 127:209–227

Simpson GG (1944) Tempo and mode in evolution. Columbia University Press, New York

Sober E (1984) The nature of selection: evolutionary theory in philosophical focus. MIT Press, Cambridge, MA

Stanley S (1975) A theory of evolution above the species level. Proc Natl Acad Sci USA 72 (2):6467–6650

Sterelny K (2001) The reality of ecological assemblages: a palaeo-ecological puzzle. Biol Philos 16:437–461

Tamas I et al (2002) 50 million years of genomic stasis in endosymbiotic bacteria. Science 296:2376–2379

Turner DD (2010) Punctuated equilibrium and species selection: what does it mean for one theory to suggest another? Theory Biosci 129:113–123

Turner DD (2011) Paleontology: a philosophical introduction. Cambridge University Press, Cambridge

Wake DN, Roth G, Wake MH (1983) On the problem of stasis in organismal evolution. J Theor Biol 101:211–224